Emerging Technologies and Marketing Strategies for Sustainable Agriculture

Jabulani Garwi
University of the Free State, South Africa

Reason Masengu
Middle East College, Oman

Option Takunda Chiwaridzo
University of Science and Technology Beijing, Beijing, China

IGI Global
PUBLISHER of TIMELY KNOWLEDGE

A volume in the Practice, Progress, and
Proficiency in Sustainability (PPPS) Book Series

Published in the United States of America by
IGI Global
Engineering Science Reference (an imprint of IGI Global)
701 E. Chocolate Avenue
Hershey PA, USA 17033
Tel: 717-533-8845
Fax: 717-533-8661
E-mail: cust@igi-global.com
Web site: http://www.igi-global.com

Library of Congress Cataloging-in-Publication Data

CIP DATA IN PROCESS

Emerging Technologies and Marketing Strategies for Sustainable Agriculture
Jabulani Garwi, Reason Masengu, Option Takunda Chiwaridzo
2024 Engineering Science Reference

ISBN: 9798369348642 (h/c)
ISBN: 9798369348659 (s/c)
eISBN: 9798369348666

This book is published in the IGI Global book series Practice, Progress, and Proficiency in Sustainability (PPPS) (ISSN: 2330-3271; eISSN: 2330-328X)

British Cataloguing in Publication Data
A Cataloguing in Publication record for this book is available from the British Library.

Practice, Progress, and Proficiency in Sustainability (PPPS) Book Series

Ayman Batisha
International Sustainability Institute, Egypt

ISSN:2330-3271
EISSN:2330-328X

Mission

In a world where traditional business practices are reconsidered and economic activity is performed in a global context, new areas of economic developments are recognized as the key enablers of wealth and income production. This knowledge of information technologies provides infrastructures, systems, and services towards sustainable development.

The **Practices, Progress, and Proficiency in Sustainability (PPPS) Book Series** focuses on the local and global challenges, business opportunities, and societal needs surrounding international collaboration and sustainable development of technology. This series brings together academics, researchers, entrepreneurs, policy makers and government officers aiming to contribute to the progress and proficiency in sustainability.

Coverage

- Global Content and Knowledge Repositories
- Outsourcing
- Green Technology
- Strategic Management of IT
- ICT and knowledge for development
- Knowledge clusters
- Sustainable Development
- Technological learning
- Environmental informatics
- E-Development

IGI Global is currently accepting manuscripts for publication within this series. To submit a proposal for a volume in this series, please contact our Acquisition Editors at Acquisitions@igi-global.com or visit: http://www.igi-global.com/publish/.

The Practice, Progress, and Proficiency in Sustainability (PPPS) Book Series (ISSN 2330-3271) is published by IGI Global, 701 E. Chocolate Avenue, Hershey, PA 17033-1240, USA, www.igi-global.com. This series is composed of titles available for purchase individually; each title is edited to be contextually exclusive from any other title within the series. For pricing and ordering information please visit http://www.igi-global.com/book-series/practice-progress-proficiency-sustainability/73810. Postmaster: Send all address changes to above address.

Titles in this Series

For a list of additional titles in this series, please visit:
http://www.igi-global.com/book-series/practice-progress-proficiency-sustainability/73810

Quality of Life and Climate Change Impacts, Sustainable Adaptation, and Social-Ecological esilience
Kasturi Shukla (Symbiosis Institute of Health Sciences, Symbiosis International University (Deemed), Pune, India) Yogesh B. Patil (Symbiosis Centre for Research and Innovation, Symbiosis International University (Deemed), Pune, India) Ronald C. Estoque (Forestry and Forest Products Research Institute, Japan) and Pedro Antonio López de Haro (Universidad Autónoma Indígena de México, Mexico)
Engineering Science Reference • © 2024 • 499pp • H/C (ISBN: 9781668498637) • US $250.00

Sustainable Practices for Agriculture and Marketing Convergence
Jabulani Garwi (University of the Free State, South Africa) Reason Masengu (Middle East College, Oman) and Option Takunda Chiwaridzo (University of Science and Technology, Beijing, China)
Engineering Science Reference • © 2024 • 324pp • H/C (ISBN: 9798369320112) • US $245.00

Sustainable Disposal Methods of Food Wastes in Hospitality Operations
Amrik Singh (Lovely Professional University, India) Pankaj Kumar Tyagi (Chandigarh University, India) and Anshul Garg (Taylor's University, Malaysia)
Engineering Science Reference • © 2024 • 370pp • H/C (ISBN: 9798369321812) • US $285.00

Sustainable Development in AI, Blockchain, and E-Governance Applications
Rajeev Kumar (Moradabad Institute of Technology, India) Abu Bakar Abdul Hamid (Infrastructure University, Kuala Lumpur, Malaysia) Noor Inayah Binti Ya'akub (Infrastructure University, Kuala Lumpur, Malaysia) Hari Om Sharan (Rama University, Kanpur, India) and Sandeep Kumar (Rightzone Technologies Pvt. Ltd., India)
Engineering Science Reference • © 2024 • 282pp • H/C (ISBN: 9798369317228) • US $285.00

Operational Research for Renewable Energy and Sustainable Environments
Joshua Thomas (University of Wollongong Malaysia, Malaysia) Gerhard Wilhelm Weber (Poznan University of Technology, Poland) Roman Rodríguez Aguilar (Universidad Panamericana, Mexico) Elias Munapo (North West University, South Africa) and Pandian Vasant (Ton Duc Thang University, Vietnam)
Engineering Science Reference • © 2024 • 350pp • H/C (ISBN: 9781668491300) • US $240.00

Sustainable Solutions for E-Waste and Development
Rajesh Kumar K. V. (Woxsen University, India) Hemachandran Kannan (Woxsen University, India) Dmitry Spodarets (Data Phoenix, USA) Parvez Alam Khan (Universiti Teknologi PETRONAS, Malaysia) and Bikash Kumar Pradhan (Woxsen University, India)
Engineering Science Reference • © 2024 • 396pp • H/C (ISBN: 9798369310182) • US $265.00

IGI Global
PUBLISHER of TIMELY KNOWLEDGE

701 East Chocolate Avenue, Hershey, PA 17033, USA
Tel: 717-533-8845 x100 • Fax: 717-533-8661
E-Mail: cust@igi-global.com • www.igi-global.com

Table of Contents

Detailed Table of Contents

It includes practices like crop management, sustainable farming, agroforestry, precision farming, and climate-resilient crop varieties. CSA improves soil health, water use efficiency, erosion, carbon sequestration, biodiversity conservation, and resource utilization. Precision agriculture tools like remote sensing and data analytics enable site-specific management decisions. Climate-resilient crop types, such as drought-tolerant, heat-tolerant, and disease-resistant, are crucial for food production under changing climatic conditions. Encouraging laws, institutional frameworks, and financial incentives are essential for CSA implementation. However, more research and collaboration are needed to promote widespread adoption of CSA practices.

Vidya Kumbhar, Symbiosis Institute of Geo-Informatics, Symbiosis International University (Deemed), India
Sneha Kumari, Symbiosis School of Economics, Symbiosis International University (Deemed), India
Pratyush Kumar, Symbiosis Institute of Geo-Informatics, Symbiosis International University (Deemed), India
Kailash Parihar, Symbiosis Institute of Geo-Informatics, Symbiosis International University (Deemed), India
Samaya Venkatesh Pillai, Symbiosis Institute of Digital and Telecom Management, Symbiosis International University (Deemed), India

In India, agriculture is a source of employment to more than 50% of the population. Despite major strengths such as geographical location, diverse agro-climatic conditions, variety of soils, and agro-ecological congeniality, Indian farmers face diverse types of risks. Climate change is one the major risks confronting Indian farmers. Variations in climatic parameters such as rainfall, temperature, humidity, etc. result in a decrease in soil moisture and reduction in soil fertility. This has adversely affected crop productivity. History reflects that unpredictable climate change has affected the standard of living of Indian farmers. This gives rise to a strong need to study the risk management techniques of the Indian farmers. The chapter explores the diverse risks faced by Indian farmers and the risk management strategies adopted by them. The purpose of the research is to examine the different hazards that farmers encounter as well as risk management techniques. Primary data has been collected from 192 farmers in Maharashtra. Purposive sampling technique was used for collecting responses from the farmers. And an open-ended interview has also been conducted for another 30 farmers. The data was analysed using QDA miner software. The research shows how farmers deal with hazards using various ways and strategies. It can be advanced by looking into cutting-edge risk management strategies as well as potential enablers and obstacles associated with it.

Syed Muhammad Jalal, Department of Management Studies, Middle East College, Oman
Iqra Arshad, Graduate School of Public Administration (GSPA), National Institute of Development Administration, Thailand
Shahid Khalil, School of Transportation and Logistics, Malaysia University of Science and Technology, Malaysia
Muhammad Khuram Khalil, Department of Management Studies, Middle East College, Oman

The negative impact that individuals have on the natural world led to the development of the concept of sustainable marketing. Increasing consumer trends regarding environmentally conscious consumers

Online shopping has increased with the development of technology in the modern era, which leads people to make unplanned purchases. The purpose of this study is to examine the factors affecting the rush to buy online in the Omani market. These factors (social shopping, convenience shopping, adventure shopping, idea shopping) motivate consumers to initiate shopping readiness and impulse buying. The research also aims to know the relationship between factors and impulsive buying via the internet and how they affect the Omani market. The data were collected from 210 online buyers belong to capital city Muscat, Oman. Structure equation modelling was employed via SmartPLS to check the hypotheses. Results endorse that out of four hypotheses, only two are accepted. Following structure equation modelling results, there is a positive relationship between adventure shopping and impulse buying. In addition, a similar relationship has been seen between value shopping and impulse buying. Results are also helpful to understand the importance and great potential of concerned factors to encourage the consumers' buying behaviour towards impulsive buying.

The broad objective of this chapter is to underscore the importance of collaborative PPP in agricultural research financing. The specific objectives are to reveal Nigeria's comparative advantage in agriculture, reveal how Nigeria's comparative advantage in agriculture could be transformed into Michael Porter's concept of competitive advantage through Albert Hirschman's unbalanced growth strategy, and initiate a triple bottom-line approach to agricultural research financing. To achieve the last objective, the United Nations Global Compact (UNGC) members' account balances of firms in Nigeria were collected from annual reports to speculate the agricultural research finance revenue realizable from the private sector. The study adopted the UNGC's financial commitment benchmark for every functional firm registered with the Corporate Affairs Commission (CAC) in Nigeria as an appropriate bottom-line commitment that private sectors should have towards agricultural research financing – if to be regarded as socially responsible in Nigeria.

Crop prediction and fertilizer recommendation are essential for optimizing agricultural practices, a crucial concern for the agricultural sector. However, accurate crop and fertilizer prediction has long been challenging, requiring innovative solutions rooted in the vast pool of available data. This work presents a comprehensive system for

Agricultural lands are swiftly transforming into residential complexes and high-rise buildings, leaving limited space for traditional gardening. A sustainable and creative venture, terrace gardening (TG) focuses on using rooftop areas for the growth of plants, vegetables, and herbs. To give people and communities access to fresh food and a green haven in urban settings, it entails converting underutilized rooftops into gardens. The authors review and present relevant literature on the topic of TG. The review identifies that papers on TG mainly cover themes like motivations to participate in TG, advantages of TG, sustainability and innovation in TG, operations and stakeholder management in TG, and limitations of TG. The review shows that time management, project management, networking, and stakeholder involvement are essential for TG. A SWOC analysis reveals that TG offers potential for business owners seeking to have a positive environmental impact while supplying the rising demand for fresh, locally grown food.

This quantitative study explores the relationships between precision agriculture, vertical farming, and bio-genetic technologies among 402 smallholder farmers in Zimbabwe. Results highlight the positive impact of these technologies on sustainable farming practices and marketing, with sustainable practices playing a mediating role. The integrated approach emphasizes the importance of combining high-tech solutions with agroecological practices to realize full benefits and meet emerging consumer demands for responsible produce. Recommendations are provided for policy interventions to promote synergistic development, ultimately enhancing productivity, sustainability, and market access for smallholder farmers in Zimbabwe.

Preface

In an era defined by the urgent imperative for sustainable solutions, the convergence of emerging technologies and innovative marketing strategies emerges as a beacon of hope for the future of agriculture. As global stakeholders rally to achieve the ambitious targets outlined by the United Nations Sustainable Development Goals (SDGs), the pivotal role of sustainable agriculture in fostering resilience, equity, and prosperity has never been more pronounced.

With this in mind, it is with great pleasure that we introduce *Emerging Technologies and Marketing Strategies for Sustainable Agriculture*, an edited reference book that represents a collaborative endeavor to explore the multifaceted landscape of sustainable agricultural practices and their implications for global development. Authored by esteemed researchers and scholars from diverse academic and professional backgrounds, each chapter within this volume offers unique insights into the intricate relationship between technological innovation, marketing dynamics, and agricultural sustainability.

Chapter 1, "Sustainable Agriculture and Sustainable Development Goals (SDGs)," penned by Anjana J. Atapattu and esteemed colleagues from the Coconut Research Institute of Sri Lanka, casts a spotlight on the transformative potential of sustainable agriculture in advancing specific SDGs. Through a meticulous synthesis of research and innovative case studies, the authors underscore the paramount importance of evidence-based approaches that harmonize productivity with environmental stewardship.

Following suit, Chapter 2, "Climate Smart Agriculture and Resilience," authored by Muhammad Safdar and collaborative partners from Pakistan and the USA, delves into the imperative of climate-smart agriculture in mitigating the adverse impacts of climate change on global food security. Advocating for sustainable farming practices and requisite institutional support, the authors chart a course towards enhancing agricultural productivity while concurrently curtailing greenhouse gas emissions.

The subsequent chapters traverse a diverse array of topics, ranging from risk management strategies in Indian agriculture to the nuanced role of sustainable marketing in elevating brand evaluation and consumer perception. With contributions spanning stakeholder perspectives on African agricultural marketing to precision irrigation techniques in Pakistan, this compendium offers a comprehensive examination of the challenges and opportunities facing sustainable agriculture on a global scale.

As editors of this volume, we extend our profound gratitude to the contributors whose unwavering dedication, expertise, and passion have enriched this collaborative endeavor. Their insights and research findings have not only deepened our understanding of the complex interplay between agriculture, technology, and marketing dynamics but have also illuminated pathways toward tangible progress.

Looking forward, we envision this book serving as a catalyst for dialogue, innovation, and actionable solutions in pursuit of a more sustainable and equitable agricultural future. By harnessing the transformative potential of emerging technologies and marketing strategies, we are poised to cultivate resilient food systems that nourish both people and the planet.

In conclusion, we extend our sincere appreciation to all those who have contributed to this book and to the broader community committed to advancing sustainability in agriculture. May the insights shared within these pages inspire meaningful change and catalyze collective efforts toward building a prosperous and harmonious world for generations to come.

Chapter 1: Sustainable Agriculture and Sustainable Development Goals (SDGs)

Authored by Anjana J. Atapattu, C. S. Ranasinghe, Tharindu Nuwarapaksha, Shashi Udumann, and Nuwandhya Dissanayaka from the Coconut Research Institute of Sri Lanka, this chapter delves into the critical role of sustainable agriculture in achieving the UN Sustainable Development Goals (SDGs) by 2030. By synthesizing research on sustainable techniques such as conservation agriculture, integrated farming systems, and organic agriculture, the authors provide an overview of how sustainable agriculture practices directly influence specific SDGs, including zero hunger, climate action, and responsible production and consumption. Through innovative case studies, they demonstrate how principles of regeneration and agroecology can drive progress on interconnected SDGs, including poverty alleviation, clean water access, gender equality, and reduced inequalities. The chapter emphasizes the importance of evidence-based, context-specific approaches that balance productivity, resilience, and environmental protection across local and global scales.

Chapter 2: Climate Smart Agriculture and Resilience

Written by Muhammad Safdar, Muhammad Adnan Shahid, Ce Yang, Fahd Rasul, Muhammad Tahir, Aamir Raza, and Rehan Mehmood Sabir from the University of Agriculture, Faisalabad, Pakistan, and the University of Minnesota, USA, this chapter explores the challenges posed by climate change to global agriculture and the necessity of adopting sustainable practices to ensure food security and mitigate environmental impacts. The authors discuss climate-smart agriculture (CSA) techniques such as crop management, sustainable farming, agroforestry, and precision farming, emphasizing their role in enhancing agricultural productivity, adaptive capacity, and greenhouse gas emissions reduction. They highlight the importance of encouraging laws, institutional frameworks, and financial incentives to facilitate widespread adoption of CSA practices while calling for further research and collaboration in this area.

Chapter 3: Sustainable Agriculture With Special Emphasis on Risk Management

Authored by Vidya Kumbhar, Sneha Kumari, Pratyush Kumar, Kailash Parihar, and Samaya Pillai from various institutes affiliated with Symbiosis International University, India, this chapter focuses on the challenges faced by Indian farmers, particularly in the context of climate change and its adverse effects on agricultural productivity. The authors examine the diverse risks encountered by Indian farmers and explore the risk management strategies adopted to mitigate these challenges. The chapter underscores the importance of understanding the hazards farmers face and the need for effective risk management techniques backed by both primary and secondary data.

Chapter 4: A Policy Framework for Sustainable Marketing Practices and Brand Evaluation Through Instagram Marketing

Penned by Syed Jalal, Iqra Arshad, Shahid Khalil, and Muhammad Khalil from various academic institutions, including Middle East College, Oman, and National Institute of Development Administration (NIDA), Thailand, this chapter delves into the concept of sustainable marketing and its impact on brand evaluation. The authors explore how sustainability marketing practices influence consumer perceptions and brand evaluations, with a specific focus on the mediating role of fashion consciousness. Drawing on data collected from 336 respondents and analyzed using SmartPLS 4.0, the chapter highlights the significant impact of sustainable marketing practices on brand evaluation while examining the nuanced role of fashion consciousness in this relationship.

Chapter 5: The Future of Sustainable Marketing: Stakeholder Perspectives on Sustainable Agricultural Marketing Practices

Authored by Sinothando Tshuma, Mercy Dube, Delight Hungwe, and Reason Masengu from various academic institutions, including Zimbabwe Open University and Midlands State University, this chapter addresses the critical need for sustainable agricultural marketing practices in African countries amidst escalating environmental degradation. The authors investigate various stakeholders' perspectives, including government, suppliers, shareholders, and customers, on embracing green agricultural products for sustainable development. Informed by Stakeholder theory, the study examines the trends in sustainable agricultural marketing and the legislative frameworks necessary to guide business operations effectively.

Chapter 6: Empowering Smallholder Farmers Through Community-Based Marketing Initiatives in Promoting Sustainable Agriculture

Crafted by Tendai Kaponda and Option Chiwaridzo from Central China Normal University and University of Science and Technology Beijing, China, respectively, this chapter focuses on Zimbabwe's agricultural sector's challenges and the potential of technology and sustainable marketing to enhance farmer-consumer linkages. Through mixed methods research, including surveys and interviews, the authors aim to understand the current situation, benefits, and challenges of technology adoption and sustainable marketing in Zimbabwe's agricultural value chain. The findings from this study will inform recommendations for enhancing transparency, equity, and sustainability in the agricultural sector.

Chapter 7: Shopping Therapy: How Social, Adventurous, Idea, and Value Shopping Behaviors Stimulate Impulse Buying Across Individuals in Oman

Penned by Zulfiqar Ali Jumani, Atheer Rashid Amur Al Salmani, and Muhammad Khalil from Khon Kaen University, Thailand, and Middle East College, Oman, this chapter explores the factors influencing impulse buying in the Omani market. The authors investigate how social, adventurous, idea, and value shopping behaviors motivate consumers to engage in impulse buying online. Through data collected from 210 online buyers in Muscat, Oman, and analyzed using structural equation modeling, the chapter highlights the positive relationship between adventure shopping, value shopping, and impulse buying, shedding light on consumer behavior in the digital marketplace.

Chapter 8: Public Private Partnership (PPP) in Agricultural Research Financing in Nigeria

Authored by Ibukunoluwa Akinrinde, Opeyemi Aromolaran, Sodiq Arogundade, and Peripaul Onuoha from various academic institutions, including North West University, South Africa, and University of Johannesburg, South Africa, this chapter underscores the importance of collaborative public-private partnerships (PPP) in agricultural research financing. The authors delve into Nigeria's comparative advantage in agriculture and explore strategies to transform this advantage into competitive advantage through innovative financing models. Drawing on data from the United Nations Global Compact (UNGC) members' financial reports, the chapter proposes a Triple Bottom-Line Approach to agricultural research financing, emphasizing the need for private sector involvement and social responsibility.

Chapter 9: Data-Driven Precision Agriculture for Crop Prediction and Fertilizer Recommendation Using Machine Learning

Crafted by Yashi Tiwari, Ayush Verma, and Manju Khari from Banasthali Vidyapith and Jawaharlal Nehru University, India, respectively, this chapter addresses the challenges of crop prediction and fertilizer recommendation in agriculture and the potential of data-driven precision agriculture to address these challenges. Through a comprehensive system leveraging Machine Learning (ML) algorithms, the authors propose a solution for predicting crops and recommending suitable fertilizers based on historical data. Their research findings highlight the accuracy of ML algorithms such as Naive Bayes and Random Forest in soil classification and underscore the potential of ML techniques to aid farmers in making informed decisions about soil management and fertilizer selection.

Chapter 10: Precision Irrigation for Sustainable Agricultural Productivity

Penned by Moazzam Mushtaq, Hasnain Ali, Aamir Raza, Sheraz Maqbool, Muhammad Safdar, Mubashir Ahmed, and Jaffar Sattar from the University of Agriculture Faisalabad, Pakistan, and Khwaja Fareed University of Engineering and Information Technology, Pakistan, this chapter provides a detailed overview of precision irrigation as a sustainable water management strategy in agriculture. The authors discuss the benefits, types, components, design, implementation, and management of precision irrigation systems, highlighting their role in enhancing crop yields, reducing water use, improving fertilizer efficiency, and minimizing environmental impact. The chapter also addresses the challenges and opportunities for adoption of precision irrigation and its future in sustainable agriculture.

Chapter 11: Fabrication of Equipment and IoT-Assisted Monitoring for Enhanced Zero-Till Farming

Authored by Rajesh P. V., Ashvin J, Harish M., Kishore G., and Nidheesh S. from Saranathan College of Engineering, India, this chapter focuses on enhancing Zero-Till Farming practices through the development of specialized equipment and Internet of Things (IoT) based monitoring systems. The authors highlight the significance of Zero-Till Farming in promoting sustainable agricultural practices and addressing global food security challenges. Through the fabrication of Non-Tilling seed Sower with solid fertilizer sprayer and IoT-based monitoring capabilities, the chapter contributes to advancing precision agriculture and promoting sustainable food production systems.

Chapter 12: Review of Terrace Gardening Literature and Suggestions for Increasing Adoption

Authored by Chitra Dey, V A Monica, Amit Karanth Gurpur, Swetha Venugopal, Lavanya M, Sai Choornika Chandana Samudrala, Aakash Raja A, and Aishwar Haris J from various academic institutions, including College of Engineering Guindy Anna University and Vellore Institute of Technology Chennai, India, this chapter explores the concept of terrace gardening (TG) as a sustainable and creative venture in urban settings. The authors review relevant literature on TG, focusing on motivations for participation, advantages, sustainability, innovation, operations, stakeholder management, and limitations. Through a SWOC analysis, they identify the potential of TG for business owners seeking to have a positive environmental impact while meeting the rising demand for fresh, locally grown food.

Chapter 13: Enhancing Farmer-Consumer Linkages Through Technology Adoption and Sustainable Marketing in Zimbabwe's Agricultural Sector

Crafted by Option Chiwaridzo, Tendai Kaponda, and Reason Masengu from various academic institutions, including University of Science and Technology Beijing and Middle East College, Oman, this chapter examines the relationships between technology adoption, sustainable marketing, and bio-genetic technologies among smallholder farmers in Zimbabwe. Through a quantitative study involving 402 farmers, the authors explore the impact of innovative technologies on sustainable farming practices and marketing, emphasizing the importance of integrating high-tech solutions with agroecological practices to meet emerging consumer demands for responsible produce. The chapter provides recommendations for policy interventions aimed at promoting synergistic development and enhancing productivity, sustainability, and market access for smallholder farmers in Zimbabwe.

In compiling this edited reference book, we aimed to provide a comprehensive overview of sustainable agricultural practices and their implications for achieving global development goals. Each chapter offers unique insights into various aspects of sustainable agriculture, ranging from climate-smart techniques to precision irrigation and innovative marketing strategies. We hope that this book serves as a valuable resource for researchers, policymakers, practitioners, and stakeholders interested in advancing sustainable agriculture and contributing to global food security, environmental conservation, and economic development.

In concluding our edited reference book, we reflect on the profound significance of sustainable agricultural practices and their intersection with effective marketing strategies. Throughout this comprehensive exploration, we have witnessed the intricate web of relationships between agricultural sustainability, market dynamics, and global development goals.

Our journey through the chapters of this book has illuminated the urgent need for transformative action in agriculture to address the myriad challenges facing our food systems. From the imperative of climate-smart agriculture to the crucial role of precision irrigation, each chapter has underscored the vital importance of adopting sustainable practices to ensure the resilience and longevity of our agricultural systems.

Moreover, our examination of sustainable marketing strategies has revealed the pivotal role of consumer behavior, stakeholder engagement, and policy frameworks in driving the adoption of sustainable agricultural practices. By highlighting the interplay between marketing dynamics and agricultural sustainability, we have uncovered opportunities for collaboration, innovation, and collective action to advance our shared goals of environmental stewardship and food security.

As editors, we are immensely grateful to the contributors who have lent their expertise, insights, and passion to this collaborative endeavor. Their dedication to advancing knowledge and driving positive change in agriculture and marketing has been instrumental in shaping this book into a valuable resource for researchers, policymakers, practitioners, and students alike.

Looking ahead, we envision this book serving as a catalyst for continued dialogue, innovation, and action in the pursuit of sustainability in agriculture and marketing. By harnessing the lessons learned and the insights gained from these pages, we can chart a course towards a more resilient, equitable, and sustainable future for our food systems and our planet.

In closing, we extend our heartfelt appreciation to all those who have contributed to this book and to the broader community committed to advancing sustainability in agriculture and marketing. May our collective efforts inspire meaningful change and usher in a new era of prosperity and harmony for people and planet alike.

Jabulani Garwi
University of the Free State, South Africa

Reason Masengu
Middle East College, Oman

Option Takunda Chiwaridzo
University of Science and Technology, Beijing, China

Chapter 1
Sustainable Agriculture and Sustainable Development Goals (SDGs)

Anjana J. Atapattu
https://orcid.org/0000-0003-4806-4647
Coconut Research Institute of Sri Lanka, Sri Lanka

C. S. Ranasinghe
Coconut Research Institute of Sri Lanka, Sri Lanka

Tharindu D. Nuwarapaksha
Coconut Research Institute of Sri Lanka, Sri Lanka

Shashi S. Udumann
https://orcid.org/0000-0002-5717-7419
Coconut Research Institute of Sri Lanka, Sri Lanka

Nuwandhya S. Dissanayaka
https://orcid.org/0000-0002-2062-1133
Coconut Research Institute of Sri Lanka, Sri Lanka

ABSTRACT

Sustainable agriculture plays a crucial role in achieving the United Nations Sustainable Development Goals (SDGs) by 2030. This chapter analyzes the linkages between sustainable agriculture practices and specific SDGs, examining the potential of eco-friendly farming techniques to address socio-economic and environmental issues. It provides an overview of prominent goals directly influenced by agriculture like zero hunger, climate action, and responsible production and consumption. Additionally, it highlights innovative case studies that demonstrate how principles of regeneration and agroecology can drive progress on interlinked SDGs spanning poverty alleviation, clean water access, gender equality, and reduced inequalities. By synthesizing research on sustainable techniques like conservation agriculture, integrated farming systems, and organic agriculture, this chapter offers actionable recommendations for agricultural stakeholders to align with the 2030 agenda for sustainable development. It emphasizes the need for evidence-based, context-specific approaches that balance productivity, resilience, and environmental protection across local and global scales.

DOI: 10.4018/979-8-3693-4864-2.ch001

INTRODUCTION

Sustainable agriculture has become an increasingly important concept in recent decades as the world grapples with the challenges of feeding a growing population while protecting the environment (Sarker et al., 2019). The concept of sustainable agriculture encompasses a range of practices and goals that aim to meet current and future societal needs for food and other agricultural products, while ensuring profitable farm incomes and supporting the sustainable use of natural resources. As illustrated in Figure 1, the three pillars of sustainability in agriculture include economic, social and environmental sustainability (Fischer et al., 2020). Achieving truly sustainable agriculture requires taking a holistic systems approach that integrates economic profitability, social equity and environmental health. It aligns agricultural production with long-term economic social and environmental sustainability. Sustainable agriculture considers the health of the entire agro-ecosystem including the people who live and work on farms. Economic sustainability in agriculture means that farms are financially viable and contribute to robust local and regional economies (Berti & Mulligan, 2016). Farmers need access to fair markets, reasonable profits and wages, financial planning tools, and support during hard times (Valkila, 2009). Value can be added through processing, marketing, agritourism and farm-to-table initiatives. Infrastructure like storage facilities, roads and broadband internet enable market access.

Social sustainability in agriculture promotes justice, equity and quality of life for all. Safe and ethical working conditions, fair wages, farmer autonomy, land tenure rights, access to education and community development are important social sustainability goals (Petrescu-Mag et al., 2019). Empowering women, indigenous groups, youth and other marginalized populations contributes to social sustainability in farming systems around the world. Environmental sustainability in agriculture involves maintaining natural resources and ecosystem services like soil fertility, clean water, pollinators, and biological pest control (Power, 2010). Practices like crop rotation, cover cropping, reduced tillage, integrated pest management, and habitat conservation support environmental sustainability (Redlich et al., 2021). Farmers aim to minimize pollution, preserve biodiversity, improve resilience to climate change, and reduce dependence on non-renewable inputs. Technologies like precision agriculture, drought-tolerant crops, and renewable energy help reduce agriculture's environmental footprint (Zhao et al., 2023). The united nation SDGs, adopted in 2015, expanded upon this concept of sustainable development by providing 17 goals and 169 targets to guide inclusive social and economic progress while protecting the environment (Halkos & Gkampoura, 2021; Leal Filho et al., 2019). Although not one of the 17 official SDGs, sustainable agriculture is considered an important component for achieving many of the goals. Countries around the world have committed to achieving the SDGs by 2030 through partnerships, policy and investment. Advancing sustainable agriculture is key to accomplishing these ambitious global sustainability targets.

This chapter aims to discuss how sustainable agriculture promotes SDGs and examines the challenges and recommendations to achieving SDGs globally. Agricultural productivity must substantially increase to provide for population and income growth, urbanization, and changing diets (Fróna et al., 2019). This must be achieved while adapting to climate change impacts and reducing the environmental footprint of agriculture. A transformation to sustainable agriculture will depend on system-level changes that integrate socioeconomic and environmental considerations across food systems (Wezel et al., 2020). This includes sustainable intensification of production and closing yield gaps through ecologically-based practices, use of well-adapted crop varieties and livestock breeds, and efficient water management. Diversified agricultural systems, such as integrated systems and agroforestry, can enhance productivity (Shennan et al., 2017). Loss and waste of food must be minimized through investments in storage, transport, pro-

Figure 1. Three pillars of sustainability for sustainable agriculture
Source: Drawn up by author with Reference to Rodriguez et al., 2002.

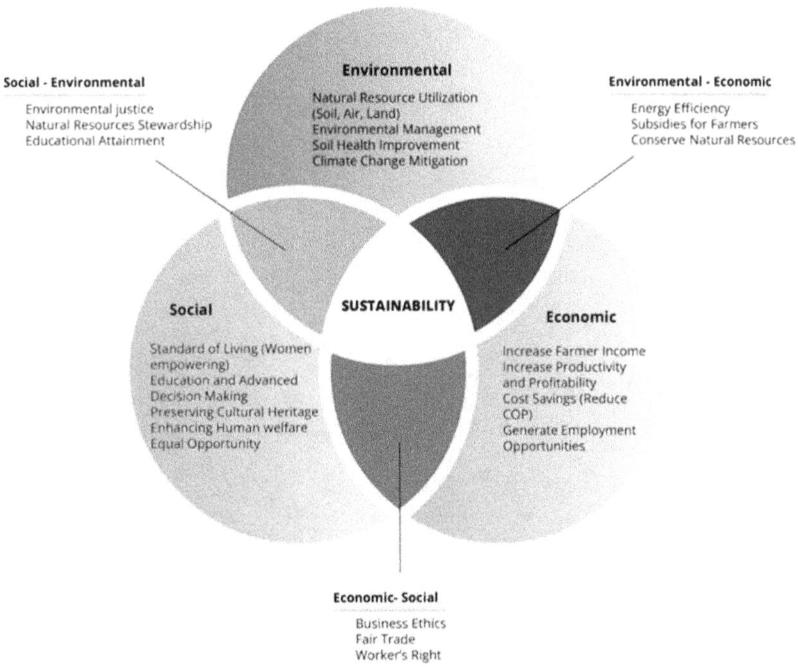

cessing, and distribution infrastructure. Dietary changes towards more plant-based diets in high-meat consuming regions can reduce pressure on land and lower greenhouse gas emissions from agriculture (Joe Millward & Garnett, 2010). Connecting farmers to value chains and economies of scale can provide pathways out of poverty while meeting food security and environmental goals (Tomich et al., 2019). However, based on above things this chapter provides a way to achieve SDGs while protecting the three pillars of sustainability through sustainable agriculture.

SUSTAINABLE AGRICULTURE TO ACHIEVING KEY SDGS

Sustainable agriculture plays a transformative role across a spectrum of SDGs, contributing to economic, social, and environmental dimensions (Figure 2). In the economic realm, it addresses SDG 1 (No Poverty) by enhancing the economic resilience of farming communities, offering income-generating opportunities through diversified and sustainable farming practices (Rahman, 2022). Goal 2 (Zero Hunger) is pursued as sustainable agriculture ensures food security by promoting resilient and diverse crop cultivation, thereby mitigating hunger and malnutrition (Arora & Mishra, 2022). Socially, sustainable agriculture supports Goal 3 (Good Health and Well-being) by reducing reliance on harmful pesticides, fostering healthier ecosystems, and improving food quality, thereby positively impacting community health (McElwee et al., 2020). It also contributes to Goal 5 (Gender Equality) by empowering women in agriculture, ensuring their equal participation, and promoting gender-inclusive farming practices. Additionally, in advancing SDG 10 (Reduced Inequality), sustainable agriculture advocates for fair trade practices and equitable

Figure 2. Visual concept of key SDGs achieving by sustainable agriculture
Source: Drawn up by author

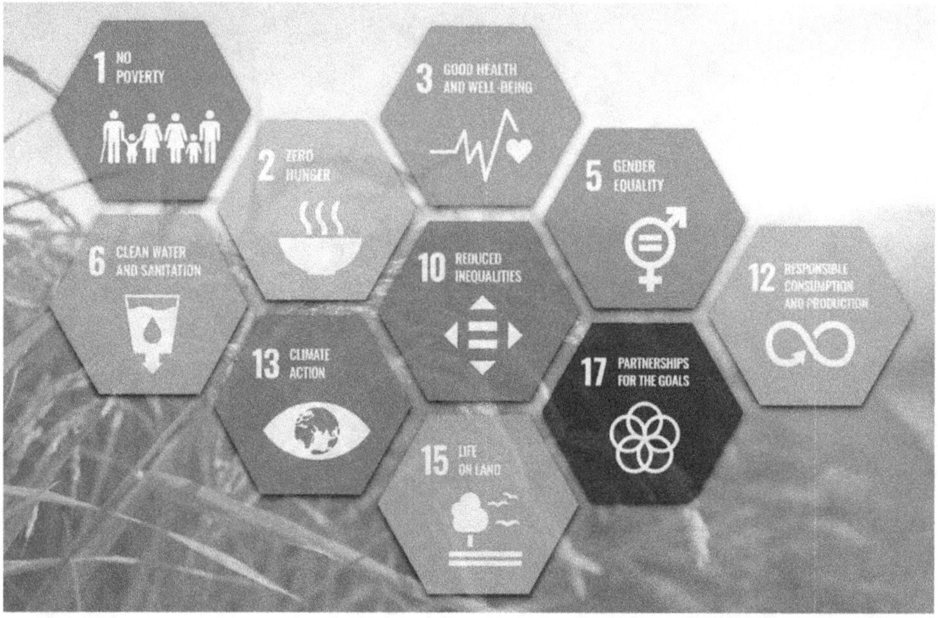

distribution of resources, addressing disparities among farming communities (Sexsmith & McMichael, 2015). Enabling responsible consumption and production patterns (SDG 12), sustainable agriculture minimizes waste, promotes efficient resource utilization, and advocates for environmentally conscious farming practices (Arora & Mishra, 2023).

In the environmental domain, sustainable agriculture supports Goal 6 (Clean Water and Sanitation) by endorsing water-efficient practices and minimizing agricultural runoff, ensuring sustainable water management (Mujtaba et al., 2024). Goal 13 (Climate Action) is addressed through climate-resilient agricultural practices, such as agroforestry and conservation agriculture, mitigating the sector's impact on climate change and fostering climate adaptation (Campbell et al., 2018). In promoting sustainable land use SDG 15 (Life on Land), agriculture becomes a vital ally in halting biodiversity loss, preserving ecosystems, and ensuring the long-term health of terrestrial environments (Octavia et al., 2022). Finally, sustainable agriculture aligns with SDG 17 (Partnerships for the Goals) by emphasizing collaboration between governments, private sectors, and local communities to enhance the scalability and impact of sustainable agricultural practices globally (Mariani et al., 2022). In essence, sustainable agriculture emerges as a cornerstone for a holistic approach to achieving these interconnected SDGs, integrating economic prosperity, social equity, and environmental stewardship within the fabric of global development.

Economically Sustainable Agriculture for SGDs

SDG 1 (No Poverty)

The implementation of sustainable agriculture practices is crucial in contributing to the achievement of the targets set forth in SDG 1, which seeks to eliminate poverty in all its manifestations (Ladha et al., 2020). Encouraging sustainable and inclusive agricultural systems has the potential to elevate rural livelihoods, bolster food security, and play a role in reducing poverty. One key aspect of sustainable agriculture in the context of SDG 1 is the promotion of agroecological practices (Feliciano, 2019). Agroecology underscores the incorporation of ecological principles into agricultural systems, aiming to boost productivity while minimizing adverse environmental effects. Techniques like diversified cropping systems, agroforestry, and integrated pest management actively contribute to fostering sustainable and resilient farming practices (Altieri et al., 2015). Adopting this approach not only enhances crop yields but also diminishes reliance on external inputs, rendering it economically viable for farmers. Empowering smallholder farmers is crucial for achieving SDG 1. Sustainable agriculture practices that focus on capacity building, education, and access to resources help enhance the skills and knowledge of farmers (Siebrecht, 2020). This includes training in climate-smart farming techniques, efficient water management, and soil conservation. By equipping farmers with the tools to adapt to changing conditions and improve productivity, sustainable agriculture contributes to poverty alleviation and improved rural livelihoods.

In the realm of sustainable agriculture, incorporating crop diversification stands out as a key strategy in working towards the objectives of SDG 1. The cultivation of a diverse array of crops allows farmers to fortify their resilience against the impacts of climate change, mitigate the risk of crop failure, and enhance overall income stability (Lasco et al., 2014). Furthermore, diversified cropping systems enhance nutrition and food security by offering a variety of crops with distinct nutritional profiles. Market access is crucial for reducing poverty, and sustainable agricultural practices contribute to this dimension of SDG 1 (Feliciano, 2019). Developing sustainable value chains and promoting fair trade practices enable smallholder farmers to access markets on more equitable terms. By connecting farmers with consumers and facilitating fair compensation for their produce, sustainable agriculture contributes to the economic well-being of rural communities. Water management is a critical component of sustainable agriculture practices aligned with SDG 1(Tsani et al., 2020). Smallholder farmers, frequently located in regions facing water scarcity, gain advantages from embracing water-efficient irrigation techniques, such as drip irrigation and rainwater harvesting. These methods not only preserve water resources but also enhance agricultural productivity, thereby improving the economic well-being of farming communities.

Gender equality is an integral part of SDG 1, and sustainable agriculture practices can play a transformative role in empowering women in rural areas (Glazebrook & Opoku, 2020). By providing equal access to resources, education, and decision-making opportunities, sustainable agriculture contributes to breaking the cycle of poverty and fostering inclusive development. Moreover, the promotion of sustainable livelihoods beyond traditional agriculture is essential for achieving SDG 1. Agro-tourism, adding value to agricultural products, and establishing non-farm income-generating activities contribute to economic diversification in rural areas (Pingali, 2010). These initiatives create additional avenues for income, reduce vulnerability to external shocks, and foster overall economic development (Williams & Vorley, 2014). Furthermore, sustainable agriculture practices are instrumental in achieving SDG 1 by addressing poverty in rural areas. Through promoting sustainable practices, we can strive to eliminate poverty and construct a more equitable and prosperous future for all.

SDG 2 (Zero Hunger)

Sustainable agriculture promotes SDG 2 targeted to end hunger, achieve food security and improved nutrition (Lartey, 2015). This objective acknowledges the imperative to tackle worldwide issues such as the increasing population, climate change, and environmental decline, all while guaranteeing universal access to an ample and nourishing food supply (Fanzo, 2019). Establishing sustainable agricultural practices is crucial for realizing SDG 2 and promoting a robust and fair food system. Within the framework of SDG 2, agroecology underscores the incorporation of ecological principles into agricultural systems, taking into account the intricate interconnections among crops, livestock, and the environment (Blesh et al., 2019). This approach prioritizes biodiversity, soil health, and efficient resource use, contributing to enhanced resilience and productivity. Agroecological practices include diversified crop rotations, agroforestry, and the use of legume cover crops, which increase nitrogen fixation, reduce soil erosion, and mitigate the impact of pests and diseases without relying on harmful chemical inputs (Dissanayaka et al., 2024; Nuwarapaksha et al., 2023a; Udumann et al., 2023).

Another pivotal aspect involves the adoption of conservation agriculture. This approach concentrates on minimizing soil disturbance, sustaining permanent soil cover, and diversifying crop rotations. Conservation agriculture enhances water retention, mitigates soil erosion, and enhances soil structure, thereby rendering agricultural systems more resilient to climate variability (Page et al., 2020). By giving precedence to soil health, farmers can achieve sustainable increases in yields and secure the long-term viability of their land for future generations. Integrated pest management stands as another sustainable agriculture strategy that aligns with SDG 2 (Rossi et al., 2023). It involves using a combination of biological, cultural, and mechanical control methods to manage pests while minimizing the use of chemical pesticides. This approach safeguards biodiversity, protects beneficial insects, and reduces the environmental impact associated with conventional pest control methods. The implementation of integrated pest management not only contributes to sustainable food production but also safeguards ecosystems and human health (Hillocks & Cooper, 2012).

Furthermore, advocating for sustainable water management practices is essential for accomplishing SDG 2. Agriculture, being a significant consumer of water resources, faces the risk of depletion and degradation of water sources through unsustainable water use. Incorporating water-efficient irrigation systems, adopting rainwater harvesting techniques, and enhancing water-use efficiency in agriculture are crucial components of sustainable water management (Nazari et al., 2018). These practices help ensure the availability of water for agriculture while minimizing negative environmental impacts. Furthermore, fostering inclusive and resilient food systems is vital for achieving SDG 2. This involves supporting smallholder farmers, particularly women, and promoting equitable access to resources and markets (Batchelor, 1999). Strengthening the empowerment of local communities and bolstering their adaptive capacity to climate change are pivotal contributions to constructing resilient food systems. These resilient systems can withstand shocks and guarantee food security for all. However, sustainable agriculture practices are integral to achieving SDG 2 by addressing the complex challenges of hunger, food security, and nutrition (Viana et al., 2022). Agroecology, conservation agriculture, integrated pest management, sustainable water management, and inclusive food systems all play crucial roles in building a resilient and sustainable future for global agriculture.

Socially Sustainable Agriculture for SGDs

SDG 3 (Good Health and Well-Being)

While SDG 3 primarily focuses on ensuring healthy lives and promoting well-being for all at all ages, sustainable agriculture practices play a significant role in achieving this goal by contributing to food security, nutrition, and public health (Lal et al., 2021). Through the adoption and promotion of sustainable agricultural methods, it becomes feasible to tackle a myriad of health-related challenges, thereby contributing to the cultivation of a healthier and more resilient global population. One essential sustainable agriculture practice for achieving SDG 3 is promoting agroecology. Agroecological approaches prioritize the integration of ecological principles into farming systems, emphasizing sustainable and environmentally friendly practices (Wezel et al., 2020). Diversified cropping systems, agroforestry, and integrated pest management are key components of agroecology. By reducing the reliance on synthetic inputs and promoting biodiversity, agroecological practices contribute to producing healthier and more nutritious food, addressing malnutrition and promoting overall well-being (Kahane et al., 2013). Crop diversification is crucial for achieving SDG 3, as it directly impacts nutrition and dietary diversity. Monoculture farming, which focuses on cultivating a single crop, can lead to not properly utilizing sunlight and land (Nuwarapaksha et al., 2022).

Sustainable agriculture practices that encourage the cultivation of a variety of crops contribute to more balanced and nutrient-rich diets. This, in turn, helps prevent malnutrition and related health problems, aligning with the goal of ensuring healthy lives. Promoting sustainable livestock practices is another essential component of achieving SDG 3(Grosso et al., 2020). Livestock farming, if not managed sustainably, can contribute to zoonotic diseases and environmental degradation. Sustainable livestock management involves practices such as rotational grazing, responsible antibiotic use, and improving animal welfare standards. These practices not only reduce the risk of disease transmission but also contribute to producing healthier and safer food products. Incorporating livestock into the underutilized pastures within lands has the potential to enhance food sources and generate additional income for the farmer, utilizing the existing resources (Nuwarapaksha et al., 2023b). Furthermore, sustainable agriculture practices contribute to mitigating air pollution, another factor impacting public health. Crop residues burning and certain agricultural practices can release pollutants into the air. Sustainable agriculture methods, such as conservation tillage and cover cropping, reduce the need for such practices, mitigating air pollution and promoting respiratory health (Pokharel et al., 2023). Education and awareness initiatives focusing on sustainable agriculture practices play a vital role in accomplishing SDG 3. It is imperative to inform farmers, communities, and consumers about the advantages of sustainable farming methods for human health (Petway et al., 2019). This includes understanding the nutritional advantages of diverse crops, the risks associated with chemical inputs, and the broader impact of sustainable agriculture on public health.

SDG 5 (Gender Equality)

Sustainable agriculture practices are essential for achieving SDG 5, which focuses on achieving gender equality and empowering all women and girls. Addressing gender disparities in various facets of agriculture is crucial for promoting sustainable development (Agarwal, 2018). Implementing sustainable agriculture practices that specifically consider and benefit women in the agricultural sector can contribute significantly to achieving SDG 5. One key aspect of sustainable agriculture aligned with SDG 5 is promoting gender-inclusive access to resources. Women constitute a significant portion of the agricultural workforce, yet they often face barriers in accessing essential resources such as land, credit, and technology (Friedemann-Sánchez, 2006). Sustainable agriculture practices should prioritize breaking down these barriers and ensuring that women have equal access to productive resources. This includes implementing policies and programs that recognize and secure women's land rights, providing financial services tailored to their needs, and promoting the use of gender-responsive agricultural technologies.

Empowering women in decision-making processes within the agricultural sector is another crucial element for achieving SDG 5 (Shahbaz et al., 2022). Sustainable agriculture practices should actively involve women in planning, implementing, and evaluating agricultural initiatives. This empowerment can be facilitated through training programs, capacity-building initiatives, and the creation of inclusive platforms that allow women to voice their concerns and contribute to shaping agricultural policies (Quisumbing et al., 2023). Recognizing and valuing the diverse roles women play in agriculture, from farm management to marketing, is essential for fostering a more equitable and inclusive sector. Promoting sustainable and gender-sensitive agricultural practices is vital for enhancing the well-being of women and girls in rural communities (Onyimadu, 2023). For instance, agroecological approaches that prioritize biodiversity, soil health, and sustainable resource use can benefit women by providing diverse and nutritious food sources. Sustainable practices, especially agroforestry, not only contribute to improved food security but also offer income-generating opportunities for women, strengthening their economic independence (Dissanayaka et al., 2023a).

Additionally, integrating women into the value chains of agricultural products is essential for achieving SDG 5. Sustainable agriculture practices should aim to create equitable market opportunities for women farmers and entrepreneurs (Agarwal, 2018). This involves addressing challenges such as market access, fair pricing, and providing support for women to engage in value-added activities like processing and marketing. Creating inclusive business models and supply chains that recognize and value the contributions of women can significantly enhance their economic empowerment within the agricultural sector (Kini, 2022). Furthermore, focusing on the intersectionality of gender with other factors such as age, ethnicity, and socioeconomic status is crucial for achieving comprehensive gender equality in agriculture (Tran et al., 2023). Sustainable agriculture practices should be sensitive to the diverse needs and experiences of women, considering the unique challenges faced by different groups. This intersectional approach ensures that interventions are tailored to address the specific barriers that women from various backgrounds may encounter in the agricultural sector.

SDG 10 (Reduced Inequalities)

The implementation of sustainable agricultural practices is essential for realizing the goals outlined in SDG 10, which seeks to decrease inequalities within and among countries. By promoting inclusive and sustainable agricultural systems, it is possible to address disparities in access to resources, income, and opportunities, fostering a more equitable and just society (Bacon et al., 2012). A pivotal sustainable agriculture practice for advancing SDG 10 involves the promotion of agroecology. Agroecological approaches center on utilizing ecological principles to boost agricultural productivity while minimizing adverse environmental and social effects. Through the adoption of agroecological practices, such as diversified cropping systems, agroforestry, and integrated farming systems, farmers can enhance yields, decrease dependence on external inputs, and foster resilience. This contributes to narrowing the gap in agricultural productivity and income among different communities (Selvan et al., 2023).

Boosting the capacity of smallholder farmers is crucial for mitigating inequalities, a central focus of SDG 10. Sustainable agriculture practices that offer training, education, and access to resources empower small-scale farmers, enabling them to adopt innovative and climate-smart farming techniques (Antwi-Agyei et al., 2021). By equipping these farmers with the necessary knowledge and skills, they can improve their productivity and income, ultimately narrowing the gap between smallholder farmers and larger, more resource-endowed agricultural enterprises. Access to markets is another critical aspect of sustainable agriculture for achieving SDG 10 (Ladha et al., 2020). By developing sustainable value chains and promoting fair trade practices, smallholder farmers can access markets on more equitable terms. This helps address the inequalities in market access and ensures that farmers receive fair compensation for their produce. The establishment of direct links between producers and consumers can further contribute to reducing the income disparities within the agricultural sector.

Crop diversification is a sustainable agriculture practice that directly aligns with SDG 10 by addressing the inequalities in access to diverse and nutritious food. Encouraging the cultivation of a variety of crops contributes to improved dietary diversity, ensuring that communities have access to a range of essential nutrients (Rajendran et al., 2017). This practice not only supports food security but also promotes a more balanced and equitable distribution of agricultural resources. Gender equality is an integral component of SDG 10, and sustainable agriculture practices can contribute to bridging gender gaps in agriculture. Providing equal access to resources, education, and decision-making opportunities for women in agriculture can help reduce gender-based inequalities. Initiatives that empower women as key contributors to sustainable agriculture foster more inclusive and resilient agricultural systems (Adefare et al., 2024). Additionally, the development of sustainable livelihoods beyond traditional agriculture is essential for achieving SDG 10. Diversifying income-generating activities, promoting rural entrepreneurship, and investing in non-farm activities contribute to reducing economic disparities in rural areas. These initiatives create alternative sources of income, reducing vulnerability and fostering overall economic development.

SDG 12 (Responsible Consumption and Production)

Achieving SDG 12, focused on fostering sustainable consumption and production patterns, heavily relies on the implementation of sustainable agriculture practices. This goal addresses the need for responsible resource use, waste reduction, and environmental conservation. Adopting sustainable agriculture practices is essential to strike a balance between meeting the world's food demands and preserving the planet's ecosystems for future generations (Muhie, 2022). Certainly, one pivotal aspect of sustainable agriculture is agroecology, an approach that underscores the integration of ecological principles into agricultural systems. This involves comprehending and collaborating with natural processes to improve soil fertility, manage pests, and enhance overall ecosystem health. Conservation tillage stands out as another crucial sustainable agriculture practice. In contrast to traditional tillage methods that involve overturning the soil, conservation tillage aims to reduce erosion, preserve soil structure, and minimize carbon emissions (Hussain et al., 2021). Conservation tillage methods, such as no-till or minimum tillage, help retain soil moisture, reduce erosion, and enhance carbon sequestration (Busari et al., 2015). Such practices actively promote long-term soil health and help mitigate the environmental impact of agricultural activities. Additionally, precision farming technologies play a crucial role in sustainable agriculture. These technologies, which may include GPS-guided tractors and drones, empower farmers to optimize resource utilization by precisely managing inputs such as water, fertilizers, and pesticides (Balyan et al., 2024). By minimizing waste and improving efficiency, precision farming contributes to achieving SDG 12's targets related to resource efficiency and pollution reduction.

Water conservation is a significant focus within sustainable agriculture practices. Implementing efficient irrigation techniques, such as drip irrigation or rainwater harvesting, helps conserve water resources and reduce the environmental impact of agriculture. Additionally, adopting drought-resistant crop varieties and improving water use efficiency contribute to achieving SDG 12's objectives of sustainable water management (Kimaro, 2019). By combining biological control, crop rotation, and resistant crop varieties, farmers can control pests while preserving beneficial insects and avoiding the negative impacts of chemical pesticides on ecosystems and human health (Baker et al., 2020). Integrated pest management aligns with SDG 12's targets by promoting responsible chemical use and reducing the ecological footprint of agriculture. Promoting sustainable farming systems, such as organic farming, is also essential for achieving SDG 12 (Ladha et al., 2020). Organic farming avoids synthetic inputs, promotes biodiversity, and enhances soil fertility through natural processes. Certifying and supporting organic farmers can incentivize the adoption of environmentally friendly practices, contributing to sustainable consumption and production patterns.

Environmentally Sustainable Agriculture for SGDs

SDG 6 (Clean Water and Sanitation)

Implementing sustainable agricultural practices is essential for attaining SDG 6, which emphasizes securing universal access to clean water and sanitation management. Agriculture is a major consumer of water resources, and unsustainable practices can lead to water scarcity, pollution, and negative impacts on ecosystems (McLaughlin & Kinzelbach, 2015). Implementing sustainable agriculture practices that prioritize water conservation, efficient water use, and environmental protection is crucial for advancing SDG 6 objectives. One key aspect of sustainable agriculture aligned with SDG 6 is the adoption of water-efficient irrigation systems (Obaideen et al., 2022). Conventional irrigation techniques frequently lead to the inefficient use of water and the deterioration of soil quality. Sustainable alternatives, such as drip irrigation and precision farming, enhance water efficiency by directing water precisely to the crop root zone, thereby reducing losses due to evaporation and runoff (Evans & Sadler, 2008). These practices contribute to water conservation, increase crop yield, and reduce the overall environmental impact of agricultural water use. Conservation agriculture is another important practice for achieving SDG 6. This method entails minimizing soil disruption, preserving continual soil cover, and diversifying crop rotations. Conservation agriculture enhances water retention in the soil by promoting better soil structure and mitigating erosion. This not only aids in preserving water but also alleviates the adverse impacts of severe weather events, fostering resilience in agricultural systems to climate fluctuations. Advocating for agroforestry practices is essential for attaining sustainable agriculture and water management objectives. Agroforestry integrates trees and shrubs into agricultural environments, offering numerous advantages, including the regulation of water (Atapattu et al., 2017; Dissanayaka et al., 2023b). Tree roots play a crucial role in soil stabilization, erosion reduction, and increased water infiltration. The existence of trees further aids in microclimate regulation, lowering water evaporation from the soil and enhancing overall water-use efficiency in agriculture.

The implementation of cover cropping is another sustainable agriculture practice that aligns with SDG 6. Cover crops are strategically planted to cover the soil between main crops, preventing erosion, improving soil structure, and reducing water runoff. These crops also contribute to weed suppression and enhance biodiversity, creating a more resilient and sustainable agroecosystem. Additionally, cover crops can capture excess nutrients, preventing them from leaching into water bodies and causing pollution (Dissanayaka et al., 2022; Udumann et al., 2023). Integrated water management approaches are essential for achieving SDG 6 in the context of agriculture. This involves considering the entire water cycle in agricultural practices, from efficient irrigation to rainwater harvesting. Sustainable agriculture should prioritize the conservation and responsible use of water resources, ensuring that water is used judiciously, and agricultural activities do not contribute to water pollution (McLaughlin & Kinzelbach, 2015). This approach safeguards water quality, protects aquatic ecosystems, and supports the overall goal of sustainable water management. Furthermore, sustainable agriculture practices should address the potential negative impacts of agricultural activities on water quality. The responsible use of fertilizers and pesticides, as well as the adoption of agroecological approaches and organic agriculture integrated pest management, can minimize water pollution risks (Jacquet et al., 2022). By promoting practices that reduce the use of harmful chemicals, sustainable agriculture contributes to maintaining clean and safe water sources, aligning with the objectives of SDG 6.

SDG 13 (Climate Action)

Addressing the challenges presented by climate change is significantly facilitated by the affecting of sustainable agricultural practices, aligning with the goals of SDG 13. This goal emphasizes the need to take urgent action to combat climate change and its impacts, recognizing the profound threats it poses to ecosystems, food security, and overall global well-being (Caron et al., 2018). Implementing sustainable agriculture practices is essential for both adaptation to climate change and mitigation of its effects. One fundamental aspect of sustainable agriculture in the context of SDG 13 is the adoption of climate-smart farming techniques (Qureshi et al., 2022). The objective of these practices is to bolster resilience and productivity amidst evolving climate conditions. Agroecological methods, including diversified cropping systems and integrated farming systems, play a role in climate resilience by advocating biodiversity and diminishing the susceptibility of agricultural systems. These approaches not only enhance crop yields but also nurture ecosystem health, providing sustained support for climate adaptation. Cover cropping and agroforestry stand out as pivotal strategies within sustainable agriculture for realizing SDG 13. Cover crops shield the soil against erosion, enhance water retention, and improve soil fertility (Dissanayaka et al., 2022; Nuwarapaksha et al., 2023a). These practices contribute to carbon sequestration, mitigating the impacts of greenhouse gas emissions. Agroforestry, which involves the integration of trees into agricultural landscapes, provides multiple benefits such as carbon sequestration, improved water management, and enhanced biodiversity, making it a powerful tool for climate change mitigation and adaptation (Atapattu et al., 2017; Raveendra et al., 2017). In addition to the aforementioned benefits, the cultivation of legume shrubs and trees in agroforestry systems can significantly enhance nitrogen fixation. Legumes possess a distinctive capability to establish symbiotic relationships with nitrogen-fixing bacteria known as rhizobia. This interaction takes place within the nodules on the roots of legumes, where the bacteria convert atmospheric nitrogen into a form accessible to plants for growth. Table 1 shows the nitrogen-fixing potential of leguminous trees and shrubs, predominantly thriving in agroforestry systems.

Livestock management is another area where sustainable agriculture practices can significantly impact SDG 13. Livestock farming is a notable contributor to greenhouse gas emissions, particularly methane (O'Mara, 2011). Practices such as rotational grazing, improved feeding regimes, and manure management contribute to reducing greenhouse gas emission emissions and enhancing the overall sustainability of livestock production (Nuwarapaksha et al., 2023c). Integrating agroecological principles into animal husbandry practices promotes biodiversity and ecosystem resilience. Furthermore, sustainable agricultural practices contribute to soil health and carbon sequestration, aligning with SDG 13 goals. Conservation tillage, cover cropping, and organic farming practices improve soil structure, reduce erosion, and enhance carbon storage in the soil. Healthy soils not only support increased agricultural productivity but also play

Table 1. Nitrogen fixing potential of legume trees and shrubs in agroforestry systems

Tree Species	N-Fixation (kg ha^{-1} $Year^{-1}$)
Leguminous trees	
Gliricidia sepium	13 – 50
Sesbania sesban	260
Leucaena leucocephala	100 – 500
Erythrina sp.	57 – 66
Acacia mearnsii	200
Albizia lebbeck	260
Faidherbia albida	20 – 50
Grevellea robusta	100 – 300
Inga edulis	35 – 40

Source: Drawn up by author with reference to Ram et al., 2017.

a crucial role in mitigating climate change by acting as carbon sinks (Rodrigues et al., 2023). Table 2 shows the soil carbon stocks in major agroforestry systems under an ecological-based approach within 50cm depth. Education and capacity-building initiatives are vital for the successful implementation of sustainable agriculture practices related to SDG 13 (Bloomfield et al., 2018). Farmers, policymakers, and communities need access to knowledge and resources that empower them to make informed decisions and implement sustainable practices. This includes promoting climate-smart agricultural techniques, supporting research and development of resilient crop varieties, and facilitating the adoption of sustainable land management practices.

SDG 15 (Life on Land)

Adopting sustainable agriculture practices is crucial for attaining the goals outlined in SDG 15. This particular Sustainable Development Goal centers on safeguarding, restoring, and encouraging the sustainable utilization of terrestrial ecosystems, with a primary emphasis on arresting biodiversity loss and combating desertification (Reith et al., 2021). Embracing environmentally responsible farming techniques presents an opportunity to actively contribute to the conservation of ecosystems, the preservation of biodiversity, and the prevention of land degradation. A pivotal sustainable agriculture practice aligned with SDG 15 is agroforestry, which integrates trees and shrubs into agricultural settings, offering a multitude of ecological advantages. Additionally, conservation tillage stands as another crucial practice in realizing SDG 15 objectives. Traditional tillage methods can result in soil erosion, compromised soil structure, and heightened susceptibility to desertification (Nunes et al., 2023). Conservation tillage, including techniques like no-till farming, helps maintain soil health by minimizing soil disturbance. This practice promotes water conservation, reduces erosion, and supports sustainable land management, aligning with the goal of preventing land degradation. Biodiversity conservation is a central theme in SDG 15, and sustainable agriculture practices directly contribute to this objective (Shahmohamadloo et al., 2022). Crop rotation, cover cropping, and diversified farming systems promote biodiversity by cre-

Table 2. Soil carbon stocks in major agroforestry systems under an ecological-based approach

Agroforestry System	Soil Carbon Stock (Mg/C/ha^{-1})
Humid lowlands	
Alley cropping systems	10 – 25
Woodlots	61 – 75
Improved fallow lands	123 – 149
Home gardens	108 – 119
Tree intercropping	27 – 78
Silvopasture	96 – 173
Shaded perennial cropping systems	21 – 235
Tropical highlands	
Silvopastoral systems	132 – 173
Shaded perennial systems	21 – 97
Arid and semiarid lowlands	
Live fences	24
Grazing lands	27 – 33
Fodder – tree dominated	25 – 80
Crop dominated	20 – 70
Forage banks	33
Fuelwood dominated	30 – 90

Source: Drawn up by author with reference to Nair et al., 2009.

ating habitats for various plant and animal species. By avoiding monoculture and incorporating diverse crops, farmers can enhance ecosystem resilience and support the survival of pollinators and beneficial organisms (Kovács-Hostyánszki et al., 2017).

Water management practices play a crucial role in achieving SDG 15, particularly in regions prone to desertification. Sustainable agriculture practices, such as rainwater harvesting and efficient irrigation techniques, contribute to soil moisture conservation and prevent water scarcity (Nikolaou et al., 2020). By promoting responsible water use in agriculture, these practices support sustainable land management and help combat desertification. Organic farming is a sustainable agriculture approach that directly supports SDG 15's goals (Purnhagen et al., 2021). Organic farming avoids the use of synthetic pesticides and fertilizers, promoting natural soil fertility and reducing the environmental impact of agriculture. Additionally, organic farming practices often incorporate agroecological principles, emphasizing the importance of working with natural ecosystems for sustainable food production. Furthermore, afforestation and reforestation initiatives in agricultural landscapes contribute significantly to SDG 15 (McElwee et al., 2020). Afforestation efforts, involving the planting of trees in degraded areas or the restoration of forests, play a vital role in combatting desertification, promoting biodiversity, and enhancing overall ecosystem health. Sustainable land management practices that incorporate trees into agricultural systems actively contribute to the restoration of degraded land, aligning with the objective of preventing land degradation outlined in SDG 15. Furthermore, education and awareness form crucial components of sustainable agriculture practices aimed at achieving SDG 15, fostering informed and responsible land use practices (Manolis & Manoli, 2021). Farmers, policymakers, and communities need to understand the importance of preserving terrestrial ecosystems, preventing land degradation, and adopting sustainable land management practices. Outreach programs and capacity-building initiatives can facilitate the adoption of environmentally responsible farming methods.

SDG 17 (Partnerships for the Goals)

The implementation of sustainable farming methods is crucial for realizing the goals set forth in SDG 17, which emphasizes the enhancement of worldwide collaborations to promote sustainable development. By promoting collaborative and inclusive approaches within the agricultural sector, sustainable practices contribute to building resilient and sustainable agricultural systems worldwide (Knickel et al., 2018). One key sustainable agriculture practice for achieving SDG 17 is the promotion of knowledge-sharing and technology transfer. Sustainable farming methods, innovative technologies, and best practices can be disseminated among farmers, agricultural stakeholders, and policymakers. Establishing international partnerships for the exchange of expertise and resources facilitates the adoption of sustainable practices globally, fostering collaboration toward shared agricultural sustainability goals (Martínez-Peláez et al., 2023). Investment in sustainable agricultural research and development is essential for achieving SDG 17. This includes supporting research initiatives that focus on climate-resilient crops, precision farming technologies, and sustainable soil management practices. Encouraging innovation in agriculture enables countries to bolster their capabilities in addressing global challenges like climate change, food security, and environmental sustainability. This aligns with the cooperative ethos of SDG 17 (Leakey et al., 2021). Encouraging sustainable supply chain management is another critical aspect of sustainable agriculture for achieving SDG 17. Strengthening partnerships along the agricultural supply chain, from farmers to consumers, contributes to the development of fair and sustainable trading practices. Promoting international cooperation in addressing food security challenges is inherent in SDG 17, and sustainable agriculture practices contribute significantly to this endeavor (Whitcraft et al., 2019).

By implementing agroecological approaches, such as diversified cropping systems and climate-smart farming, countries can enhance their resilience to climate change and improve local food production capacities. Sustainable agriculture practices that prioritize the conservation of natural resources align with SDG 17's focus on environmental sustainability (Delabre et al., 2020). Promoting fair and inclusive agricultural policies is essential for achieving SDG 17, particularly in terms of creating an enabling environment for sustainable agriculture. Governments and international organizations can work together to formulate policies that provide incentives for the adoption of environmentally friendly farming practices, offer support to smallholder farmers, and ensure equitable access to resources (Makate, 2019). This collaborative policy approach contributes to the establishment of a global agricultural framework that adheres to the principles of SDG 17. Moreover, constructing resilient agricultural systems capable of withstanding shocks and stresses is paramount for achieving SDG 17. Sustainable agriculture practices, including crop diversification and sustainable water management, play a role in enhancing the resilience of farming communities in the face of climate change, extreme weather events, and other challenges. Resilient agricultural systems not only support the well-being of farmers but also contribute to the overall stability of global food production (Hurni et al., 2015). Education and capacity-building initiatives are integral components of sustainable agriculture practices aimed at achieving SDG 17. Offering farmers, policymakers, and communities the necessary knowledge and skills to implement sustainable practices enables countries to fortify their agricultural sectors. This, in turn, contributes to the overall resilience and sustainability of global food systems.

Affected Challenges and Strategic Recommendations for Sustainable Agriculture

Sustainable agriculture, a linchpin for achieving SDGs, encounters a myriad of challenges and limitations that complicate its seamless integration into broader developmental frameworks. A primary challenge revolves around the resource-intensive nature of sustainable agricultural practices (Chartres & Noble, 2015). While these practices, such as organic farming and agroecology, contribute to environmental conservation, they often demand significant investments of time, labor, and financial resources. This poses a considerable hurdle for small-scale farmers in developing regions who may lack the necessary means to adopt these sustainable methodologies. Technological barriers further impede the widespread adoption of sustainable agriculture (Anibaldi et al., 2021). Cutting-edge technologies, such as precision agriculture, hold the key to optimizing resource use and minimizing environmental impact.

Precision agriculture employs advanced technologies to optimize farm management practices, contributing to SDGs (Pandey & Pandey, 2023). Global positioning system mapping and remote sensing provide detailed data on soil conditions, crop health, and yield potential, enabling site-specific input application. Sensor networks monitor real-time environmental factors, facilitating precise irrigation and reducing water waste. Predictive analytics guide judicious use of fertilizers and pesticides, minimizing environmental pollution. Variable rate technology applies inputs at optimal levels across fields, boosting yields while conserving resources. Overall, precision agriculture improves productivity, profitability, and resource-use efficiency, mitigating environmental impacts and supporting goals like responsible consumption, climate action, and sustainable food production (Balasundram et al., 2023). However, the accessibility and affordability of these technologies remain significant hurdles, particularly in regions where the digital divide exacerbates disparities in technological access (Bezuidenhout et al., 2017). Closing this divide necessitates united endeavors to guarantee that farmers, regardless of their scale or location, can avail themselves of and apply these advancements.

A critical limitation lies in the lack of knowledge and education about sustainable farming practices. Successful implementation of sustainable agriculture depends on farmers' understanding and adoption of innovative techniques. Insufficient awareness, coupled with a dearth of training opportunities, poses a substantial barrier, particularly in regions where traditional farming methods have deep-rooted practices (Slayi et al., 2023). Market dynamics also play a pivotal role in hindering the progress of sustainable agriculture. The current market structure often favors conventional farming, which may receive subsidies that create an uneven playing field. For sustainable agriculture to thrive, there is a need for policies that level the economic playing field, ensuring that environmentally friendly practices are not overshadowed by conventional methods due to economic incentives. Policy and regulatory challenges constitute another significant impediment to sustainable agriculture. Inconsistent or inadequate policies hinder the adoption of sustainable practices, as the regulatory framework may not align with the principles of environmental conservation and resource efficiency. Policymakers must play a proactive role in creating and enforcing regulations that incentivize and reward sustainable agricultural practices (Pacheco et al., 2010).

Climate change represents a significant peril to global agricultural systems. Increasing temperatures, shifts in rainfall patterns, and a heightened occurrence of extreme weather events are causing disruptions to crop yields and livestock production (Henry et al., 2018). This is especially challenging for smallholder farmers in developing regions who lack adaptive capacity. Several barriers impede climate change adaptation in agriculture. Insufficient access to climate information and early warning systems also hinders preparedness and planning (Basher, 2006). Weak institutional capacity in many countries results in inadequate policies and support services for building resilience. Additionally, limited technical knowledge and access to technologies like drought-tolerant seeds inhibits uptake of adaptive practices. Overcoming these barriers requires a multi-pronged approach. Increased investment is needed to make climate-smart technologies affordable for vulnerable communities (Partey et al., 2018). Improving climate monitoring and forecasting systems can bolster preparedness and adaptation. Institutional capacity building through training programs and agricultural extension services can expand knowledge on climate resilience practices. Policy reforms should target enhancing access to credit, insurance and social safety nets to strengthen farmer livelihoods against climate disruptions (Davies et al., 2009).

The multifaceted challenges and limitations confronting sustainable agriculture, coupled with its intricate alignment with the SDGs, underscore the complexity inherent in addressing global issues (Kwilinski et al., 2023). The task of overcoming these obstacles necessitates a comprehensive and cooperative approach, engaging not only governments but also international organizations, local communities, and individuals. The effort requires creating inventive solutions that carefully take into account the economic, social, and environmental aspects, thus guaranteeing that the pursuit of sustainability is not only all-encompassing but also inclusive, resilient, and able to adapt to the various challenges faced by humanity (Martínez-Peláez et al., 2023). Although the road ahead is fraught with intricacies, the compelling imperative to forge a sustainable and equitable future compels persistent efforts to surmount these challenges and bring to fruition the vision encapsulated in the SDGs.

ACHIEVING SDGS IN DEVELOPED, EMERGING, AND LEAST DEVELOPED COUNTRIES: CASE STUDIES

The case study of SDGs within European Union developed countries reveals underperformance in areas concerning responsible consumption and production, climate action, and biodiversity Conversely, these countries exhibit stronger performance in socio-economic goals such as SDG 1 (No poverty), SDG 3 (Good health and well-being), and SDG 6 (Clean water and sanitation) (Kluza et al., 2021). The research also highlights a twofold relationship between greenhouse gas emissions and SDGs, encompassing both direct and indirect connections. Findings suggest that the implementation of SDG 7 (Affordable and clean energy) and SDG 5 (Gender equality), SDG 4 (Quality education) and SDG 17 (Partnership for the goals) has the most significant impact on greenhouse gas emissions. Indirect relationships are observed in the improvement of SDG 4 (Quality education) and SDG 17 (Partnership for the goals) through the interplay of SDG 1 (No poverty), SDG 3 (Good health and well-being), SDG 8 (Decent work and economic growth), and SDG 15 (Life on land).

A study was conducted by Ali et al. (2018) concluded BRICS countries as emerging countries have not been adopting the defined SDGs in its entirety. China shows more concern towards the SDGs when compared to South Africa, India, Russia, and Brazil. The most highlighted and adopted sustainable goals are SDG 16 (Peace justice and strong institutions), and SDG 8 (Decent work and economic growth. The results also depict that BRICS countries do not place importance on SDG 4 (Quality education), SDG 13 (Climate action), and SDG 14 (Life below water). Implementation of these goals cannot be done by a single company or sector; partnerships are critical and necessary. This study highly recommends that BRICS countries pay more attention to sustainable goals that are pivotal to quality of life. SDG 13 (Climate action), SDG 4 (Quality education), and SDG 14 (Life below water) are important and highlighted by the United Nations, and these issues should also be addressed in the vision and mission statements of BRICS companies.

A case study in Afghanistan as least developed country, focused on improving the quality of life of rural communities by meeting their basic daily needs, aimed to achieve the following SDGs (Danish et al., 2021). The objectives were to foster self-reliance among local people for sustainable long-term development, promote a peaceful environment to achieve SDG 16 (Peace, justice, and strong institutions), reduce greenhouse gas emissions to achieve SDG 13 (Climate action), and protect the natural environment to achieve SDG 14 (Life below water). By directly addressing the goals outlined in the SDG 2030 Agenda as a developing country, this initiative demonstrates a commitment to collaborative efforts aimed at ensuring sustainability. Sustaining this positive trajectory requires long-term operational strategies and innovative approaches at national and regional levels, fostering cooperation between states and within national frameworks.

CONCLUSION

Sustainable agriculture stands as a pivotal strategy for achieving the ambitious and interlinked objectives encompassed in the SDGs. As highlighted throughout this analysis, agricultural practices that integrate environmental health, economic viability, and social wellbeing have the capacity to advance multiple SDGs simultaneously. By adopting agroecological approaches, improving resource efficiency, building climate resilience, empowering smallholder farmers, and strengthening value chains, food systems can provide economic opportunity and nutrition while restoring ecosystems. However, profound challenges remain that require coordinated action across sectors, economies, and societies. Enabling environments shaped by supportive policies, regulations, and incentives must be established by governments to accelerate the transition to sustainability. International cooperation through technology transfers, research networks, and funding mechanisms can boost capacities for innovation and adoption of sustainable techniques. Partnerships along entire supply chains are needed to ensure practices that protect environmental resources are also economically viable for farmers at the production level.

Sustainable agriculture offers a roadmap toward food security and environmental sustainability but reaching this vision requires active collaboration, financing, and accountability mechanisms. High-income countries must provide funding for sustainability initiatives particularly in low-income regions which lack resources but are often rich in biodiversity. Global agreements such as the Paris agreement and the convention on biological diversity should be incorporated into the national policy frameworks. Continued monitoring using platforms like the SDG indicators can track progress and gaps. Achieving the SDGs will only be possible through unprecedented levels of partnership, investment, and political will. Sustainable agriculture provides a model for the cooperation and systemic, ecological thinking needed to balance human development with environmental protection. Much work remains to overcome technical, capacity and policy constraints. However, the synergies between sustainable agriculture and the SDGs highlight the immense potential to address urgent global priorities simultaneously with sound management of precious natural resources. With diligent commitment, adaptive management, and equitable sharing of knowledge and technologies, a sustainable future for all may be within reach.

REFERENCES

Adefare, T., Adeola, O., Mogaji, E., Nguyen, N. P., & Mogaji, S. A. (2024). Empowering women agriculture entrepreneurs: Banks' role in achieving sustainable development goals. *International Journal of Bank Marketing*. Advance online publication. doi:10.1108/IJBM-03-2023-0128

Agarwal, B. (2018). Gender equality, food security and the sustainable development goals. *Current Opinion in Environmental Sustainability, 34*, 26–32. doi:10.1016/j.cosust.2018.07.002

Ali, S., Hussain, T., Zhang, G., Nurunnabi, M., & Li, B. (2018). The Implementation of Sustainable Development Goals in "BRICS" Countries. *Sustainability (Basel), 10*(7), 2513. doi:10.3390/su10072513

Altieri, M. A., Nicholls, C. I., Henao, A., & Lana, M. A. (2015). Agroecology and the design of climate change-resilient farming systems. *Agronomy for Sustainable Development, 35*(3), 869–890. doi:10.1007/s13593-015-0285-2

Anibaldi, R., Rundle-Thiele, S., David, P., & Roemer, C. (2021). Theoretical Underpinnings in Research Investigating Barriers for Implementing Environmentally Sustainable Farming Practices: Insights from a Systematic Literature Review. *Land (Basel), 10*(4), 386. doi:10.3390/land10040386

Antwi-Agyei, P., Abalo, E. M., Dougill, A. J., & Baffour-Ata, F. (2021). Motivations, enablers and barriers to the adoption of climate-smart agricultural practices by smallholder farmers: Evidence from the transitional and savannah agroecological zones of Ghana. *Regional Sustainability, 2*(4), 375–386. doi:10.1016/j.regsus.2022.01.005

Arora, N. K., & Mishra, I. (2022). Current scenario and future directions for sustainable development goal 2: A roadmap to zero hunger. *Environmental Sustainability, 5*(2), 129–133. doi:10.1007/s42398-022-00235-8 PMID:37521584

Arora, N. K., & Mishra, I. (2023). Responsible consumption and production: A roadmap to sustainable development. *Environmental Sustainability, 6*(1), 1–6. doi:10.1007/s42398-023-00266-9

Atapattu, A. A. A. J., Senarathne, S. H. S., Raveendra, S. A. S. T., Egodawatte, W. C. P., & Mensah, S. (2017). Effect of short-term agroforestry systems on soil quality in marginal coconut lands in Sri Lanka. *Agricultural Research Journal, 54*(3), 324. doi:10.5958/2395-146X.2017.00060.6

Bacon, C. M., Getz, C., Kraus, S., Montenegro, M., & Holland, K. (2012). The Social Dimensions of Sustainability and Change in Diversified Farming Systems. *Ecology and Society, 17*(4), 41. doi:10.5751/ES-05226-170441

Baker, B. P., Green, T. A., & Loker, A. J. (2020). Biological control and integrated pest management in organic and conventional systems. *Biological Control, 140*, 104095. doi:10.1016/j.biocontrol.2019.104095

Balasundram, S. K., Shamshiri, R. R., Sridhara, S., & Rizan, N. (2023). The Role of Digital Agriculture in Mitigating Climate Change and Ensuring Food Security: An Overview. *Sustainability (Basel), 15*(6), 5325. doi:10.3390/su15065325

Balyan, S., Jangir, H., Tripathi, S. N., Tripathi, A., Jhang, T., & Pandey, P. (2024). Seeding a Sustainable Future: Navigating the Digital Horizon of Smart Agriculture. *Sustainability (Basel), 16*(2), 475. doi:10.3390/su16020475

Basher, R. (2006). Global early warning systems for natural hazards: systematic and people-centred. *Philosophical Transactions of the Royal Society A: Mathematical, Physical and Engineering Sciences, 364*(1845), 2167–2182. doi:10.1098/rsta.2006.1819

Batchelor, C. (1999). Improving water use efficiency as part of integrated catchment management. *Agricultural Water Management, 40*(2–3), 249–263. doi:10.1016/S0378-3774(98)00125-5

Berti, G., & Mulligan, C. (2016). Competitiveness of Small Farms and Innovative Food Supply Chains: The Role of Food Hubs in Creating Sustainable Regional and Local Food Systems. *Sustainability (Basel), 8*(7), 616. doi:10.3390/su8070616

Bezuidenhout, L. M., Leonelli, S., Kelly, A. H., & Rappert, B. (2017). Beyond the digital divide: Towards a situated approach to open data. *Science & Public Policy, 44*(4), 464–475. doi:10.1093/scipol/scw036

Blesh, J., Hoey, L., Jones, A. D., Friedmann, H., & Perfecto, I. (2019). Development pathways toward "zero hunger.". *World Development, 118*, 1–14. doi:10.1016/j.worlddev.2019.02.004

Bloomfield, G., Bucht, K., Martínez-Hernández, J. C., Ramírez-Soto, A. F., Sheseña-Hernández, I., Lucio-Palacio, C. R., & Ruelas Inzunza, E. (2018). Capacity building to advance the United Nations sustainable development goals: An overview of tools and approaches related to sustainable land management. *Journal of Sustainable Forestry, 37*(2), 157–177. doi:10.1080/10549811.2017.1359097

Busari, M. A., Kukal, S. S., Kaur, A., Bhatt, R., & Dulazi, A. A. (2015). Conservation tillage impacts on soil, crop and the environment. *International Soil and Water Conservation Research, 3*(2), 119–129. doi:10.1016/j.iswcr.2015.05.002

Campbell, B. M., Hansen, J., Rioux, J., Stirling, C. M., & Twomlow, S., & Wollenberg, E. (2018). Urgent action to combat climate change and its impacts (SDG 13): Transforming agriculture and food systems. *Current Opinion in Environmental Sustainability, 34*, 13–20. doi:10.1016/j.cosust.2018.06.005

Caron, P., Ferrero y de Loma-Osorio, G., Nabarro, D., Hainzelin, E., Guillou, M., Andersen, I., Arnold, T., Astralaga, M., Beukeboom, M., Bickersteth, S., Bwalya, M., Caballero, P., Campbell, B. M., Divine, N., Fan, S., Frick, M., Friis, A., Gallagher, M., Halkin, J.-P., ... Verburg, G. (2018). Food systems for sustainable development: Proposals for a profound four-part transformation. *Agronomy for Sustainable Development, 38*(4), 41. doi:10.1007/s13593-018-0519-1 PMID:30956691

Chartres, C. J., & Noble, A. (2015). Sustainable intensification: Overcoming land and water constraints on food production. *Food Security, 7*(2), 235–245. doi:10.1007/s12571-015-0425-1

Danish, M. S. S., Senjyu, T., Urasaki, N., Rahmany, N. A., Ershad, A. M., Sabory, N. R., Zarabie, A. K., Anwarzai, M. A., Karimy, H., & Zaheb, H. (2021). Developing nations as a foremost allotment for achieving 2030 SDGs - A case study. *Journal of Engineering and Technology Revolution, 2*(1), 1–10. doi:10.37357/1068/jetr.2.1.01

Davies, M., Guenther, B., Leavy, J., Mitchell, T., & Tanner, T. (2009). Climate Change Adaptation, Disaster Risk Reduction and Social Protection: Complementary Roles in Agriculture and Rural Growth? *IDS Working Papers, 2009*(320), 1–37. doi:10.1111/j.2040-0209.2009.00320_2.x

Delabre, I., Alexander, A., & Rodrigues, C. (2020). Strategies for tropical forest protection and sustainable supply chains: Challenges and opportunities for alignment with the UN sustainable development goals. *Sustainability Science, 15*(6), 1637–1651. doi:10.1007/s11625-019-00747-z

Dissanayaka, D., Nuwarapaksha, T., Udumann, S., Dissanayake, D., & Atapattu, A. J. (2022). A sustainable way of increasing productivity of coconut cultivation using cover crops: A review. *Circular Agricultural Systems, 2*(1), 1–9. doi:10.48130/CAS-2022-0007

Dissanayaka, D. M. N. S., Dissanayake, D. K. R. P. L., Udumann, S. S., Nuwarapaksha, T. D., & Atapattu, A. J. (2023a). Agroforestry—a key tool in the climate-smart agriculture context: A review on coconut cultivation in Sri Lanka. *Frontiers in Agronomy, 5*, 1162750. Advance online publication. doi:10.3389/fagro.2023.1162750

Dissanayaka, D. M. N. S., Udumann, S. S., Nuwarapaksha, T. D., & Atapattu, A. J. (2024). *Harnessing the potential of Mucuna cover cropping: a comprehensive review of its agronomic and environmental benefits*. Circular Agricultural Systems. doi:10.48130/cas-0024-0001

Dissanayaka, N. S., Udumann, S. S., Nuwarapaksha, T. D., & Atapattu, A. J. (2023b). Agroforestry: An Avenue for Resilient and Productive Farming through Integrated Crops and Livestock Production. In *Transitioning to Zero Hunger*. MDPI Books. https://www.mdpi.com/books/edition/1311-transitioning-to-zero-hunger

Evans, R. G., & Sadler, E. J. (2008). Methods and technologies to improve efficiency of water use. *Water Resources Research, 44*(7). Advance online publication. doi:10.1029/2007WR006200

Fanzo, J. (2019). Healthy and Sustainable Diets and Food Systems: The Key to Achieving Sustainable Development Goal 2? *Food Ethics, 4*(2), 159–174. doi:10.1007/s41055-019-00052-6

Feliciano, D. (2019). A review on the contribution of crop diversification to Sustainable Development Goal 1 "No poverty" in different world regions. *Sustainable Development (Bradford), 27*(4), 795–808. doi:10.1002/sd.1923

Fischer, D., Brettel, M., & Mauer, R. (2020). The Three Dimensions of Sustainability: A Delicate Balancing Act for Entrepreneurs Made More Complex by Stakeholder Expectations. *Journal of Business Ethics, 163*(1), 87–106. doi:10.1007/s10551-018-4012-1

Friedemann-Sánchez, G. (2006). Assets in Intrahousehold Bargaining Among Women Workers in Colombia's Cut-Flower Industry. *Feminist Economics, 12*(1–2), 247–269. doi:10.1080/13545700500508551

Fróna, D., Szenderák, J., & Harangi-Rákos, M. (2019). The Challenge of Feeding the World. *Sustainability (Basel), 11*(20), 5816. doi:10.3390/su11205816

Glazebrook, T., & Opoku, E. (2020). Gender and Sustainability: Learning from Women's Farming in Africa. *Sustainability (Basel), 12*(24), 10483. doi:10.3390/su122410483

Grosso, G., Mateo, A., Rangelov, N., Buzeti, T., & Birt, C. (2020). Nutrition in the context of the Sustainable Development Goals. *European Journal of Public Health, 30*(Supplement_1), i19–i23. doi:10.1093/eurpub/ckaa034 PMID:32391903

Halkos, G., & Gkampoura, E.-C. (2021). Where do we stand on the 17 Sustainable Development Goals? An overview on progress. *Economic Analysis and Policy, 70*, 94–122. doi:10.1016/j.eap.2021.02.001

Henry, B. K., Eckard, R. J., & Beauchemin, K. A. (2018). Review: Adaptation of ruminant livestock production systems to climate changes. *Animal, 12*, s445–s456. doi:10.1017/S1751731118001301 PMID:30092851

Hillocks, R. J., & Cooper, J. E. (2012). Integrated Pest Management – Can it Contribute to Sustainable Food Production in Europe with Less Reliance on Conventional Pesticides? *Outlook on Agriculture, 41*(4), 237–242. doi:10.5367/oa.2012.0107

Hurni, H., Giger, M., Liniger, H., Mekdaschi Studer, R., Messerli, P., Portner, B., Schwilch, G., Wolfgramm, B., & Breu, T. (2015). Soils, agriculture and food security: The interplay between ecosystem functioning and human well-being. *Current Opinion in Environmental Sustainability, 15*, 25–34. doi:10.1016/j.cosust.2015.07.009

Hussain, S., Hussain, S., Guo, R., Sarwar, M., Ren, X., Krstic, D., Aslam, Z., Zulifqar, U., Rauf, A., Hano, C., & El-Esawi, M. A. (2021). Carbon Sequestration to Avoid Soil Degradation: A Review on the Role of Conservation Tillage. *Plants, 10*(10), 2001. doi:10.3390/plants10102001 PMID:34685810

Jacquet, F., Jeuffroy, M.-H., Jouan, J., Le Cadre, E., Litrico, I., Malausa, T., Reboud, X., & Huyghe, C. (2022). Pesticide-free agriculture as a new paradigm for research. *Agronomy for Sustainable Development, 42*(1), 8. doi:10.1007/s13593-021-00742-8

Joe Millward, D., & Garnett, T. (2010). Plenary Lecture 3 Food and the planet: Nutritional dilemmas of greenhouse gas emission reductions through reduced intakes of meat and dairy foods. *The Proceedings of the Nutrition Society, 69*(1), 103–118. doi:10.1017/S0029665109991868 PMID:20003639

Kahane, R., Hodgkin, T., Jaenicke, H., Hoogendoorn, C., & Hermann, M., Keatinge, J. D. H., d'Arros Hughes, J., Padulosi, S., & Looney, N. (2013). Agrobiodiversity for food security, health and income. *Agronomy for Sustainable Development, 33*(4), 671–693. doi:10.1007/s13593-013-0147-8

Kimaro, J. (2019). A Review on Managing Agroecosystems for Improved Water Use Efficiency in the Face of Changing Climate in Tanzania. *Advances in Meteorology, 2019*, 1–12. doi:10.1155/2019/9178136

Kini, J. (2022). Gender-aware inclusive value chain: A theoretical perspective. *Frontiers in Sustainable Food Systems, 6*, 1047190. Advance online publication. doi:10.3389/fsufs.2022.1047190

Kluza, K., Zioło, M., Bąk, I., & Spoz, A. (2021). Achieving Environmental Policy Objectives through the Implementation of Sustainable Development Goals. The Case for European Union Countries. *Energies, 14*(8), 2129. doi:10.3390/en14082129

Knickel, K., Redman, M., Darnhofer, I., Ashkenazy, A., Calvão Chebach, T., Šūmane, S., Tisenkopfs, T., Zemeckis, R., Atkociuniene, V., Rivera, M., Strauss, A., Kristensen, L. S., Schiller, S., Koopmans, M. E., & Rogge, E. (2018). Between aspirations and reality: Making farming, food systems and rural areas more resilient, sustainable and equitable. *Journal of Rural Studies*, *59*, 197–210. doi:10.1016/j.jrurstud.2017.04.012

Kovács-Hostyánszki, A., Espíndola, A., Vanbergen, A. J., Settele, J., Kremen, C., & Dicks, L. V. (2017). Ecological intensification to mitigate impacts of conventional intensive land use on pollinators and pollination. *Ecology Letters*, *20*(5), 673–689. doi:10.1111/ele.12762 PMID:28346980

Ladha, J. K., Jat, M. L., Stirling, C. M., Chakraborty, D., Pradhan, P., Krupnik, T. J., Sapkota, T. B., Pathak, H., Rana, D. S., Tesfaye, K., & Gerard, B. (2020). Achieving the sustainable development goals in agriculture: The crucial role of nitrogen in cereal-based systems. doi:10.1016/bs.agron.2020.05.006

Lal, R., Bouma, J., Brevik, E., Dawson, L., Field, D. J., Glaser, B., Hatano, R., Hartemink, A. E., Kosaki, T., Lascelles, B., Monger, C., Muggler, C., Ndzana, G. M., Norra, S., Pan, X., Paradelo, R., Reyes-Sánchez, L. B., Sandén, T., Singh, B. R., ... Zhang, J. (2021). Soils and sustainable development goals of the United Nations: An International Union of Soil Sciences perspective. *Geoderma Regional*, *25*, e00398. doi:10.1016/j.geodrs.2021.e00398

Lartey, A. (2015). End hunger, achieve food security and improved nutrition and promote sustainable agriculture. *UN Chronicle*, *51*(4), 6–8. doi:10.18356/5940d90a-en

Lasco, R. D., Delfino, R. J. P., & Espaldon, M. L. O. (2014). Agroforestry systems: Helping smallholders adapt to climate risks while mitigating climate change. *Wiley Interdisciplinary Reviews: Climate Change*, *5*(6), 825–833. doi:10.1002/wcc.301

Leakey, R., Mabhaudhi, T., & Gurib-Fakim, A. (2021). African Lives Matter: Wild Food Plants Matter for Livelihoods, Justice, and the Environment—A Policy Brief for Agricultural Reform and New Crops. *Sustainability (Basel)*, *13*(13), 7252. doi:10.3390/su13137252

Leal Filho, W., Tripathi, S. K., Andrade Guerra, J. B. S. O. D., Giné-Garriga, R., Orlovic Lovren, V., & Willats, J. (2019). Using the sustainable development goals towards a better understanding of sustainability challenges. *International Journal of Sustainable Development and World Ecology*, *26*(2), 179–190. doi:10.1080/13504509.2018.1505674

Makate, C. (2019). Effective scaling of climate smart agriculture innovations in African smallholder agriculture: A review of approaches, policy and institutional strategy needs. *Environmental Science & Policy*, *96*, 37–51. doi:10.1016/j.envsci.2019.01.014

Manolis, E. N., & Manoli, E. N. (2021). Raising awareness of the Sustainable Development Goals through Ecological Projects in Higher Education. *Journal of Cleaner Production*, *279*, 123614. doi:10.1016/j.jclepro.2020.123614

Mariani, L., Trivellato, B., Martini, M., & Marafioti, E. (2022). Achieving Sustainable Development Goals Through Collaborative Innovation: Evidence from Four European Initiatives. *Journal of Business Ethics*, *180*(4), 1075–1095. doi:10.1007/s10551-022-05193-z PMID:35855697

Martínez-Peláez, R., Ochoa-Brust, A., Rivera, S., Félix, V. G., Ostos, R., Brito, H., Félix, R. A., & Mena, L. J. (2023). Role of Digital Transformation for Achieving Sustainability: Mediated Role of Stakeholders, Key Capabilities, and Technology. *Sustainability (Basel), 15*(14), 11221. doi:10.3390/su151411221

McElwee, P., Calvin, K., Campbell, D., Cherubini, F., Grassi, G., Korotkov, V., Le Hoang, A., Lwasa, S., Nkem, J., Nkonya, E., Saigusa, N., Soussana, J., Taboada, M. A., Manning, F., Nampanzira, D., & Smith, P. (2020). The impact of interventions in the global land and agri-food sectors on Nature's Contributions to People and the UN Sustainable Development Goals. *Global Change Biology, 26*(9), 4691–4721. doi:10.1111/gcb.15219 PMID:32531815

McLaughlin, D., & Kinzelbach, W. (2015). Food security and sustainable resource management. *Water Resources Research, 51*(7), 4966–4985. doi:10.1002/2015WR017053

Muhie, S. H. (2022). Novel approaches and practices to sustainable agriculture. *Journal of Agriculture and Food Research, 10*, 100446. doi:10.1016/j.jafr.2022.100446

Mujtaba, G., Shah, M. U. H., Hai, A., Daud, M., & Hayat, M. (2024). A holistic approach to embracing the United Nation's Sustainable Development Goal (SDG-6) towards water security in Pakistan. *Journal of Water Process Engineering, 57*, 104691. doi:10.1016/j.jwpe.2023.104691

Nair, P. K. R., Nair, V. D., Kumar, B. M., & Haile, S. G. (2009). Soil carbon sequestration in tropical agroforestry systems: A feasibility appraisal. *Environmental Science & Policy, 12*(8), 1099–1111. doi:10.1016/j.envsci.2009.01.010

Nazari, B., Liaghat, A., Akbari, M. R., & Keshavarz, M. (2018). Irrigation water management in Iran: Implications for water use efficiency improvement. *Agricultural Water Management, 208*, 7–18. doi:10.1016/j.agwat.2018.06.003

Nikolaou, G., Neocleous, D., Christou, A., Kitta, E., & Katsoulas, N. (2020). Implementing Sustainable Irrigation in Water-Scarce Regions under the Impact of Climate Change. *Agronomy (Basel), 10*(8), 1120. doi:10.3390/agronomy10081120

Nunes, A. N., Gonçalves, J. P., & Figueiredo, A. (2023). Soil Erosion in Extensive versus Intensive Land Uses in Areas Sensitive to Desertification: A Case Study in Beira Baixa, Portugal. *Land (Basel), 12*(8), 1591. doi:10.3390/land12081591

Nuwarapaksha, T., Udumann, S., Dissanayaka, D., Dissanayake, D., & Atapattu, A. J. (2022). Coconut based multiple cropping systems: An analytical review in Sri Lankan coconut cultivations. *Circular Agricultural Systems, 2*(1), 1–7. doi:10.48130/CAS-2022-0008

Nuwarapaksha, T. D., Dissanayaka, N. S., Udumann, S. S., & Atapattu, A. J. (2023a). Gliricidia as a Beneficial Crop in Resource-Limiting Agroforestry Systems in Sri Lanka. *Indian Journal of Agroforestry, 25*, 12–18.

Nuwarapaksha, T. D., Rajapaksha, U. N., Ekanayake, J., Weerasooriya, S. A., & Atapattu, A. J. (2023b). Exploring the Economic Viability of Integrating Jamnapari Goat into Underutilized Pastures under Coconut Cultivations in Coconut Research Institute, Sri Lanka. *IECAG, 27*, 27. Advance online publication. doi:10.3390/IECAG2023-14990

Nuwarapaksha, T. D., Udumann, S. S., Dissanayaka, N. S., & Atapattu, A. J. (2023c). Coconut-Based Livestock Farming: A Sustainable Approach to Enhancing Food Security in Sri Lanka. In *Transitioning to Zero Hunger*. MDPI Books. https://www.mdpi.com/books/edition/1311-transitioning-to-zero-hunger

O'Mara, F. P. (2011). The significance of livestock as a contributor to global greenhouse gas emissions today and in the near future. *Animal Feed Science and Technology*, *166–167*, 7–15. doi:10.1016/j.anifeedsci.2011.04.074

Obaideen, K., Yousef, B. A. A., AlMallahi, M. N., Tan, Y. C., Mahmoud, M., Jaber, H., & Ramadan, M. (2022). An overview of smart irrigation systems using IoT. *Energy Nexus*, *7*, 100124. doi:10.1016/j.nexus.2022.100124

Octavia, D., Suharti, S., Murniati, Dharmawan, I. W. S., Nugroho, H. Y. S. H., Supriyanto, B., Rohadi, D., Njurumana, G. N., Yeny, I., Hani, A., Mindawati, N., Suratman, Adalina, Y., Prameswari, D., Hadi, E. E. W., & Ekawati, S. (2022). Mainstreaming Smart Agroforestry for Social Forestry Implementation to Support Sustainable Development Goals in Indonesia: A Review. *Sustainability (Basel)*, *14*(15), 9313. doi:10.3390/su14159313

Onyimadu, C. O. (2023). Climate change adaptation and wellbeing among smallholder women farmers in Gwagwalada and Kokona, Nigeria. *Futures*, *153*, 103238. doi:10.1016/j.futures.2023.103238

Pacheco, D. F., Dean, T. J., & Payne, D. S. (2010). Escaping the green prison: Entrepreneurship and the creation of opportunities for sustainable development. *Journal of Business Venturing*, *25*(5), 464–480. doi:10.1016/j.jbusvent.2009.07.006

Page, K. L., Dang, Y. P., & Dalal, R. C. (2020). The Ability of Conservation Agriculture to Conserve Soil Organic Carbon and the Subsequent Impact on Soil Physical, Chemical, and Biological Properties and Yield. *Frontiers in Sustainable Food Systems*, *4*, 31. Advance online publication. doi:10.3389/fsufs.2020.00031

Pandey, P. C., & Pandey, M. (2023). Highlighting the role of agriculture and geospatial technology in food security and sustainable development goals. *Sustainable Development (Bradford)*, *31*(5), 3175–3195. doi:10.1002/sd.2600

Partey, S. T., Zougmoré, R. B., Ouédraogo, M., & Campbell, B. M. (2018). Developing climate-smart agriculture to face climate variability in West Africa: Challenges and lessons learnt. *Journal of Cleaner Production*, *187*, 285–295. doi:10.1016/j.jclepro.2018.03.199

Petrescu-Mag, R. M., Petrescu, D. C., & Reti, K.-O. (2019). My land is my food: Exploring social function of large land deals using food security–land deals relation in five Eastern European countries. *Land Use Policy*, *82*, 729–741. doi:10.1016/j.landusepol.2019.01.003

Petway, J. R., Lin, Y.-P., & Wunderlich, R. F. (2019). Analyzing Opinions on Sustainable Agriculture: Toward Increasing Farmer Knowledge of Organic Practices in Taiwan-Yuanli Township. *Sustainability (Basel)*, *11*(14), 3843. doi:10.3390/su11143843

Pingali, P. (2010). Agriculture renaissance: making "Agriculture for development" work in the 21st century. Handbook of agricultural economics, 4, 3867-3894. doi:10.1016/S1574-0072(09)04074-2

Pokharel, A., Hennessy, D. A., & Wu, F. (2023). Health burden associated with tillage-related PM2.5 pollution in the United States, and mitigation strategies. *The Science of the Total Environment*, *903*, 166161. doi:10.1016/j.scitotenv.2023.166161 PMID:37574060

Power, A. G. (2010). Ecosystem services and agriculture: Tradeoffs and synergies. *Philosophical Transactions of the Royal Society of London. Series B, Biological Sciences*, *365*(1554), 2959–2971. doi:10.1098/rstb.2010.0143 PMID:20713396

Purnhagen, K. P., Clemens, S., Eriksson, D., Fresco, L. O., Tosun, J., Qaim, M., Visser, R. G. F., Weber, A. P. M., Wesseler, J. H. H., & Zilberman, D. (2021). Europe's Farm to Fork Strategy and Its Commitment to Biotechnology and Organic Farming: Conflicting or Complementary Goals? *Trends in Plant Science*, *26*(6), 600–606. doi:10.1016/j.tplants.2021.03.012 PMID:33893049

Quisumbing, A., Cole, S., Elias, M., Faas, S., Galiè, A., Malapit, H., Meinzen-Dick, R., Myers, E., Seymour, G., & Twyman, J. (2023). Measuring Women's Empowerment in Agriculture: Innovations and evidence. *Global Food Security*, *38*, 100707. doi:10.1016/j.gfs.2023.100707 PMID:37752898

Qureshi, M. R. N. M., Almuflih, A. S., Sharma, J., Tyagi, M., Singh, S., & Almakayeel, N. (2022). Assessment of the Climate-Smart Agriculture Interventions towards the Avenues of Sustainable Production–Consumption. *Sustainability (Basel)*, *14*(14), 8410. doi:10.3390/su14148410

Rahman, M. (2022). Is co-management a double-edged sword in the protected areas of Sundarbans mangrove? *Biology & Philosophy*, *37*(1), 4. doi:10.1007/s10539-022-09836-3

Rajendran, S., Afari-Sefa, V., Shee, A., Bocher, T., Bekunda, M., dominick, I., & Lukumay, P. J. (2017). Does crop diversity contribute to dietary diversity? Evidence from integration of vegetables into maize-based farming systems. *Agriculture & Food Security*, *6*(1), 50. doi:10.1186/s40066-017-0127-3

Ram, A., Dev, I., Uthappa, A. R., Kumar, D., Kumar, N., Chaturvedi, O. P., Dotaniya, M. L., & Meena, B. P. (2017). *Reactive Nitrogen in Agroforestry Systems of India. In The Indian Nitrogen Assessment.* Elsevier. doi:10.1016/B978-0-12-811836-8.00014-8

Raveendra, S. A. S. T., Atapattu, A. A. A. J., Senarathne, S. H. S., Ranasinghe, C. S., & Weerasinghe, K. W. L. K. (2017). Evaluation of the carbon sequestration potential of intercropping systems under coconut in Sri Lanka. *International Journal of Geology, Earth & Environmental Sciences*, *7*, 1–7.

Redlich, S., Martin, E. A., & Steffan-Dewenter, I. (2021). Sustainable landscape, soil and crop management practices enhance biodiversity and yield in conventional cereal systems. *Journal of Applied Ecology*, *58*(3), 507–517. doi:10.1111/1365-2664.13821

Reith, J., Ghazaryan, G., Muthoni, F., & Dubovyk, O. (2021). Assessment of Land Degradation in Semiarid Tanzania—Using Multiscale Remote Sensing Datasets to Support Sustainable Development Goal 15.3. *Remote Sensing (Basel)*, *13*(9), 1754. doi:10.3390/rs13091754

Rodrigues, C. I. D., Brito, L. M., & Nunes, L. J. R. (2023). Soil Carbon Sequestration in the Context of Climate Change Mitigation: A Review. *Soil Systems*, *7*(3), 64. doi:10.3390/soilsystems7030064

Rodriguez, S. I., Roman, M. S., Sturhahn, S. C., & Terry, E. H. (2002). *Sustainability assessment and reporting for the University of Michigan's Ann Arbor Campus*. Center for Sustainable Systems, Report No. CSS02-04. University of Michigan.

Rossi, V., Caffi, T., Salotti, I., & Fedele, G. (2023). Sharing decision-making tools for pest management may foster implementation of Integrated Pest Management. *Food Security*, *15*(6), 1459–1474. doi:10.1007/s12571-023-01402-3

Sarker, A., Bornman, J., & Marinova, D. (2019). A Framework for Integrating Agriculture in Urban Sustainability in Australia. *Urban Science (Basel, Switzerland)*, *3*(2), 50. doi:10.3390/urbansci3020050

Selvan, T., Panmei, L., Murasing, K. K., Guleria, V., Ramesh, K. R., Bhardwaj, D. R., Thakur, C. L., Kumar, D., Sharma, P., Digvijaysinh Umedsinh, R., Kayalvizhi, D., & Deshmukh, H. K. (2023). Circular economy in agriculture: Unleashing the potential of integrated organic farming for food security and sustainable development. *Frontiers in Sustainable Food Systems*, *7*, 1170380. Advance online publication. doi:10.3389/fsufs.2023.1170380

Sexsmith, K., & McMichael, P. (2015). Formulating the SDGs: Reproducing or Reimagining State-Centered Development? *Globalizations*, *12*(4), 581–596. doi:10.1080/14747731.2015.1038096

Shahbaz, P., ul Haq, S., Abbas, A., Batool, Z., Alotaibi, B. A., & Nayak, R. K. (2022). Adoption of Climate Smart Agricultural Practices through Women Involvement in Decision Making Process: Exploring the Role of Empowerment and Innovativeness. *Agriculture*, *12*(8), 1161. doi:10.3390/agriculture12081161

Shahmohamadloo, R. S., Febria, C. M., Fraser, E. D. G., & Sibley, P. K. (2022). The sustainable agriculture imperative: A perspective on the need for an agrosystem approach to meet the United Nations Sustainable Development Goals by 2030. *Integrated Environmental Assessment and Management*, *18*(5), 1199–1205. doi:10.1002/ieam.4558 PMID:34821459

Shennan, C., Krupnik, T. J., Baird, G., Cohen, H., Forbush, K., Lovell, R. J., & Olimpi, E. M. (2017). Organic and Conventional Agriculture: A Useful Framing? *Annual Review of Environment and Resources*, *42*(1), 317–346. doi:10.1146/annurev-environ-110615-085750

Siebrecht, N. (2020). Sustainable Agriculture and Its Implementation Gap—Overcoming Obstacles to Implementation. *Sustainability (Basel)*, *12*(9), 3853. doi:10.3390/su12093853

Slayi, M., Zhou, L., & Jaja, I. F. (2023). Constraints Inhibiting Farmers' Adoption of Cattle Feedlots as a Climate-Smart Practice in Rural Communities of the Eastern Cape, South Africa: An In-Depth Examination. *Sustainability (Basel)*, *15*(20), 14813. doi:10.3390/su152014813

Tomich, T. P., Lidder, P., Coley, M., Gollin, D., Meinzen-Dick, R., Webb, P., & Carberry, P. (2019). Food and agricultural innovation pathways for prosperity. *Agricultural Systems*, *172*, 1–15. doi:10.1016/j.agsy.2018.01.002

Tran, P. T., Nguyen, T. Q., Van Huynh, C., Pham, T. H., & Schinkel, U. (2023). A Nuanced Analysis on Livelihood Resilience of Vietnamese Upland Households: An Intersectional Lens of Ethnicity and Gender. *Sustainability (Basel)*, *15*(4), 3510. doi:10.3390/su15043510

Tsani, S., Koundouri, P., & Akinsete, E. (2020). Resource management and sustainable development: A review of the European water policies in accordance with the United Nations' Sustainable Development Goals. *Environmental Science & Policy*, *114*, 570–579. doi:10.1016/j.envsci.2020.09.008

Udumann, S. S., Dissanayaka, N. S., Nuwarapaksha, T. D., Thelwadana, E. P., & Atapattu, A. J. (2023). Assessing the growth potential of Sunn hemp (Crotalaria juncea L.) as a cover crop for major coconut-growing soils. *Trends in Horticulture*, *6*(2), 3579. doi:10.24294/th.v6i2.3579

Valkila, J. (2009). Fair Trade organic coffee production in Nicaragua — Sustainable development or a poverty trap? *Ecological Economics*, *68*(12), 3018–3025. doi:10.1016/j.ecolecon.2009.07.002

Viana, C. M., Freire, D., Abrantes, P., Rocha, J., & Pereira, P. (2022). Agricultural land systems importance for supporting food security and sustainable development goals: A systematic review. *The Science of the Total Environment*, *806*, 150718. doi:10.1016/j.scitotenv.2021.150718 PMID:34606855

Wezel, A., Herren, B. G., Kerr, R. B., Barrios, E., Gonçalves, A. L. R., & Sinclair, F. (2020). Agroecological principles and elements and their implications for transitioning to sustainable food systems. A review. *Agronomy for Sustainable Development*, *40*(6), 40. doi:10.1007/s13593-020-00646-z

Whitcraft, A. K., Becker-Reshef, I., Justice, C. O., Gifford, L., Kavvada, A., & Jarvis, I. (2019). No pixel left behind: Toward integrating Earth Observations for agriculture into the United Nations Sustainable Development Goals framework. *Remote Sensing of Environment*, *235*, 111470. doi:10.1016/j.rse.2019.111470

Williams, N., & Vorley, T. (2014). Economic resilience and entrepreneurship: Lessons from the Sheffield City Region. *Entrepreneurship and Regional Development*, *26*(3–4), 257–281. doi:10.1080/08985626.2014.894129

Zhao, J., Liu, D., & Huang, R. (2023). A Review of Climate-Smart Agriculture: Recent Advancements, Challenges, and Future Directions. *Sustainability (Basel)*, *15*(4), 3404. doi:10.3390/su15043404

Chapter 2
Climate Smart Agriculture and Resilience

Muhammad Safdar
https://orcid.org/0009-0006-1779-6967
University of Agriculture, Faisalabad, Pakistan

Muhammad Adnan Shahid
https://orcid.org/0000-0001-6403-0290
University of Agriculture, Faisalabad, Pakistan

Ce Yang
University of Minnesota, USA

Fahd Rasul
https://orcid.org/0000-0002-7813-2981
University of Agriculture, Faisalabad, Pakistan

Muhammad Tahir
University of Minnesota, USA

Aamir Raza
https://orcid.org/0009-0001-1867-2660
University of Agriculture, Faisalabad, Pakistan

Rehan Mehmood Sabir
https://orcid.org/0009-0007-4711-8304
University of Agriculture, Faisalabad, Pakistan

ABSTRACT

Climate change poses significant challenges to global agriculture, necessitating sustainable practices to ensure food security and mitigate environmental impacts. Climate-smart agriculture (CSA) is a technique that aims to boost agricultural productivity, improve adaptive ability, and reduce greenhouse gas emissions. It includes practices like crop management, sustainable farming, agroforestry, precision farming, and climate-resilient crop varieties. CSA improves soil health, water use efficiency, erosion, carbon sequestration, biodiversity conservation, and resource utilization. Precision agriculture tools like remote sensing and data analytics enable site-specific management decisions. Climate-resilient crop types, such as drought-tolerant, heat-tolerant, and disease-resistant, are crucial for food production under changing climatic conditions. Encouraging laws, institutional frameworks, and financial incentives are essential for CSA implementation. However, more research and collaboration are needed to promote widespread adoption of CSA practices.

DOI: 10.4018/979-8-3693-4864-2.ch002

Figure 1. Conceptual framework for climate-smart agriculture and expected outputs (Khatri et al., 2017).

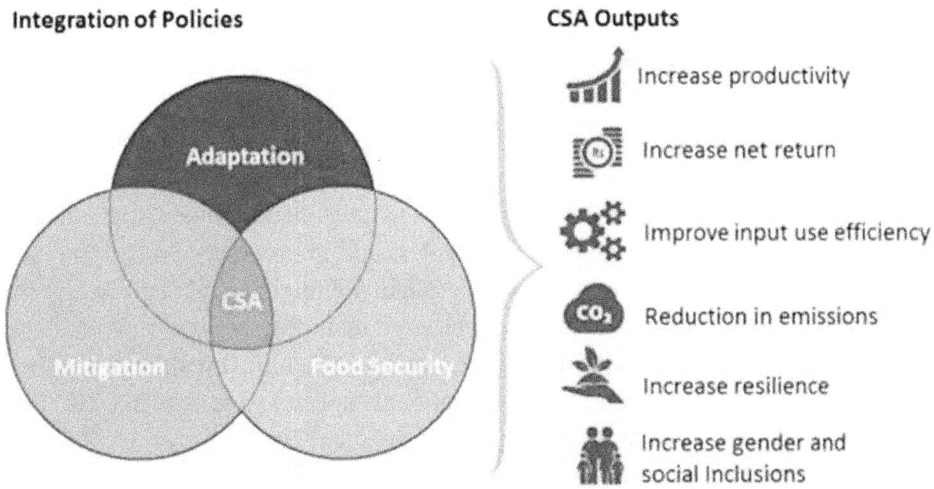

1. INTRODUCTION

1.1. Climate Smart Agriculture

Climate-smart agriculture (CSA) is a strategy aimed at enhancing sustainable food production by addressing food security and climate change. It boosts agricultural productivity and incomes by improving resource efficiency, reducing post-harvest losses, and minimizing losses. CSA also helps farmers adapt to climate change by minimizing losses and ensuring food security. It reduces greenhouse gas emissions through improved manure management and fertilizer use. CSA is a locally adapted approach to managing the agricultural system, considering interactions between crops, livestock, soil, water, and climate. (Matteoli et al., 2021). It involves farmers and stakeholders in decision-making to ensure culturally appropriate, economically viable, and technically feasible practices. CSA practices include crop diversification, agroforestry, integrated pest management, water-saving irrigation techniques, and improved livestock management. They aim to reduce crop failure, improve soil fertility, reduce greenhouse gas emissions, conserve water, and improve animal health. Overall, Figure 1 highlights the potential of CSA practices to contribute to sustainable agriculture and positive environmental and social impacts. (Khatri et al., 2017).

1.2. Overview of the Impact of Climate Change on Agriculture

'Climate change is significantly affecting agriculture worldwide, posing significant challenges to food security (Lipper et al., 2014) and rural livelihoods (Acharyya, 2022). Climate change is posing significant challenges to agriculture worldwide, including increased temperatures, reduced crop yields, heat stress on livestock, changes in precipitation patterns, droughts, floods, extreme weather events, sea level rise, salinization of agricultural land, loss of coastal agricultural land, changes in pests and diseases, and the emergence of new pests and diseases. These impacts could reduce agricultural productivity, increase food insecurity, and disrupt rural livelihoods. While some regions may experience positive impacts like longer growing seasons and increased crop yields, the overall negative impacts are expected to outweigh the positives.

1.3. Importance of Building Resilience in the Agricultural Sector

Resilience in the agricultural sector is crucial for food security, livelihood protection, rural development, ecosystem services, and climate change mitigation. Climate change, population growth, and resource scarcity threaten global food security, requiring farmers to adapt and maintain stable food production. Resilient systems create jobs, stimulate local economies, and improve access to food and resources. Ecosystems are vital for agriculture, providing services like pollination, soil fertility, and water regulation. Mitigating climate change involves reducing emissions and using climate-smart technologies. Key strategies include diversifying crops and livestock, improving soil health, investing in water management, using climate-smart technologies, and building institutional capacity (Ben et al., 2021).

Successful resilience-building initiatives include climate-smart agriculture projects, farmer-managed natural regeneration (FMNR), and community seed banks. These initiatives help farmers adapt to climate change and improve their productivity, while also preserving local crop varieties that are more resilient to climate change than commercial varieties. By investing in research, development, and extension, we can help farmers adapt to climate change and ensure a sustainable future for agriculture.

2. CLIMATE-SMART AGRICULTURE PRINCIPLES AND COMPONENTS

2.1. Principles Guiding Climate-Smart Practices

CSA is a sustainable approach to managing agricultural systems that aims to address food security and climate change. It focuses on three key objectives: sustainable intensification of agricultural production, adaptation and building resilience to climate change and reducing greenhouse gas emissions. CSA practices enhance yields, improve resource use efficiency, and reduce post-harvest losses, leading to increased food production without negative environmental impacts. It also helps farmers adapt to changing weather patterns, minimizing losses and ensuring food security. CSA is not a single technology or practice but a framework for identifying and implementing context-specific solutions, building upon existing knowledge and principles (FAO, 2013; FAO, 2016; Lipper, et al. 2014). Embrace adaptation and mitigation: Reduce emissions while preparing for changing weather, guided by science and collaboration. Prioritize efficiency and resilience: Maximize output with minimal resources, building systems that thrive in a hotter, drier future. Figure 2 vividly represents the principles of Climate-Smart Agriculture (CSA). It showcases a dynamic and healthy agricultural landscape, highlighting the diverse aspects and practices of CSA.

2.1.1. Food Security

Climate-smart agriculture (CSA) aims to increase agricultural productivity, reduce food loss and waste, and support sustainable food systems. It operates in four main domains: evidence generation, strength-

Figure 2. Climate-smart agriculture principles (authors' conceptualization)

ening institutional efficacy, promoting policy coherence between agriculture and climate change, and linking climate and agriculture finance. CSA focuses on flexible, situation-specific solutions, innovative financial and policy efforts, and lowering agricultural production risks and vulnerabilities for the world's farmers. It supports sustainable food systems by highlighting diversity, equity, and shock and stress resistance. The goal is to preserve food security in the face of climate change (Lipper et al., 2014).

2.1.2. Adaptation and Resilience

CSA practices, which include crop diversification, drought-resistant planting, and water-saving irrigation techniques, help farmers manage climate risks like floods, droughts, and extreme weather events. By fostering biodiversity, bolstering local institutions, and creating social capital, they also support ecological and social resilience. Reliable climatic data and tools for making decisions are necessary for farmers to successfully modify their agricultural operations.

2.1.3. Mitigation and Emissions Reduction

The goal of CSA techniques is to lower greenhouse gas emissions from agriculture, including carbon dioxide from degrading soil, nitrous oxide from fertilizers, and methane from cattle. By reducing climate change, improving carbon storage, and applying compost to crops, they also improve soil health. By supporting energy-efficient technology and renewable energy sources in agriculture processing and production, CSA further advances energy efficiency.

2.1.4. Equity and Inclusivity

CSA policies ought to guarantee that all farmers, irrespective of their gender, ethnicity, or financial background, have fair access to resources and technologies. Programs for assistance and targeted interventions are necessary for marginalized groups. It is important to give local communities the authority to make decisions and carry out climate-smart initiatives. Equal access to resources, technologies, and decision-making processes should be granted to women.

2.1.5. Continuous Learning and Innovation

To find and use efficient technologies and methods in a variety of settings, CSA constantly conducts research and development. To accelerate adoption, information sharing among scientists, farmers, and other stakeholders is essential. Consistent observation and assessment of CSA procedures guarantees ongoing enhancement and adjustment to evolving situations, encouraging ongoing enhancement and adjustment.

2.2. Components of Climate-Smart Agriculture

A sustainable farming method called community supported agriculture (CSA) seeks to lower greenhouse gas emissions while boosting food security, productivity, and climate change resilience. It entails techniques that raise crop yields, maintain healthy soil, and adjust to shifting weather patterns. Crop systems, including crops, precision agriculture, integrated pest control, soil, water, and animal management are all included in CSA. Technology is also included. Figure 3 illustrates the many components of the Climate-Smart Agriculture Prioritization Framework process and the consequent outcomes.

2.2.1. Sustainable Intensification

The component aims to increase agricultural output while minimizing environmental impact, focusing on improved crop and animal management, integrated soil and water management, and diversification of production systems. Key tactics include precision farming, optimal livestock nutrition, and high-yielding crop varieties. Water conservation and reduced dependence on external resources are achieved through integrated soil and water management techniques (FAO, 2016).

2.2.2. Adaptation to Climate Change

The text emphasizes the importance of enhancing agricultural systems' resistance to climate change effects. Key strategies include selecting livestock and crops that can withstand climate change, improving water management through rainwater collection and drought-resistant infrastructure, expanding revenue through ecotourism and agroforestry, and implementing early warning systems for climate-informed decisions. This will reduce farmers' dependence on vulnerable crops and provide them with timely weather information and tools (World Bank, 2023).

Figure 3. Components of the climate-smart agriculture prioritization framework process and subsequent results (Campbell et al., 2016).

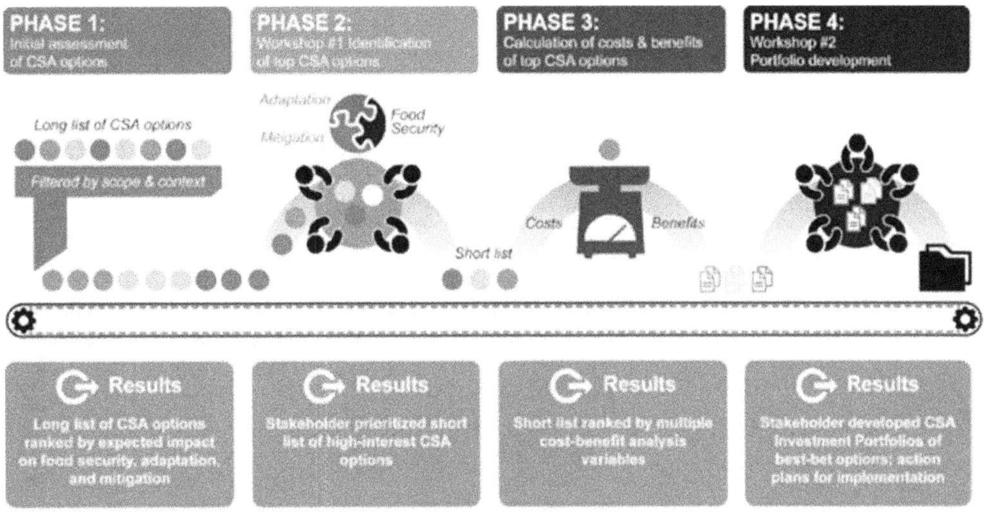

2.2.3. Mitigation of Greenhouse Gas Emissions

The component aims to decrease greenhouse gas emissions from agricultural activities by implementing strategies like improved manure management, efficient fertilizer use, conservation agriculture, and agroforestry and afforestation. Biogas digesters and composting capture methane emissions, while optimized fertilizer application rates reduce nitrous oxide emissions. Conservation agriculture practices enhance soil carbon sequestration (IPCC, 2022). The three components of climate-smart agriculture are interconnected and mutually reinforced. Sustainable intensification practices can support adaptation and mitigation efforts, while adaptation strategies can create conditions for sustainable intensification. Mitigation efforts can positively impact productivity and resilience. Implementing effective climate-smart agriculture requires a holistic approach.

3. CLIMATE CHANGE IMPACTS ON AGRICULTURE

Figure 4 highlights the effects of climate change on agriculture, including damage to crops, livestock, and soil health, resulting in reduced food production, security, and economic problems. Extreme weather events like droughts, floods, and heatwaves also harm crops and livestock, and changes in pest and disease patterns make it difficult for farmers to protect their crops and livestock. The poster urges raising awareness and taking action to mitigate the effects of climate change.

3.1. Changes in Temperature and Precipitation

Climate change is causing significant changes in temperature and precipitation patterns, which are crucial for agricultural productivity. These changes can lead to reduced soil moisture, water stress, crop failure, increased wildfire risk, physical damage, soil erosion, and reduced water infiltration and

Figure 4. Visualization of climate change's impact on agriculture (Mall et al., 2023)

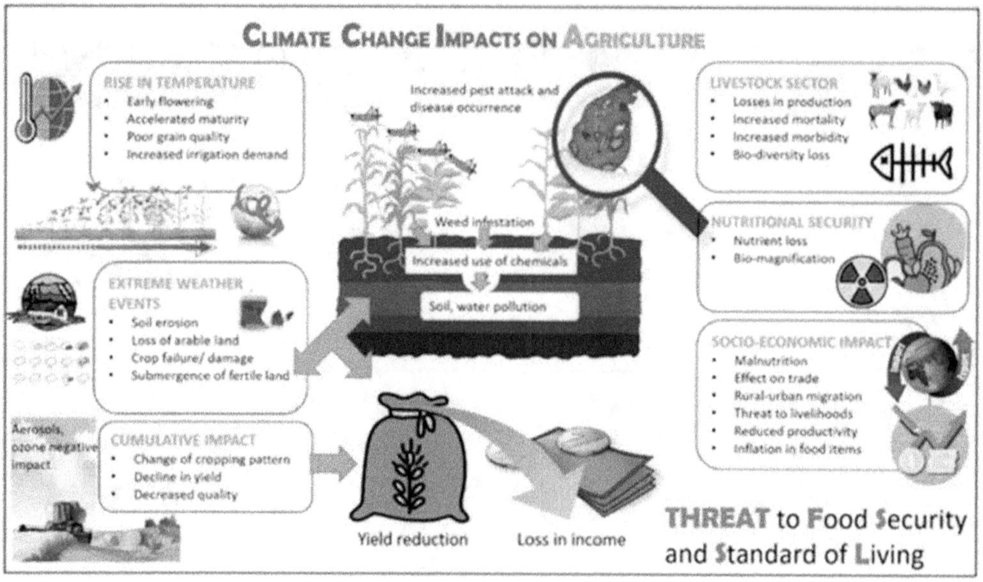

runoff. Temperatures increase evaporation and evapotranspiration, causing water scarcity and stress on agricultural systems. Changes in precipitation patterns, such as droughts and floods, reduce surface runoff and groundwater recharge, leading to declining water levels in aquifers and reservoirs. Increased competition for water resources among sectors, soil salinization, and floods can also impact crop growth and soil health.

3.1.1. Impact On Crop Yield

Increased temperatures can have both negative and positive effects on crops. Negative effects include reduced yields of sensitive crops like wheat, rice, and maize, increased evapotranspiration leading to water stress, shortened growing seasons, and increased pest and disease outbreaks. Positive effects include extended growing seasons in temperate regions. The impact of temperature and precipitation changes on crop yields is multifaceted and influenced by factors like crop type, location, and climate change scenario, with a projected negative trend causing significant global food production reductions (Lesk, et al., 2016). Climate change is posing significant challenges to Pakistan's main crops, including sugarcane, wheat, rice, and maize. A study from 1989 to 2015 found that maximum temperature negatively impacts wheat yield, and rainfall reduces crop yield except for wheat. The study suggests developing high-yielding, heat-, and drought-resistant varieties is crucial for reducing climate change's negative effects and ensuring Pakistan's food security (Ali et al., 2017).

3.1.2. Impact on Water Resources

Temperature and precipitation patterns interact, causing significant impacts on water resources. Increased temperatures worsen drought, causing water stress and scarcity. Changes in precipitation intensity can cause flash floods, soil erosion, and infrastructure damage. These combined effects threaten agricultural

Figure 5. Projected impact of climate change on agricultural yield (Cline, 2008).

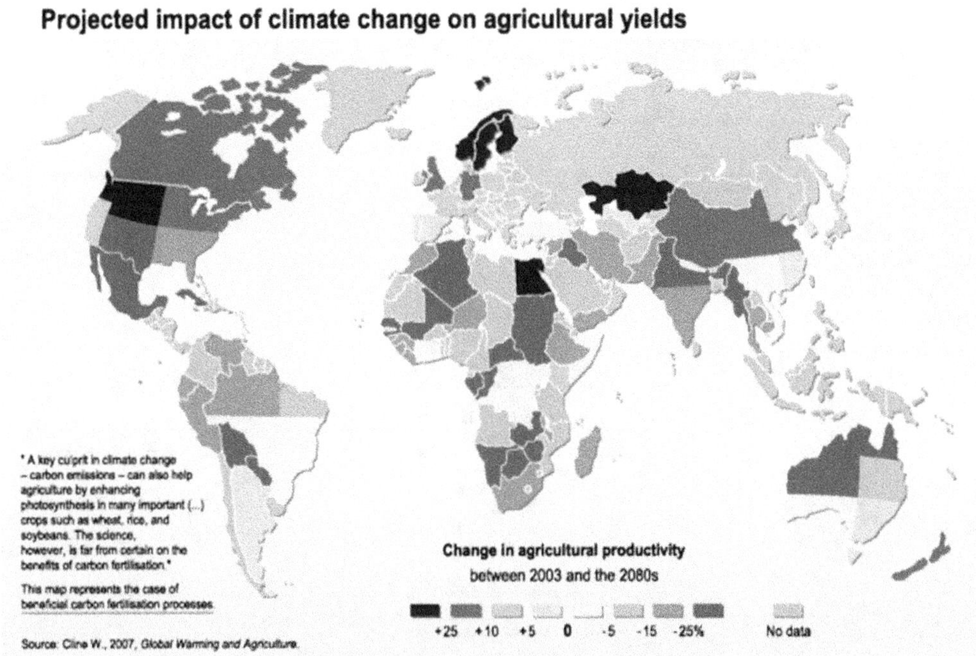

productivity and food security, especially in vulnerable regions. Climate change can significantly impact water resources, including reduced availability for irrigation, increased salinization and pollution, and disruptions in water distribution systems. Temperature and precipitation changes can lead to reduced water availability, making agricultural production unsuitable. Additionally, changes in precipitation patterns and glacial melt can cause water shortages in certain areas (IPCC, 2022). Water conservation through techniques like rainwater harvesting and efficient water allocation systems, drought-resistant crop development, and upgrading water infrastructure can enhance agricultural resilience. Addressing climate change through policies and measures can stabilize temperature and precipitation patterns, requiring adaptation and mitigation strategies for water security and agricultural sustainability. These measures are crucial for addressing the significant challenges posed by climate change.

3.2. Extreme Weather Events

Climate change is causing an increase in extreme weather events, including heatwaves, droughts, floods, wildfires, heavy rain, and extreme precipitation events. Heatwaves cause increased heat stress, reduced agricultural productivity, and increased risk of wildfires. Droughts cause water scarcity, reduced hydro-electric power generation, soil erosion, and dust storms. Floods cause land inundation, loss of life and property damage, disruptions to infrastructure, water contamination, and waterborne diseases. Wildfires spread rapidly through vegetation, causing air pollution, respiratory problems, damage to forests and ecosystems, and increased soil erosion. Heavy rain results in flash floods and landslides, causing loss of life and property damage, infrastructure damage, transportation disruptions, soil erosion, and water

Figure 6. Conceptual framework on climate change, its repercussions, and CSA practices (Agbenyo et al., 2022)

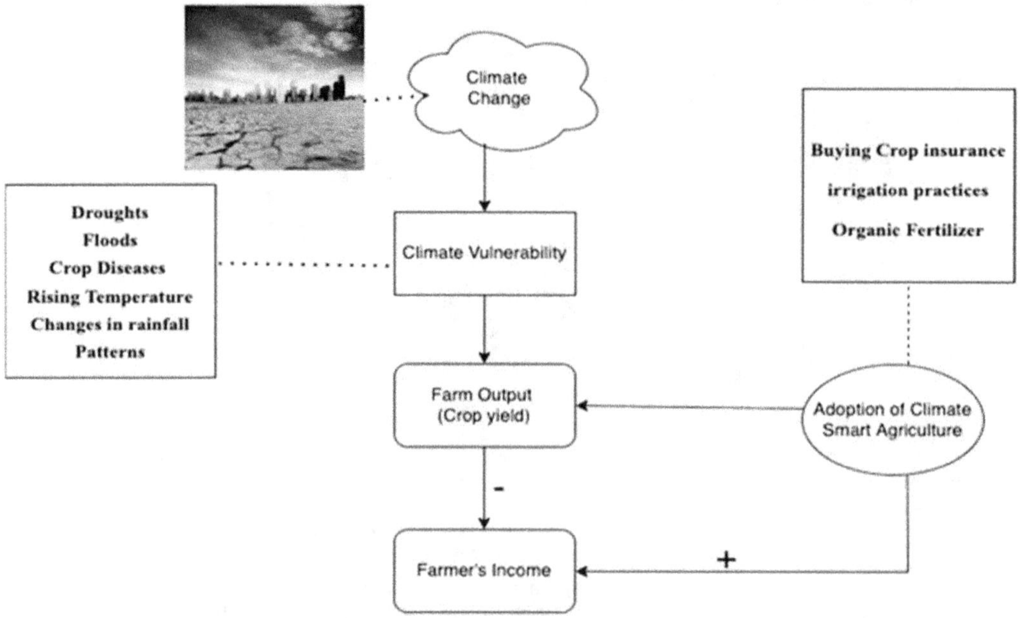

pollution. Understanding these impacts and implementing adaptation and mitigation strategies is crucial for building resilience to climate change challenges.

3.2.1. Droughts and Water Scarcity Impact on Crop Production

Droughts and water scarcity pose significant environmental challenges that affect crop production and food security globally. These issues include reduced water availability, altered soil conditions, increased pest and disease pressure, economic and social impacts, and global food security concerns. Droughts can lead to reduced yields, stunted plant growth, and even crop failure. Groundwater depletion and reduced surface water flow exacerbate water scarcity, while dry soil conditions can hinder root growth and nutrient uptake. Increased pest and disease pressure can result in crop losses and reduced yields. Economic and social consequences for farmers include income losses, debt, and food security issues (Lesk et al., 2016). The growing global population and increasing food demand exacerbate global food security concerns. To address these issues, sustainable water management practices and drought-resistant crop varieties are essential. Strategies include improving water management, implementing rainwater harvesting techniques, adopting water-saving technologies, developing drought-resistant crop varieties, sustainable land management, and developing water pricing policies (IPCC, 2022).

3.2.2. Floods and Their Consequences on Crop Production

Floods, caused by overflowing water, pose a significant threat to global agricultural production and food security. They cause physical damage to crops, soil erosion, and nutrient loss, leading to yield losses and crop failure. Seedlings and young plants are particularly vulnerable. Floodwaters also carry away fertile topsoil, causing soil erosion and decreased fertility. Nutrients essential for plant growth can be washed away by floodwaters, further impacting crop yield. Saltwater inundation in coastal areas can also reduce crop yields (Lipper et al., 2014). The study examines Pakistan's rainfall patterns over the past seven decades, focusing on the monsoon months of June to September under a warming scenario. It predicts extreme rainfall in the near and late 21st century compared to the historical period (1990-2014). The study finds that the 5-day maximum rainfall in Sindh and Baluchistan is 75% more intense than without climate warming by 1.2°C (Muzammil et al., 2023). Pakistan experienced a catastrophic flood in August 2022, affecting 33 million people and increasing food insecurity. A method for assessing productivity losses was developed using multi-sensor satellite data. 1.1 million hectares of agriculture were submerged, affecting 18% of Sindh's total area. The investigation predicted 88% of cotton, rice, and sugarcane output losses, aiding in developing evidence-based policies for rehabilitation, livelihood restoration, and post-flood recovery (Qureshi et al., 2022).

Waterlogging due to prolonged flooding can deprive roots of oxygen, leading to stunted growth and plant death. Disruptions in agricultural operations, such as planting, harvesting, and transportation, can result in delays and increased crop losses. Floods can also spread pests and diseases, posing additional health risks. To mitigate the impact of floods on crop production, strategies such as floodplain management, flood protection levees, restoration of natural floodplains, planting flood-tolerant crops, improving agricultural practices, early warning systems, and financial assistance and insurance programs can be implemented (IPCC, 2022).

4. BUILDING AGRICULTURAL RESILIENCE

Agricultural resilience is the ability of agricultural systems to maintain productivity and function amidst disturbances like climate change, extreme weather events, pests, diseases, and market fluctuations. Building resilient agricultural systems is crucial for food security and sustainable livelihoods for future generations. Key approaches include crop diversification, sustainable soil management, integrated nutrient management, improved irrigation systems, rainwater harvesting, climate-smart agriculture, and building institutions and governance. Crop diversification involves planting different crops with different maturity dates and resistance to stresses, while livestock diversification raises different livestock species. Building institutions and governance involves investing in research, strengthening farmer organizations, providing extension services, and developing risk management tools. Figure 6 depicts the essential solutions for developing agricultural resilience, with each part specifically addressing a crucial aspect of this notion (Azadi et al., 2021).

4.1. Resilience in the Context of Agriculture

4.1.1. Agricultural Resilience

Agricultural resilience refers to the farming system's ability to sustainably produce, function, and adapt to unforeseen events like droughts, floods, and insect outbreaks. It involves absorbing and recovering from these occurrences, adapting to long-term climate, market conditions, and customer preferences,

Figure 7. Concept of building agricultural resilience (authors' conceptualization)

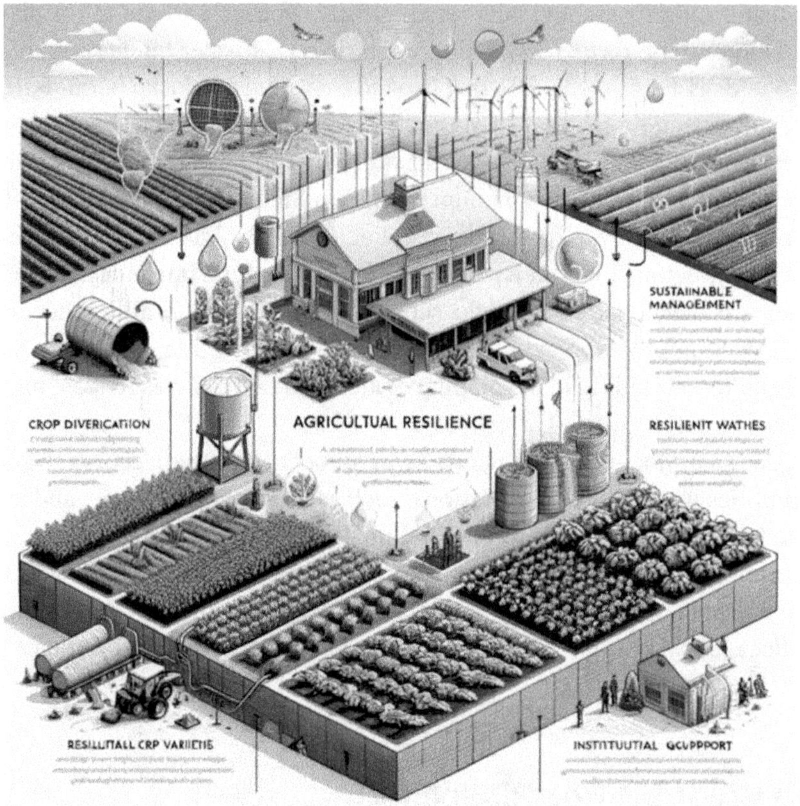

and developing new technology. This resilience is crucial for long-term sustainability and food security in an unpredictable future. (FAO, 2013).

4.1.2. Importance of Resilient Farming Systems

Resilient farming systems are crucial for food security, environmental sustainability, economic stability, and social well-being. They ensure steady output, reduce disturbances, and prevent food shortages, especially for disadvantaged populations. Resilient strategies include soil conservation, water management, and diversification. Healthy ecosystems provide pollination and pest control, while resilient farming improves farmers' standard of living and income. These communities also promote social cohesion, enhancing the quality of life and reducing vulnerability for families and communities reliant on agriculture. (World Bank, 2023). An equitable, sustainable, and safe future can be achieved by making investments in resilient farming systems. In addition to being essential for terrestrial ecosystems, food and fiber production, soil can lower greenhouse gas emissions. Still, it's early in the deployment process, and monitoring emissions is still difficult. Soil could be incorporated into a more comprehensive greenhouse gas policy thanks to recent scientific and technological advancements. The results, mitigation techniques, and suggestions for knowledge gaps are highlighted in this article. (Paustian et al., 2016).

4.2. Climate-Resilient Crop Varieties

environment-resilient crop varieties are essential for sustaining agricultural output and food security in a changing environment. Examples of these crop types include drought-tolerant maize, salt-tolerant rice, heat-tolerant wheat, submergence-tolerant rice, and pest-resistant cassava. Investments in research, effective seed delivery networks, reasonable prices, farmer education, and laws encouraging their usage are all necessary for the adoption of these crops. In the face of climate change, this encourages a more robust and sustainable agricultural system that will guarantee food security for coming generations.

4.2.1. Breeding for Drought Resistance

A major challenge to food security and agricultural productivity worldwide is drought. Crops that have been bred for drought resilience are essential for ensuring sustained food production. This incorporates genomic selection, molecular markers, and traditional breeding techniques. Backcrossing, pedigree breeding, and phenotypic selection are examples of conventional breeding techniques. QTL mapping, genomic selection, and marker-assisted selection are used in molecular markers and genomic selection. Precise breeding makes use of cutting-edge technology like genome editing and genomic-assisted breeding platforms. (Blum, 2011; Cooper et al., 2020).

Case studies demonstrate that heat-tolerant wheat, salt-tolerant rice, and drought-resistant maize can all survive in arid areas, flourish in saline water, and continue to produce well. Prospects for the future encompass creating novel mechanisms for withstanding drought, merging breeding and agronomy, and cooperating amongst academic institutions, seed businesses, and agricultural producers. Prospects for the future encompass creating novel mechanisms for enduring drought, refining agricultural techniques, and cultivating public-private collaborations amongst universities, seed firms, and farmers. We can guarantee global food security and sustainable food production by tackling these issues. (Xu et al., 2020). These are just a few examples, and the research on breeding for drought resistance is constantly evolving. By harnessing advances in genetics, genomics, and breeding technologies, we can develop more resilient.

4.2.2. Developing Flood-Tolerant Crops

Floods pose a significant threat to agricultural production worldwide, causing significant yield losses and economic damage. Researchers are developing flood-tolerant crops that can withstand temporary submergence and maintain productivity. These crops have morphological adaptations such as aerial roots, elongated internodes, and specialized stomata, as well as physiological adaptations like anaerobic respiration, enhanced antioxidant activity, and enhanced ethylene and gibberellic acid sensitivity. Breeding approaches include conventional breeding, cross-crossing flood-tolerant varieties with high-yielding varieties, using backcrossing to introduce specific flood-tolerance genes from wild or unadopted relatives, marker-assisted selection (MAS), and genome editing using CRISPR-Cas9. Examples of flood-tolerant crops include deepwater rice, flash flood-tolerant maize, and waterlogging-tolerant soybean (Lee et al., 2019).

Future directions include exploring and utilizing genetic diversity, developing multi-stress tolerance, improving field-based evaluation, and promoting farmer participation in breeding programs and variety selection. These efforts aim to develop crops that can withstand not only flooding but also other environmental stressors like drought and salinity. Additionally, researchers are focusing on enhancing field-based evaluation and involving farmers in breeding programs to ensure that developed technologies meet their needs and preferences (Bantilan et al., 2020). These references provide a valuable starting point for further research and development in breeding flood-tolerant crops. By continuing to invest in this crucial effort, we can improve food security and farmer livelihoods in flood-prone regions worldwide.

4.2.3. Importance of Crop Diversity

Crop diversity is crucial for sustainable agriculture, ensuring food security, maintaining healthy ecosystems, and promoting sustainable practices. It provides resilience against natural disasters, improves nutritional value, and reduces reliance on specific crops. It also promotes environmental sustainability by preserving biodiversity. Economic stability is achieved through increased market access, improved risk management, and rural livelihoods. Diverse crop varieties have varying tolerance to climate stressors, allowing farmers to adapt their systems. They also possess a wider range of genes that can resist pests and diseases, reducing pesticide reliance. Crop diversity also enhances resource use efficiency, leading to more sustainable agricultural practices (FAO, 2019). By recognizing and preserving crop diversity, we can build more resilient, sustainable, and equitable agricultural systems that can provide food security and environmental benefits for generations to come. A study in Karnal, India, demonstrated that Climate-smart Agriculture Practices (CSAP) can improve the energy efficiency of the traditional rice-wheat system. CSAP increased net energy, energy productivity, and energy use efficiency, but required more non-renewable and renewable energy inputs. It also improved economic efficiency, net farm revenue, and biomass yield, making it a potential substitute for increasing farm profitability, eco-efficiency, and energy use efficiency. (Kakraliya et al., 2022).

4.3. Sustainable Water Management

Sustainable water management is a vital strategy for ensuring water security for future generations. It involves a comprehensive approach that considers environmental, social, and economic aspects of water use. Key principles include integrated water resource management, demand management, water quality protection, supply augmentation, and stakeholder participation. Benefits include water security, environmental protection, economic prosperity, and social equity. Challenges include climate change, population growth, water pollution, and limited resources. Solutions include climate-resilient water management strategies, investment in infrastructure, and stricter regulations. Examples of sustainable water management practices include precision agriculture, water conservation programs, rainwater harvesting, wastewater treatment, and leakage reduction. Advances in technology and international cooperation are also crucial for addressing global water challenges.

4.3.1. Efficient Irrigation Practices

Efficient irrigation practices are crucial for sustainable agriculture and water use efficiency, especially as water resources become scarce due to population growth and climate change. Key practices include precision irrigation, improved irrigation systems, weather-based scheduling, soil moisture monitoring, water conservation practices, and technological advancements like IoT, AI, and drones. Precision irrigation uses soil moisture sensors to monitor soil moisture levels in real time, while variable rate irrigation applies to water based on specific needs. Remote sensing uses aerial imagery and satellite data to assess crop health and water stress, while improved irrigation systems include drip irrigation, sprinkler irrigation, and subsurface irrigation. Irrigation scheduling uses weather data and evapotranspiration models, while soil moisture monitoring prevents overwatering. Water conservation practices include mulching, covering crops, and land leveling (FAO, ICID).

4.3.2. Rainwater Harvesting

Rainwater harvesting is a sustainable and cost-effective method for collecting and storing rainwater for domestic, agricultural, and industrial use. It offers environmental benefits like reduced reliance on surface and groundwater resources, improved water quality, and soil erosion prevention. Techniques include rooftop harvesting, surface runoff harvesting, and groundwater recharge. Benefits include reduced water bills, increased water security, improved water quality, environmental sustainability, and economic benefits. However, challenges include initial investment, maintenance, and water treatment. Successful rainwater harvesting programs include Tamil Nadu in India, Curitiba in Brazil, and Kenya in rural communities. These initiatives have improved water security, agricultural productivity, water consumption reduction, and sanitation, contributing to overall health and well-being (FAO). By promoting rainwater harvesting and investing in innovative technologies, we can ensure a sustainable future for our water resources and contribute to a healthier planet.

4.3.3. Managing Water Scarcity in Agriculture

Water scarcity is a significant issue in agriculture, threatening food security and rural livelihoods. To tackle this, a multi-pronged approach is needed, including technological advancements, sustainable management practices, and policy changes. Technological advancements include precision irrigation, drought-resistant crop varieties, desalination and water reuse technologies, remote sensing and data analytics, and diversification of agricultural systems. Sustainable management practices include improved irrigation scheduling, soil conservation techniques, and efficient water management infrastructure. Policy changes include water pricing and allocation reforms, public-private partnerships, farmer education and extension services, and international cooperation. These measures aim to encourage water conservation and prioritize agricultural water use. Additionally, public-private partnerships promote collaboration between government, private sector, and research institutions to develop innovative solutions. Finally, farmer education and extension services provide training and resources for adopting water-saving technologies and management practices. By implementing a combination of these strategies, we can mitigate the impact of water scarcity on agriculture, ensure food security for future generations, and promote sustainable water use practices.

The study explores the use of remote-sensing-based plant stress indicators in agriculture to improve crop water stress and meet global food production demands. It examines the relationship between water content, equivalent water thickness, evapotranspiration, and sun-induced chlorophyll content and emphasizes the need for continuous research and development to enhance agricultural productivity and water stress management (Safdar, et al., 2023)

5. CLIMATE-SMART FARMING PRACTICES

Climate-smart farming (CSF) is a sustainable farming approach that aims to increase productivity, reduce greenhouse gas emissions, and improve soil health and biodiversity. Key practices include crop diversification, cover crops, integrated pest management, precision agriculture, soil management, water management, livestock management, technological advancements, and drought-resistant crop varieties. Crop diversification improves soil health, reduces pest and disease outbreaks, and adapts to changing weather patterns. Precision agriculture uses sensors, data analysis, and automation to optimize fertilizer and water use. Soil management involves conservation tillage, organic matter additions, and agroforestry for improved soil structure and carbon sequestration. Water management includes efficient irrigation and rainwater harvesting. Livestock management involves improved grazing, feed optimization, and proper manure management. Policy and market incentives can encourage CSF practices and create a supportive environment for sustainable agriculture. Figure 7 encapsulates climate-smart farming practices and key aspects of sustainable agriculture, highlighting Conservation Agriculture, Agroforestry, and Precision Agriculture.

5.1. Conservation Agriculture

Conservation Agriculture (CA) is a sustainable farming system that promotes the maintenance of permanent soil cover, minimal soil disturbance, and plant species diversification. It aims to improve soil health, conserve water resources, increase crop yields, and enhance sustainability. Key principles of CA include using cover crops, mulching, or crop residues to protect soil from erosion and improve nutrient cycling. Minimizing tillage practices reduces soil erosion, improves soil structure, and enhances biodiversity. Implementing crop rotations and incorporating diverse cover crops improves soil health, suppresses weeds, and reduces pest and disease outbreaks. Benefits of CA include improved soil health, reduced soil erosion, increased water use efficiency, enhanced biodiversity, increased crop yields, and reduced greenhouse gas emissions. However, transitioning to CA practices may require adjustments to existing farming systems, investments in equipment or technology, weed management, and training for farmers. Success stories of CA include Brazil's large-scale adoption, Ethiopia's improved soil fertility and crop yields, and Kenya's adaptation to climate change and improved food security. By embracing CA principles and practices, farmers can contribute to a more sustainable future for agriculture. Improved soil health, increased productivity, and reduced environmental impact will benefit farmers, consumers, and the planet for generations to come.

5.1.1. Minimal Tillage

Minimal tillage is a sustainable farming practice that reduces soil disturbance by reducing tillage passes compared to conventional methods. This approach offers numerous benefits for soil health, crop yields, and environmental sustainability. Key principles include reduced tillage passes, crop residue retention, and precision agriculture using technologies like GPS and sensors. Benefits include improved soil health, reduced soil erosion, increased water use efficiency, enhanced biodiversity, reduced greenhouse gas emissions, and cost savings. Practices include no-till farming, strip-till farming, mulch tilling, and cover cropping. Challenges include weed control, the initial transition period, and the knowledge and skills needed for effective adoption. Successful implementations include the United States, Australia, and Canada. By implementing minimal tillage, farmers can reduce carbon dioxide emissions, improve soil fertility, and reduce their environmental impact (FAO). By embracing minimal tillage principles and practices, farmers can contribute to a more sustainable future for agriculture. Improved soil health,

Figure 8. Climate-smart farming practices and key aspects (authors' conceptualization)

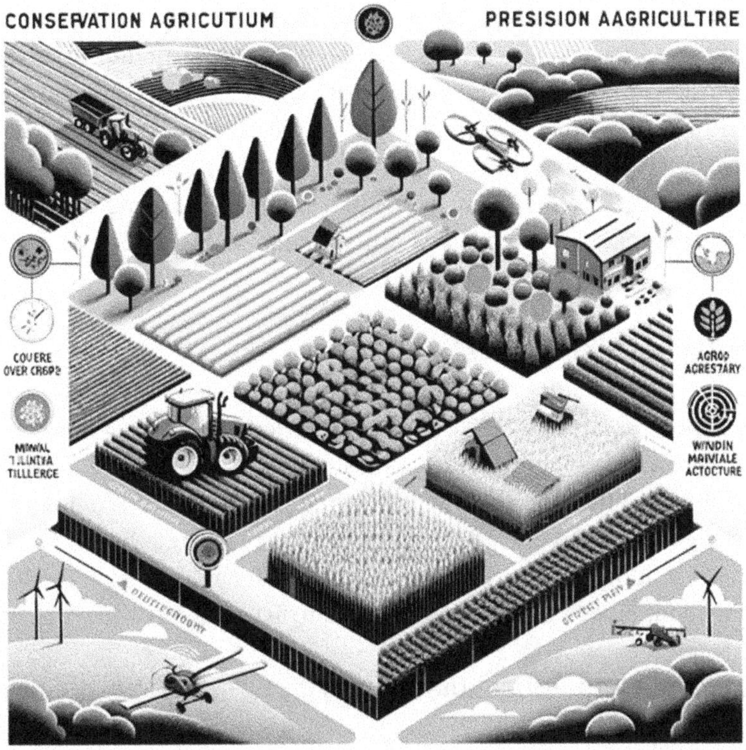

increased yields, reduced environmental impact, and lower costs will benefit farmers, consumers, and the planet for generations to come.

5.1.2. Cover Cropping

Cover cropping is a sustainable practice that involves planting crops to cover the soil during periods when the main cash crop is not growing. This practice offers numerous benefits for soil health, crop productivity, and overall agricultural sustainability. Benefits include improved soil health, enhanced microbial activity, reduced soil erosion, enhanced crop productivity, improved soil fertility through nitrogen fixation, enhanced water infiltration and utilization, reduced environmental impact, improved water quality, and enhanced biodiversity. Cover crops include grasses, legumes, brassicas, and non-leguminous broadleaves. Management involves choosing the right cover crop, planting and termination methods, seeding rate and planting depth, and fertilization. Success stories from the Midwest United States, Australia, and Brazil demonstrate the benefits of cover cropping practices (FAO). By embracing cover cropping, farmers can contribute to a more sustainable agricultural future. Improved soil health, increased crop yields, reduced environmental impact, and improved soil resilience will benefit farmers, consumers, and the planet for generations to come.

5.1.3. Crop Rotation

Crop rotation is a sustainable farming technique that involves growing different crops in the same field over a sequence of seasons, improving soil health, crop yields, and agricultural sustainability. It reduces erosion, enhances nutrient cycling, increases organic matter content, and promotes biodiversity. Principles include diversity, legumes, deep-rooted crops, cash crops, and cover crops. Planning and implementing crop rotations involves soil analysis, climate considerations, crop needs, market demands, and developing a multi-year rotation plan. Challenges include initial planning, adapting existing practices, potential short-term yield reductions, and knowledge and skills needed to design and manage effective rotations (FAO, SARE).

5.2. Agroforestry and Windbreaks

Agroforestry involves incorporating trees and shrubs into agricultural landscapes to boost crop yields, soil health, biodiversity, carbon sequestration, and income for farmers. However, challenges include resource competition, high establishment costs, and long-term farmer commitment. Windbreaks, planted to protect crops from wind damage, can reduce wind speed by up to 80%, increase crop yields by 20%, improve air quality, and enhance wildlife habitat. Trees should be tall enough for wind protection, dense enough to slow down wind without blocking sunlight, and wind resistant enough to withstand strong winds.

5.2.1. Integrating Trees Into Agricultural Landscapes

Agroforestry is a sustainable practice that involves planting and managing trees and shrubs alongside traditional crops and livestock. It benefits the environment and agricultural production by increasing crop yields, improving soil health, enhancing biodiversity, carbon sequestration, income, and water quality. Agroforestry systems can be designed in various ways, such as silvopasture, alley cropping, windbreaks, and forest farming. However, challenges include resource competition, initial investment costs, long-term farmer commitment, and specific knowledge and skill requirements. Trees can compete with crops for water, sunlight, and nutrients, necessitating careful planning and management. Integrating trees into agricultural landscapes offers a promising approach to enhancing farm productivity, environmental sustainability, and farmer livelihoods. By adopting agroforestry practices, farmers can contribute significantly to building a more resilient and sustainable food system (Montagnini et al., 2018).

5.2.2. Windbreaks for Preventing Soil Erosion

Wind erosion is a major threat to agricultural productivity and soil health, particularly in arid and semi-arid regions. It can lead to decreased crop yields and land degradation. Windbreaks are a proven method to mitigate wind erosion and conserve soil resources. They offer benefits such as reduced wind speed, increased soil moisture, improved soil structure, enhanced organic matter content, and reduced dust generation. There are various types of windbreaks, including single-row, multi-row, porous, and vegetative filter strips. The effectiveness of windbreaks depends on factors such as maturity, wind resistance, drought tolerance, fast-growing species, and non-invasive species. Trees and shrubs should be tall, dense, adapted to the local climate, and fast-growing to provide wind protection sooner (USDA, 2022).

5.3. Precision Agriculture

Precision agriculture (PA) is a growing field that uses advanced technologies and information systems to manage agricultural operations with greater precision and efficiency. By collecting and analyzing data on factors like soil conditions, weather, and crop health, farmers can make informed decisions about resource allocation, leading to increased yields, reduced environmental impact, and improved profitability. In the 1980s, precision agriculture began with remote sensing applications for soil organic matter. Modern electromagnetic wavelengths, ranging from ultraviolet to microwave, enable thermal spectroscopy, fluorescence spectroscopy, and LiDAR. Hyperspectral remote sensing, with its narrower spectrum, allows more effective analysis of chemicals, molecular interactions, crop stress, and biophysical properties, sparking interest in real-time crops, insects, and soil management (Mulla, 2013). Key features of PA include data collection through sensors, drones, satellites, yield monitors, data analysis using advanced tools, variable-rate technology, and GPS-equipped precision equipment. Benefits of PA include increased yields, reduced environmental impact, improved profitability, enhanced decision-making, and improved labor efficiency. However, challenges include cost, data management, and technology adoption. Examples of PA technologies include yield monitors, soil sensors, remote sensing, and Geographic Information Systems (GIS).

5.3.1. Use of Technology for Resource Optimization

Technology plays a crucial role in optimizing resource use across various industries, including agriculture, energy production, transportation, and manufacturing. Key examples include precision agriculture, vertical farming, renewable energy sources, smart grids, electric vehicles, industrial automation, and the circular economy. Precision agriculture uses sensors, drones, and data analytics to collect real-time data on soil conditions, crop health, and weather patterns. Vertical farming minimizes land and water requirements. Energy production uses technologies like solar panels, wind turbines, and geothermal power plants. Smart grids reduce energy waste and improve power grid efficiency. Transportation uses electric vehicles and industrial automation uses robots for productivity and waste reduction. (Beck et al., 2019)

5.3.2. Data-Driven Decision-Making in Farming

Data-driven decision-making (DDDM) is revolutionizing the agricultural industry by providing farmers with data insights to optimize operations and yields. By collecting, analyzing, and interpreting data from various sources, farmers can improve yields, profitability, and resource efficiency while minimizing environmental impact. Benefits include increased yields, improved profitability, enhanced sustainability, and better risk management. DDDM technologies include field sensors, drones and satellites, yield monitors, weather data, and farm management software. Implementing DDDM can be challenging due to costs, data management, security, and technological literacy. (Gebbers et al., 2019).

5.4. Climate-Smart Agriculture Artificial Pollination

Climate-smart agriculture (CSA) is a strategy aimed at enhancing agricultural sustainability and resilience in the face of climate change. Artificial pollination, a key component of CSA, can mitigate climate risks by ensuring pollination under adverse weather conditions or in areas with scarce natural pollinators. This resilience is crucial for maintaining food security and livelihoods, addressing food security challenges exacerbated by climate change. Artificial pollination can also support food production by increasing crop productivity and protecting biodiversity. Embracing artificial pollination is an innovative adaptation strategy for farmers facing climate-related challenges. An integrated approach to CSA includes agroforestry, soil conservation, water management, and crop diversification. By integrating artificial pollination with other climate-smart practices, farmers can enhance the resilience, productivity, and sustainability of their agricultural systems. Although artificial pollination may not address the root causes of climate change, it promotes sustainable agricultural practices that benefit both farmers and the environment (Klein et al., 2007).

5.5. Climate-Smart Agriculture Bio Bee Biological Systems

Bio bee biological systems are a new approach to climate-smart agriculture that uses beneficial organisms like bees for improved crop productivity, reduced reliance on synthetic pesticides, and biodiversity conservation. Bees, including managed honeybees and native pollinators, play a crucial role in pollinating crops, ensuring fruit set and seed production. Certain bee species, like solitary bees and bumblebees, contribute to natural pest control by preying on insect pests or parasitizing their eggs or larvae, reducing the need for chemical insecticides. Bio bee systems are used in orchards, vegetable fields, and greenhouse operations to enhance pollination of crops like apples, almonds, strawberries, and tomatoes while controlling pests like aphids, thrips, and caterpillars. This reduces chemical inputs, improves crop yields, and promotes ecosystem health. (Goulson et al., 2015)

6. POLICY AND INSTITUTIONAL SUPPORT

Climate change threatens global food security, necessitating urgent action to adapt agricultural practices and build resilience. Climate-smart agriculture (CSA) is a promising approach that promotes productivity, resource efficiency, and reduced greenhouse gas emissions. Implementing CSA requires robust policy and institutional support, including research, financial incentives, capacity building, market development, data collection, international cooperation, and institutional support. Benefits include enhanced food security, reduced emissions, improved resource efficiency, increased farmer livelihoods, and environmental sustainability. Challenges include a lack of resources, policy coherence, robust monitoring systems, and capacity building.

Governments can promote climate-smart practices (CSPs) to enhance agricultural productivity, and resource efficiency, and reduce greenhouse gas emissions. Incentives can be financial, technical, or informational. Regulations can promote agricultural resilience by establishing land use planning, water management, seed quality control, pesticide and fertilizer regulations, and building agricultural structures. Additional measures include investing in research, strengthening early warning systems, building social safety nets, and fostering international cooperation. Government policies are crucial for ensuring food security for future generations.

Climate-smart agriculture requires collaboration from various institutions beyond government policies, including research institutions, extension services, financial institutions, farmer organizations, NGOs, private sector companies, and media outlets. These institutions invest in climate-smart agriculture projects, support capacity-building programs, and facilitate access to carbon markets. Effective collaboration and coordination between these diverse institutions are crucial for a successful transition towards climate-smart agriculture. This can be achieved through establishing multi-stakeholder platforms, developing joint initiatives, sharing knowledge and resources, and monitoring progress. Strengthening these institutions' capacity can create an environment for farmers to adopt climate-smart practices, build resilience to climate change, and contribute to a more sustainable food system.

7. CHALLENGES AND OPPORTUNITIES FOR CLIMATE-SMART AGRICULTURE

7.1. Challenges in Implementing Climate-Smart Agriculture

Climate-smart practices can be challenging for smallholder farmers due to high upfront costs, limited access to financing, and market uncertainties. Investments in new technologies and infrastructure can be costly, especially in developing countries. Additionally, farmers may face uncertainties regarding fair prices and stable demand due to the evolving market for climate-smart agricultural products.

Climate change is expected to significantly impact smallholder farmers' livelihoods, potentially leading to a 28% drop in household income and reduced investment in cash crops and economic migration. However, a combination of risk-transfer mechanisms and modest cash transfers could boost adoption, increase community incomes, and reduce inequality. (Choquette et al., 2021).

CSA is often overlooked by farmers due to limited knowledge and inadequate extension services. In many regions, these services cannot provide adequate training and support. Additionally, traditional farming practices and cultural beliefs can hinder the adoption of new technologies and approaches.

7.2. Opportunities for Advancement in Technological Innovations

Advancements in breeding, precision agriculture, and renewable energy sources are transforming agriculture by improving productivity, resource efficiency, and climate resilience. As these technologies mature and scale, their costs are expected to decrease, making them more accessible to farmers. Digitalization and data analytics can also provide farmers with valuable insights, enabling data-driven decision-making for resource optimization and climate change adaptation.

Climate-smart agriculture can be promoted through collaboration between nations and organizations, with South-South cooperation enabling developing nations to benefit from each other's experiences. Financial and technical assistance from donor organizations and international organizations can support climate-smart agriculture programs. Governments can promote sustainable practices through laws and regulations, creating robust value chains for climate-smart agricultural products, and implementing training initiatives for successful adoption and long-term viability.

7. CASE STUDIES

Initiatives for climate-smart agriculture are becoming more popular throughout Asia; two case studies demonstrate this. First, the Vietnamese government and foreign NGOs implemented the System of Rice Intensification (SRI) in Vietnam. It seeks to increase climate change resistance, lower greenhouse gas emissions, and increase rice output. Innovative crop management practices, like sporadic flooding,

Figure 9. Role of existing and emerging technologies in advancing climate-smart agriculture through innovative techniques: A review (Gupta et al., 2022)

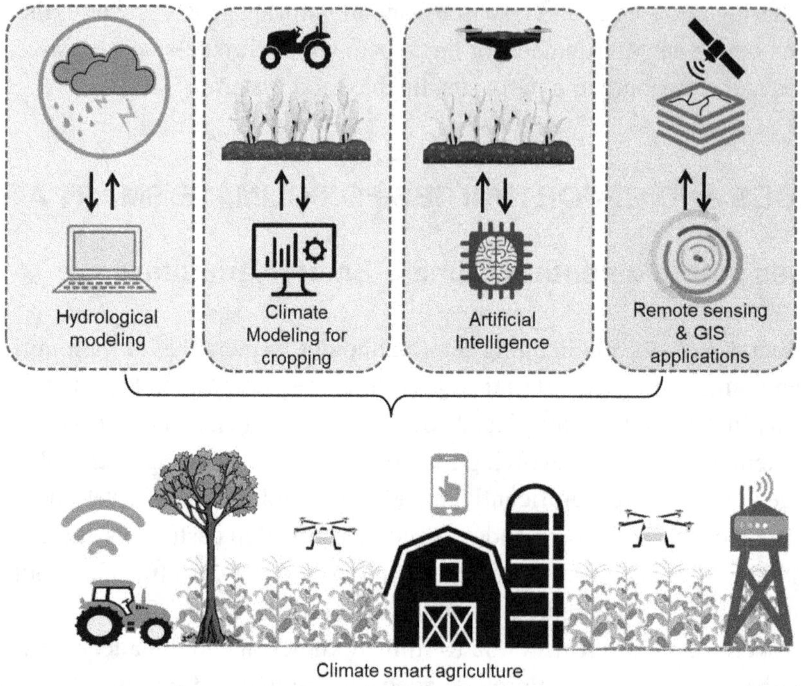

lower seedling rates, and organic fertilization, are taught to farmers. The outcomes indicate higher yields and better soil health. However, due to low knowledge and restricted access to resources, scaling up acceptance continues to be difficult. To encourage farmers to implement climate-smart techniques and incorporate them into traditional agricultural programs, policy reforms are required.

The Netherlands is implementing precision farming methods to improve sustainability and reduce climate change effects. The Precision Agriculture for Climate Resilience (PACR) initiative uses advanced sensor technologies to control inputs like herbicides, fertilizers, and water. The results show improved climate resilience, lower emissions, and increased efficiency. However, access to technology, concerns about data privacy, and policy support pose challenges. The French government should encourage investment in precision farming through grants, regulatory frameworks, and subsidies. Crop diversification and agroforestry are also being promoted.

With the help of government agencies, agricultural groups, and farmers, Iowa State (USA) is promoting climate-smart agriculture through the Iowa Cover Crop Initiative (ICCI). The effort seeks to strengthen resilience to climatic unpredictability, decrease nutrient loss, and increase soil health. To spread knowledge about cover crops and best practices, it offers instructional materials, workshops, and field days. The initial costs of establishing cover crops are reduced by financial incentives. Expanding policy backing via conservation initiatives and crop insurance benefits may promote farmers' broader use of cover crops.

Australia is utilizing precision agriculture technologies like data analytics, drones, and sensors to optimize water and nutrient use. These tools provide farmers with real-time data on crop health, soil moisture, and weather, enabling them to allocate resources effectively. This has led to increased water use efficiency, increased agricultural productivity, and reduced greenhouse gas emissions. These case studies demonstrate the success of climate-smart agriculture methods, encouraging others to adopt similar approaches.

8. CONCLUSION

CSA is a promising approach to address food security, climate change, and environmental sustainability. It promotes increased productivity, resource efficiency, and reduced greenhouse gas emissions, contributing to a more resilient and sustainable agricultural sector. However, widespread adoption faces challenges such as economic constraints, lack of awareness, and cultural barriers. To promote CSA, governments should develop policies, regulations, and financial mechanisms to incentivize farmers to adopt CSA practices.

Research institutions should invest in climate-smart technologies and practices, financial institutions should offer innovative financial solutions, extension services should provide adequate training and technical assistance, farmers should embrace continuous learning, and consumers should support products grown through sustainable practices.

Future directions in CSA include integrating digital technologies, scaling up successful CSA practices, strengthening community engagement and social capital, investing in climate-resilient crops and livestock, and expanding support systems and social safety nets to help vulnerable communities cope with climate-related challenges. Through collective effort, innovation, and continuous learning, we can harness the potential of CSA and build a more resilient and sustainable agricultural system for future generations.

REFERENCES

Acharyya, A. (2022). Climate-Smart Agriculture in Developing Economies. *Environmental Economics in Developing Countries: Issues and Challenges*, 231.

Agbenyo, W., Jiang, Y., Jia, X., Wang, J., Ntim-Amo, G., Dunya, R., Siaw, A., Asare, I., & Twumasi, M. A. (2022). Does the Adoption of Climate-Smart Agricultural Practices Impact Farmers' Income? Evidence from Ghana. *International Journal of Environmental Research and Public Health*, *19*(7), 3804. doi:10.3390/ijerph19073804 PMID:35409488

Ali, S., Liu, Y., Ishaq, M., Shah, T., Abdullah, A., Ilyas, A., & Din, I. (2017). Climate change and its impact on the yield of major food crops: Evidence from Pakistan. *Foods*, *6*(6), 1–19. doi:10.3390/foods6060039 PMID:28538704

Azadi, H., Moghaddam, S. M., Burkart, S., Mahmoudi, H., Van Passel, S., Kurban, A., & Lopez-Carr, D. (2021). Rethinking resilient agriculture: From climate-smart agriculture to vulnerable-smart agriculture. *Journal of Cleaner Production*, *319*, 128602. doi:10.1016/j.jclepro.2021.128602

Bantilan, M. C. S., Brar, D. S., & Fukao, T. (2020). Breeding crops for improved abiotic stress tolerance: Achievements and future directions. *Frontiers in Plant Science*, *11*, 1107486.

Beck, B., Zimmermann, J., McBratney, A., Whelan, B., & O'Kennedy, M. (2019). Digital soil mapping: A review of methods and applications. *European Journal of Soil Science*, *70*(6), 1164–1185.

Ben Nasr, J., Chaar, H., Bouchiba, F., & Zaibet, L. (2021). Assessing and building climate change resilience of farming systems in Tunisian semi-arid areas. *Environmental Science and Pollution Research International*, *28*(34), 46797–46808. doi:10.1007/s11356-021-13089-0 PMID:33646545

Blum, A. (2011). Drought resistance in plants and its improvement for crop production: a comprehensive overview. *Plant stress tolerance: methods and protocols*, 741-772.

Campbell, B. M., Vermeulen, S. J., Aggarwal, P. K., Corner-Dolloff, C., Girvetz, E., Loboguerrero, A. M., Ramirez-Villegas, J., Rosenstock, T., Sebastian, L., Thornton, P. K., & Wollenberg, E. (2016). Reducing risks to food security from climate change. *Global Food Security*, *11*, 34–43. doi:10.1016/j.gfs.2016.06.002

Choquette-Levy, N., Wildemeersch, M., Oppenheimer, M., & Levin, S. A. (2021). Risk transfer policies and climate-induced immobility among smallholder farmers. *Nature Climate Change*, *11*(12), 1046–1054. doi:10.1038/s41558-021-01205-4

Cline, W.R. (2008). Global Warming and Agriculture. *Finance and Development*.

Cooper, M., Messina, C. D., Podlich, D. W., Toenniessen, G. H., & Varshney, R. K. (2020). Breeding for drought tolerance: A wide array of potential approaches. *Plant Breeding Reviews*, *44*, 29–111.

FAO. (2016). *Building a common vision for sustainable food and agriculture - Principles and approaches*. FAO.

FAO. (n.d.). *Rainwater Harvesting for Agriculture and Water Supply.* https://www.fao.org/water/en/

Food and Agriculture Organization of the United Nations (FAO). (2013). *Climate-Smart Agriculture Sourcebook*. https://www.fao.org/3/i3325e/i3325e.pdf

Food and Agriculture Organization of the United Nations (FAO). (2019). *The State of the World's Biodiversity for Food and Agriculture*. FAO.

Gebbers, R., Adamchuk, V. I., Dobermann, A., Seidel, J., & Zeinstra, A. (2019). Quantitative approaches to site-specific crop management. *Advances in Agronomy, 154*, 1-50. https://pubmed.ncbi.nlm.nih.gov/31480709/

Goulson, D., Nicholls, E., Botías, C., & Rotheray, E. L. (2015). Bee Declines Driven by Combined Stress from Parasites, Pesticides, and Lack of Flowers. *Science*, *347*(6229), 1255957. doi:10.1126/science.1255957 PMID:25721506

Gupta, D., Gujre, N., Singha, S., & Mitra, S. (2022). Role of existing and emerging technologies in advancing climate-smart agriculture through modeling: A review. *Ecological Informatics*, *71*, 101805. doi:10.1016/j.ecoinf.2022.101805

Intergovernmental Panel on Climate Change (IPCC). (2023). *Impact of the 2022 floods on agriculture in Pakistan's Sindh Province.* International Commission on Irrigation and Drainage (ICID): https://icid-ciid.org/

IPCC. (2022). *Sixth Assessment Report*. https://www.ipcc.ch/report/sixth-assessment-report-working-group-ii/

Kakraliya, S. K., Jat, H. S., Singh, I., Gora, M. K., Kakraliya, M., Bijarniya, D., Sharma, P. C., & Jat, M. L. (2022). Energy and economic efficiency of climate-smart agriculture practices in a rice–wheat cropping system of India. *Scientific Reports*, *12*(1), 1–14. doi:10.1038/s41598-022-12686-4 PMID:35610271

Khatri-Chhetri, A., Poudel, B., & Shirsath, P. (2017). *Assessment of Climate-Smart Agriculture (CSA)*. Options in Nepal.

Klein, A. M., Vaissière, B. E., Cane, J. H., Steffan-Dewenter, I., Cunningham, S. A., Kremen, C., & Tscharntke, T. (2007). Importance of Pollinators in Changing Landscapes for World Crops. *Proceedings. Biological Sciences*, *274*(1608), 303–313. doi:10.1098/rspb.2006.3721 PMID:17164193

Lee, J. T., Li, G., Zhang, Y., Yang, Z., Xu, Y., & Gao, C. (2019). Genome editing of crops for improved flood tolerance. *Plant Biotechnology Journal*, *17*(10), 1907–1917.

Lesk, C., Rowhani, P., & Ramankutty, N. (2016). Influence of extreme weather disasters on global crop production. *Nature*, *529*(7584), 84–87. doi:10.1038/nature16467 PMID:26738594

Lipper, L., Thornton, P., Campbell, B. M., Baedeker, T., Braimoh, A., Bwalya, M., Caron, P., Cattaneo, A., Garrity, D., Henry, K., Hottle, R., Jackson, L., Jarvis, A., Kossam, F., Mann, W., McCarthy, N., Meybeck, A., Neufeldt, H., Remington, T., ... Torquebiau, E. (2014). Climate-smart agriculture for food security. *Nature Climate Change*, *4*(12), 1068–1072. doi:10.1038/nclimate2437

Mall, R. K., Singh, N., Patel, S., Singh, S., Arora, A., Bhatla, R., & Srivastava, P. K. (2023). Climate Changes over the Indian Subcontinent: Scenarios and Impacts. In *Science, Policies and Conflicts of Climate Change: An Indian Perspective* (pp. 27–52). Springer International Publishing.

Matteoli, F., Schnetzer, J., & Jacobs, H. (2021). Climate-Smart Agriculture (CSA): An Integrated Approach for Climate Change Management in the Agriculture Sector. In J. M. Luetz & D. Ayal (Eds.), *Handbook of Climate Change Management*. Springer. doi:10.1007/978-3-030-57281-5_148

Montagnini, F., Ewel, J. J., & Rheingans, R. (2018). *Agroforestry for soil management in Latin America*. CRC Press.

Mulla, D. J. (2013). Twenty-five years of remote sensing in precision agriculture: Key advances and remaining knowledge gaps. *Biosystems Engineering*, *114*(4), 358–371. doi:10.1016/j.biosystemseng.2012.08.009

Muzammil, H., Zaman, M., Shahid, M. A., Safdar, M., Majeed, M. D., & Sabir, R. M. (2023). The Impacts of Climate Change on Monsoon Flood Situations in Pakistan. *Environmental Sciences Proceedings*, *25*(1), 63.

Paustian, K., Lehmann, J., Ogle, S., Reay, D., Robertson, G. P., & Smith, P. (2016). Climate-smart soils. *Nature*, *532*(7597), 49–57. doi:10.1038/nature17174 PMID:27078564

Qureshi, A., Sarwar, A., & Qureshi, A. S. (2022). Water Management in the Indus Basin in Pakistan. *Challenges and Opportunities Water Management in the Indus Basin in Pakistan: Challenges and Opportunities.*, *31*, 252–260.

Safdar, M., Shahid, M. A., Sarwar, A., Rasul, F., Majeed, M. D., & Sabir, R. M. (2023). Crop Water Stress Detection Using Remote Sensing Techniques. *Environmental Sciences Proceedings*, *25*(1), 20.

Shafiq, M. N., Gillani, S., & Shafiq, S. (2021). Climate Change and Agricultural Production in Pakistan. *IRASD Journal of Energy & Environment*, *2*(2), 47–54. doi:10.52131/jee.2021.0202.0016

Sustainable Agriculture Research & Education (SARE). (n.d.). https://www.sare.org/

USDA Natural Resources Conservation Service. (2022, April). *Windbreaks and hedge rows for erosion control*. United States Department of Agriculture.

World Bank. (2021). *Climate-smart agriculture: A call to action*. https://www.worldbank.org/en/topic/climate-smart-agriculture

World Bank. (2023). *Climate-Smart Agriculture*. https://www.worldbank.org/en/topic/climate-smart-agriculture

Xu, L., Liu, F., Fu, Z., Jin, J., & Zhang, H. (2020). Breeding for crop drought resistance: Strategies and progress. *Plant Methods*, *16*(1), 1–21. PMID:33292390

Chapter 3
Sustainable Agriculture With Special Emphasis on Risk Management

Vidya Kumbhar
Symbiosis Institute of Geo-Informatics, Symbiosis International University (Deemed), India

Kailash Parihar
Symbiosis Institute of Geo-Informatics, Symbiosis International University (Deemed), India

Sneha Kumari
Symbiosis School of Economics, Symbiosis International University (Deemed), India

Samaya Venkatesh Pillai
Symbiosis Institute of Digital and Telecom Management, Symbiosis International University (Deemed), India

Pratyush Kumar
Symbiosis Institute of Geo-Informatics, Symbiosis International University (Deemed), India

ABSTRACT

In India, agriculture is a source of employment to more than 50% of the population. Despite major strengths such as geographical location, diverse agro-climatic conditions, variety of soils, and agro-ecological congeniality, Indian farmers face diverse types of risks. Climate change is one the major risks confronting Indian farmers. Variations in climatic parameters such as rainfall, temperature, humidity, etc. result in a decrease in soil moisture and reduction in soil fertility. This has adversely affected crop productivity. History reflects that unpredictable climate change has affected the standard of living of Indian farmers. This gives rise to a strong need to study the risk management techniques of the Indian farmers. The chapter explores the diverse risks faced by Indian farmers and the risk management strategies adopted by them. The purpose of the research is to examine the different hazards that farmers encounter as well as risk management techniques. Primary data has been collected from 192 farmers in Maharashtra. Purposive sampling technique was used for collecting responses from the farmers. And an open-ended interview has also been conducted for another 30 farmers. The data was analysed using QDA miner software. The research shows how farmers deal with hazards using various ways and strategies. It can be advanced by looking into cutting-edge risk management strategies as well as potential enablers and obstacles associated with it.

DOI: 10.4018/979-8-3693-4864-2.ch003

INTRODUCTION

Indian economy is heavily reliant on agriculture (Johnston and Mellor,1961). Indian farmer households rely solely on revenue from agriculture (73.33%), but 26.66% of farmers also make money from dairy, commerce, labour, and livestock. (Kaur, 2009). According to the Indian Economy report, share of agriculture in the national income has come down due to multiple reasons such as variation in climate, loss of soil fertility and reduction in crop yield (Dev, 2008; Dwivedy, 2011; Kumar and Nain, 2013; Mahendra, 2014; Joshi, 2015; Srinivasa et al., 2015; Vasu et al., 2016). India's agricultural output is influenced by climate factors like temperature, precipitation, and how much water the soil evaporates. Crop productivity in water-scarce regions has been impacted by the uneven distribution of rainfall and the lack of adequate irrigation infrastructure (Sinha et al., 1998; Kumar et al., 2004; Kumari et al., 2017; Mall et al., 2004; Mall et al., 2006). Another limiting factor that directly impacts crop productivity is the loss of soil moisture due to climatic changes. (Sarker and Biswas, 1978; Virmani et al., 1978; Biggs et al., 2008). Today, it is evident that there is a strong connection between agriculture, climate change, and water (Piao et al., 2010; Kumari et al., 2017). The three variables are linked and influence one another either directly or indirectly. A slight disturbance in one factor disturbs the whole cycle (Kumari and Patil, 2017). These factors raise the probability of agriculture risks in India. There are several risks faced by the farmers which need to be explored along with techniques that can manage these risk. These techniques not only lead to reduction in number of suicides but also minimise the risks in agriculture. A more sophisticated understanding of risk and risk management techniques has aided farmers in the risky situations and has helped policymakers assess the efficacy of various types of risk protection tools in the post-Farm Act environment, which has moved away from heavy government intervention. (Heifner et al., 1999). According to Velandia et al. (2009), the percentage of owned acres, off-farm income, education, age, and degree of business risk are the main factors influencing the adoption of risk management methods in agriculture. Information on such aspects can be used by educators, extension agents, and other information multipliers to improve risk management (Mishra and El-Osta, 2002). The majority of the risk encountered in the agriculture sector, according to Firas, 2011, is caused by drought. The rainfall fluctuations and drought conditions cannot be changed. However, farmers can undergo several adaptation strategies to sustain in agriculture. These strategies include diversification, crop rotation, creating other sources of income, adopting beneficial agriculture schemes, reducing intermediaries and creating proper storage. Farmers' response to risks pertaining to production, price or climate lead to better decision making. Farmers choose new crops or varieties depending on risk management factors such as water availability, cost-return analysis, resistance to drought and flooding, market demand, profit, and uncertain conditions. The decision making ability varies from small farmers to big farmers. Many farmers adapt multiple cropping system to mitigate the risk. There is a need of research for developing risk management strategies for farmers (Kumar et al., 2017).

Enterprise diversification, vertical integration, production and marketing contracts, insurance, and hedging in futures and options contracts are all ways to control risk. Kimura et al. (2010) investigated a range of risk management techniques, such as diversification, insurance strategy, government initiatives, and other useful technologies, in order to protect agriculture in the face of climate change. (Bhatta and Aggarwal, 2016; Sneha, 2017) . We feel that there is a need for further research in the area of risk management by farmers. Driven by the need, the research proceeds with an aim

- To explore the climate disaster risk management by farmers for sustaining agriculture
- To explore the role of government and other stakeholders in climate disaster risk management process
- To study the factors affecting farmers' disaster risk management

Researchers have investigated the different sorts of risks that farmers encounter and the managing strategies used in agriculture based on the current situation of the farmer. The study was carried out in Maharashtra. Regarding risk management in agriculture, the farmers have provided the primary data. In order to investigate the risk management strategies used by farmers for sustainable agriculture, both primary and secondary data have been gathered.

LITERATURE REVIEW

The comprehensive report on "Climate Change and Food Systems: Global Assessments" paints a detailed picture of climate change's impact on food production, security, and nutrition globally. It analyses regional vulnerabilities, explores mitigation strategies like reducing greenhouse gas emissions from agriculture, and presents adaptation options for different food systems, like diversifying crops or improving water management. "Agriculture and Climate Change: An Introduction" (World Bank): This introductory guide serves as a foundational read, defining key terms, explaining climate change mechanisms, and exploring potential impacts on agriculture, from temperature shifts to changing precipitation patterns. It delves into mitigation strategies like improved soil management and adaptation options like using drought-resistant crops. "Food Security, Nutrition and Climate Change" (IFPRI): This report tackles the intricate relationship between climate change, food security, and nutrition, particularly for vulnerable populations. It analyzes potential impacts on food availability, access, utilization, and stability, highlighting challenges like malnutrition and missed meals. "Climate Change and Agriculture: Addressing the Challenges" (UNFCCC) dives into the challenges and opportunities presented by climate change in agriculture, focusing on national and international responses. It discusses policy approaches like emissions reduction targets, technology development for climate-smart agriculture, and financial support for adaptation in developing countries. "Climate Change Impacts on Global Food Production" (FAO) projects the impacts of climate change on various aspects of global food production, including crops, livestock, fisheries, and aquaculture. It estimates yield losses and analyzes regional vulnerabilities under different climate scenarios. "Agriculture: Key to a Sustainable Future" (WRI): This report emphasizes the pivotal role of agriculture in achieving a sustainable future. It explores transformative changes needed in agricultural practices, like agroforestry, regenerative agriculture, and reducing food waste, to address climate change, conserve resources, and improve food security. "The State of Food Security and Nutrition in the World 2022" (FAO): This annual report provides a global overview of food security and nutrition, including dedicated sections on the impacts of climate change on food production and access. It analyzes trends, highlights challenges like rising food prices and conflicts, and emphasizes the need for urgent action and coordinated solutions. Consider focusing on specific regions or populations mentioned and researching specific challenges they face due to climate change. "Building Climate-Resilient Food Systems: Actions for Transforming Our World" (FAO): This report outlines concrete actions needed to build climate-resilient food systems across various components. It details diversified crop production, climate-smart agriculture practices, improved water management, and knowledge empowerment for farmers. "Food

Systems, Land Use, and Climate Change: Special Report of the IPCC": This comprehensive report by the IPCC analyzes the interactions between food systems, land use, and climate change from a scientific perspective. It highlights mitigation and adaptation strategies across different scenarios, emphasizing the need for sustainable land management and dietary shifts. "Global Food Policy Report 2023" (IFPRI): This report analyzes current trends and challenges in global food policy, including sections on climate change impacts on food systems and potential policy responses.

Agriculture risks are classified as production risk, price risk, place risk, financial risk, institutional risk and promotional risk (Kumari et al., 2023). The production risk is one of the major risks where farmers are challenged to protect their crops during climate disasters like floods, cyclones or drought. The pricing risk comprises the risk that farmers bear in getting a lower price for their produce. Improper application of fertilizers, pesticides and management practices also result in production risk. The high number of intermediaries in agriculture leads to a lower price for the farmers. Farmers also suffer from promotional risk in the form of branding the agriculture quality produce. Access to loans and credit from the banks leads to financial risks (Cromwell et al., 1992). Most of the farmers sell their products at the current price in the fear that price may decline in future. Storage of the products does not help because agriculture products deteriorate with time (Blackburn and Scudder, 2009). The farmers do not have appropriate storage conditions (Chatterjee et al., 2010; Rao et al., 2015). Improper storage conditions result in deterioration of the agriculture produce Farmers try to manage most of the risks at their level. The literature review shows that agriculture risks are a concern in the era of climate change. However there has been a research gap in highlighting the agriculture risks from the farmers' perspective. To address this research gap, the study attempts to undergo farmers' perspective on the risks faced in agriculture and discuss different risk management strategies in agriculture. The study will have a significant contribution in addressing the risks in agriculture.

METHODOLOGY

Maharashtra has been selected as the study area since it is a drought prone area. The state has high agriculture land coverage and farmers suicides have been high in the state (Farmers' suicide in India, 2018; https://en.wikipedia.org/wiki/Farmers%27_suicides_in_India). The study is based on both quantitative and qualitative data. Quantitative data has been analyzed through spss and qualitative data has been analyzed through QDA miner. The structured questionnaire has been designed to collect the data about risk management techniques used by farmers. Quantitative data has been collected from about 192 farmers across Maharashtra and interviews were also conducted with 30 farmers. Purposive sampling technique has been used for collecting response from the farmers.

RESULT AND DISCUSSION

Quantitative Data Analysis

The study shows that there is a variation in the climate of the study area and 94% farmers experience that their crop is affected by weather while only 6% of the farmers were of the opinion that crop is not affected by weather .Farmers typically use weather forecasts to manage the risk of poor weather since it is a climate risk (Carleton, 2017). Farmers typically use weather forecasts to manage the risk of poor weather since it is a climate risk (Carleton, 2017). 11% farmers think that seed and fertilizers have major effect on the crop yield while 89% of the farmers were of the opinion that seed and fertilizers did not affect crop yield much. High quality good seeds are obtained by the farmers from Seed Corporation centers to minimize the production risk. 73.44% farmers are aware of government schemes while 27% are not. Government schemes have been helpful to mitigate the Institutional risks (Kumari and Patil, 2018). Therefore, it has been found that majority of the farmers are aware of the government schemes. The only 6% farmers have well equipped farm machinery. Most of the farmers do not have well equipped machinery because, they are very costly. According to the farmers, machinery do not have a major role in managing risks for agriculture. 16% farmers think that electricity is one of the major factors which affects the crop yield for the study area. 30.73% farmers have availed crop loans from different banks and credit cooperative societies. Crop loans has been taken by farmers who are well aware of the banking procedures and documents. Most of the farmers borrow money from their relatives or friends. This is because, they do not want to undergo heavy load of paper work, documentation, security and standing in a queue. According to many farmers, a lot of time is invested in banks. Only 32.81% of farmers have crop insurance, despite the fact that farmers are aware of several government programs. The quantitative data shows that there are very few farmers who manage price and institutional risk. The correlation between variables and crop loan was done through chi square test. A large group of farmers are able to manage the climatic risk. Most of the farmers have been using modern techniques like drip irrigation, diversification of crops and contour bunding to meet the adversity of floods and drought. National Bank for Agriculture and Rural Development is working on the rural infrastructure and socio-economic index to fulfil the requirements of farmers. Ground water recharge and solar micro-irrigation programs by National Bank for Agriculture and Rural Development (NABARD) have ensured food security and enhanced resilience. NABARD has launched several programs and projects to promote sustainable agriculture and rural growth in country. To further extend the study to risk management, we have conducted qualitative study from 30 farmers.

Weather and Crop Loan

Weather and crop loan both affect the crop yield. The result shows that farmers affected adversely by weather calamity do not go for crop loan. Around 68 percent of the farmers opting weather as a major factor that affect crop yield disagree with the statement that crop loan affects the crop yield. Farmers claim that when weather calamity affects their crop, they do not go for crop loan (Chand et al., 2015; Chand, 2017). The farmers' income decreases due to weather calamity. However, 31% of the farmers affected by low yield go for crop loan. Such farmers believe that when weather destroys the crop, the loan becomes a major factor that reduces the yield as farmers do not have enough money to utilize other techniques for crop growth. Table 2 shows the Pearson chi square value to be 0.321. This means that there is a very weak insignificant relationship between crop loan and weather as major factors for affecting crop yield. Weather is not associated with the crop loan for increasing yield of farmers.

Disease and Crop Loan

Crops are easily affected by disease and pests. Table 3 and 4 shows that 30% of the farmers who said that disease is a major factor that affects crop yield also said that crop loan affects the crop yield. There

Table 1. Relationship between crop loan and weather

Crosstab					
			CropLoan		Total
			No	Yes	
Weather		Count	7	2	9
		% within Weather	77.8%	22.2%	100.0%
		% within Crop Loan	5.3%	3.4%	4.7%
		% of Total	3.6%	1.0%	4.7%
	Y	Count	126	57	183
		% within Weather	68.9%	31.1%	100.0%
		% within Crop Loan	94.7%	96.6%	95.3%
		% of Total	65.6%	29.7%	95.3%
Total		Count	133	59	192
		% within Weather	69.3%	30.7%	100.0%
		% within Crop Loan	100.0%	100.0%	100.0%
		% of Total	69.3%	30.7%	100.0%

is an association between crop loan and disease. Farmers believe that crops affected by disease leads to crop loan by farmers in order to earn their livelihood.

Insect and Crop Loan

Crops affected by insects result in low yield. Table 5 and Table 6 shows that 32% of the farmers believe that insects and crop loan both are the major factor affecting the crop yield. There is a strong association between crop loan and insects.

Qualitative Data Analysis

Qualitative study has been conducted on data collected from 30 farmers. Open-ended questions were asked through the interview. The qualitative data has been analyzed using QDA miner. The results of the qualitative data are as below.

Table 2. Chi-square test for weather and crop loan

	Value	df	Asymp. Sig. (2-Sided)
Pearson Chi-Square	.321[a]	1	.571
Continuity Correction[b]	.039	1	.844
Likelihood Ratio	.340	1	.560
Fisher's Exact Test			
N of Valid Cases	192		

Table 3. Relationship between crop loan and disease

Crosstab					
			Crop Loan		**Total**
			No	**Yes**	
Disease		Count	4	3	7
		% within Disease	57.1%	42.9%	100.0%
		% within Crop Loan	3.0%	5.1%	3.6%
		% of Total	2.1%	1.6%	3.6%
	Y	Count	129	56	185
		% within Disease	69.7%	30.3%	100.0%
		% within Crop Loan	97.0%	94.9%	96.4%
		% of Total	67.2%	29.2%	96.4%
Total		Count	133	59	192
		% within Disease	69.3%	30.7%	100.0%
		% within Crop Loan	100.0%	100.0%	100.0%
		% of Total	69.3%	30.7%	100.0%

Savings

Most of the farmers keep their savings in the form of cash or in Agri-shops. Few of them neither keep any savings nor have any other sources of income (Table 7). For managing risks, it has been found that farmers keep an amount for future use. It has been observed that very few farmers keep their savings in bank. This is because of minimum balance requirement in bank account throughout the year. In few of the banks, minimum amount requirement is too high. This makes it difficult for the farmers to keep their savings in banks.

Credit Sources

Farmers claim that banks want a lot of paper paperwork and a security deposit. As a result, their sources of income are either their family, friends, or agricultural inputs (Cromwell et al., 1992) (Table 8). This is a better and simpler method for farmers to control their financial risk. However, there are financial

Table 4. Chi-square test for crop loan and disease

	Value	**df**	**Asymp. Sig. (2-Sided)**
Pearson Chi-Square	.502[a]	1	.047
Continuity Correction[b]	.085	1	.077
Likelihood Ratio	.475	1	.091
Fisher's Exact Test			
N of Valid Cases	192		

Table 5. Relationship between crop loan and insect

			Crop Loan		Total
			No	Yes	
Insect		Count	16	4	20
		% within Insect	80.0%	20.0%	100.0%
		% within Crop Loan	12.0%	6.8%	10.4%
		% of Total	8.3%	2.1%	10.4%
	Y	Count	117	55	172
		% within Insect	68.0%	32.0%	100.0%
		% within Crop Loan	88.0%	93.2%	89.6%
		% of Total	60.9%	28.6%	89.6%
Total		Count	133	59	192
		% within Insect	69.3%	30.7%	100.0%
		% within Crop Loan	100.0%	100.0%	100.0%
		% of Total	69.3%	30.7%	100.0%

institutions and banking plans that have developed a number of services for farmers. Actually, just a few farmers make use of this. The number of farmers using the banking facility, however, has grown over time.

Strategies

Several farmers have adopted diversification of crops and drip irrigation to cope up with unfavourable conditions (Table 9). Farmers are adopting diversification in crops so that they can earn in any of crop (Kumar et al., 2004). Many of them manage the crop, livestock and fisheries for sustaining their livelihood.

Table 6. Chi-square test for crop loan and insect

	Value	df	Asymp. Sig. (2-Sided)
Pearson Chi-Square	1.207[a]	1	.272
Continuity Correction[b]	.710	1	.399
Likelihood Ratio	1.297	1	.255
Fisher's Exact Test			
N of Valid Cases	192		

Table 7. Form of savings

Code	Count	% Codes	% Cases
Agrishop	5	25.00%	50.00%
Don't keep savings	7	35.00%	50.00%
Cash	8	40.00%	50.00%

Table 8. Main credit sources available in your area

Code	Count	% Codes	% Cases
Agricultural inputs	13	31.00%	50.00%
Relatives and Friends	6	14.30%	50.00%
Bank loan	3	7.10%	50.00%

Table 9. Strategies used to deal with revenue losses brought on by the aforementioned danger circumstances

Code	Count	% Codes	% Cases
Diversification	10	16.70%	50.00%
Irrigation	8	13.30%	50.00%

We further added the qualitative data of farmers in QDA Miner as variables for Dendogram analysis. Therefore, 9 variables were created from 30 cases. These 9 variables were the questions asked during the interview and the cases were the number of farmers who were interviewed. Then we created 58 codes for all the variables. Using these codes, we performed content analysis in Wordstat (QDA Miner). We created clusters using keywords in whole documents using Jaccord's Coefficient Index and created the dendrogram. The dendrogram had the 3 clusters where Single word clusters had been removed.

Cluster Mapping, Similarity Index, and Distribution of Keywords

Figure 1, 2, 3 and 4 show the result for qualitative data analysis through QDA miner. After qualitative analysis of the interview conducted with the farmers, five clusters have been found as given below:

Figure 1. Cluster map

Figure 2. Agglomeration index

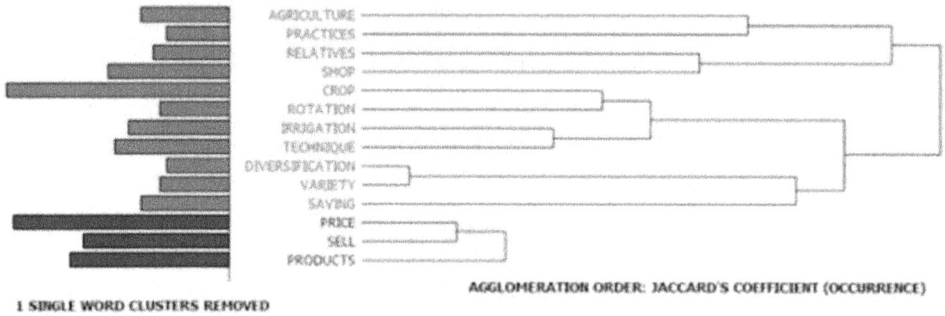

Figure 3. Similarity index

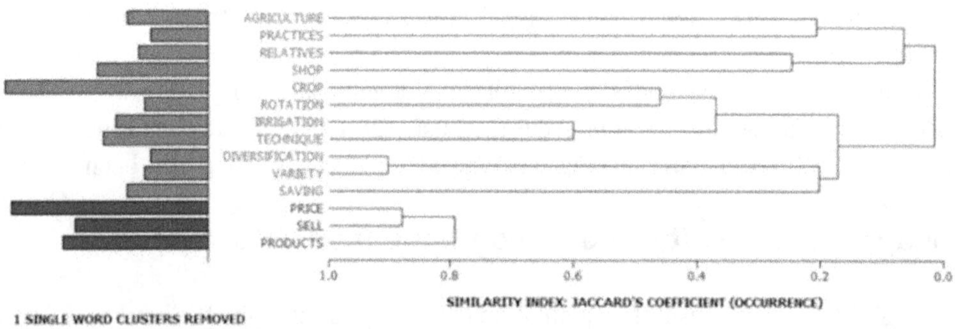

Figure 4. Distribution of keywords

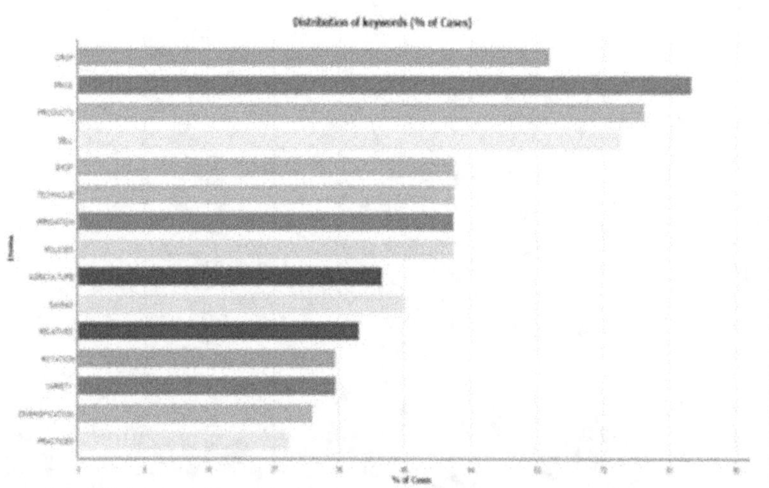

Crop Rotation and Crop

Crop and rotation were discovered to be in a cluster as a consequence of the interview. This implies that farmers cannot adequately manage risk by simply cultivating crops. Growing several crops on a plot of land is known as crop rotation. In addition to decreasing soil erosion, it also improves soil fertility (Mehar et al., 2016). According to the season and cropping pattern, crop rotation is planned (Beddington et al., 2012; Venkateswarlu, 2011). In order to boost the soil's micronutrient content, many farmers combine leguminous crops with other crops. Fish farming is used by farmers in water-prone areas to raise paddy. They earn more money as a result. (Acharya, 2006; Asplund et al.,1989).

Agriculture and Practices

Second cluster found after qualitative study is agriculture and practices. Agriculture is the cultivation of several crops (Milestad and Darnhofer, 2003). This process would increase with practices.

Variety and Diversification

Third cluster is variety and diversification. Crop diversity is the variance in the genetic and phenotypic characteristics (Kumbhar et al., 2017). This is important for risk management as it ensures food security. Several varieties resistant to disease and pests, adapted to climate change and prevent environmental degradation, and can be obtained through diversification.

Irrigation and Technique

Irrigation and technique were grouped into one as farmers usually face water scarcity. In order to meet the challenges of water scarcity, several techniques like drip irrigation and sprinkle irrigation has been practiced.

Price and Selling

Price and selling were grouped into one cluster as farmers should try to avoid intermediary channels in order to obtain right price. Direct selling by farmers often results in better pricing for the farmers. This is also a measure to reduce risk by the farmers.

Price Climate Disaster Risk Management

Most of the farmers sell their products at the current price in the fear that price may decline in future. Storage of the products does not help because agriculture products deteriorate with time (Blackburn and Scudder, 2009). The farmers do not have appropriate storage conditions. Therefore they either try to sell the product with the current price or few of them go for other places where they can sell. Price risks cannot be managed by most of the farmers due to scarcity of infrastructure and storage conditions. There have been researches to directly link the farmers with the consumers so that the farmers can manage price risk. Government has also supported farmers with Minimum Support Prices so that the farmers instead of loss can receive the minimum amount. According to the farmers there should be an increase in the minimum support prices for every crop.

Production and Yield Climate Disaster Risks Management

Farmers have been utilising different strategies and tactics over time to manage production and yield risk. To boost the production, the farmers utilise agro manures, compost, insecticides, and fertilisers. The farmers receive accurate information from Krishi Vigyan Kendra and other agro centers. The farmers aim to use high-quality seeds to reduce production risk and take part in various awareness programmes.

Institutional Climate Disaster Risk Management

All the farmers were of the opinion that the government policies have no adverse impact on agriculture. Government schemes as discussed in the literature review have supported the farmers in agriculture. The policies do affect the channel intermediaries. Most of the farmers are unaware of the government policies.

Complicated Procedures of Formal Credit Structures

Farmers face a great problem dealing with the complicated procedures of documentation. Lots of paper work, verification and security leads to overexertion of the farmers. Many farmers are not eligible for taking loan from banks due to fewer documents. In banks, the formal credit system takes a lot of time. The government has established better risk management measures than the current set of regulations. According to the farmers, more awareness campaigns should be made regarding several risk management techniques. Subsidies on fertilizers, agri inputs and machinery have been provided by the government. Insurance claims should not be time taking. New technology and methods must be introduced at Panchayat and block level. Weather forecast is very beneficial for the farmers to manage the climate risk.

Improvement of the Strategies to Cope With Climate Disaster Risks

According to the farmers it is better to change the crop when the yield is adversely affected. It is better to grow the crops as per the Socio-climatic zones. Farmers can manage their risks accordingly through weather forecast inputs. Advice from the agriculture centres are of great help (Chatterjee et al., 2010; Rao et al., 2015). Farmers try to manage most of the risks at their level. Right use of fertilizers, pesticides and management practices results in better risk management. To manage the problem of water scarcity, most of the farmers have adopted drip irrigation.

CONCLUSION AND FURTHER RESEARCH DIRECTION

Risk Management is vital to farmers in order to sustain agriculture related risks such asClimate, Price, Yield, Credit and Production (Wall and Smith, 2005). Over the time, farmers have improvised methods of farming and tried to sustain agriculture. In addition to the increase in minimum support prices, government has taken several measures for the management of institutional risk of the farmers (Sharma, 2011). Several schemes have been introduced for the welfare of farmers. Paramparagat Krishi Vikas Yojna has been introduced to promote organic farming. The Pradhan Mantri Krishi SinchaiYojna has been launched to create assured irrigation. For soil health management a scheme has been launched which issues soil health card to all the farmers. Government has also created several portals on crop insurance. Several Krishi channels have also aided in creating awareness about various issues. The authors have discussed the current types of risks and the management strategies used by the farmers. Stabilization of prices and National Agriculture Market which overcomes the impediments of agricultural products and provides better price are some of the important measures in agricultural risk management. The findings present the methods and strategies used by the farmers to cope up with risks. However, the study is based on cross sectional data, improved results can be gained by longitudinal data. The trends of the risk management techniques by farmers can help in understanding and comparison of the changing scenario. The authors feel that the research only focus on farmers' views and strategies. It can be extended by studying new risk management strategies and the upcoming barriers and enablers pertaining to them. The study can be extended with a wider sample size, secondary data and the management strategies followed by the farmers in other parts of the country.

Conflict of Interest: There is no conflict of interest among the authors

Funding: NA

Ethical approval: Not applicable

REFERENCES

Acharya, S. S. (2006). Agricultural marketing and rural credit for strengthening. *Indian Agriculturist.* INRM Policy Brief No. 3,Asian Development BankbIndia Resident Mission (INRM), https://casi.sas.upenn.edu/sites/default/files/iit/ADB%20-%20INRM%20Policy%20Brief%20No.%203.pdf. Accessed on 29 March 2024

Agriculture and Climate Change: An Introduction. (2012). World Bank. https://openknowledge.worldbank.org/server/api/core/bitstreams/93d709b7-4b38-548b-bd0a-d5431b34f296/content. Accessed on March 28 2024

Agriculture: Key to a Sustainable Future. (2021). The World Resources Institute (WRI). https://www.wri.org/research/agriadapt-strategies-participatory-design-and-development-climate-risk-tool. Accessed on 28 March 2024

Asplund, N. M., Forster, D. L., & Stout, T. T. (1989). Farmers' use of forward contracting and hedging. *Review of Futures Markets, 8*(1), 24–37.

Beddington, J. R., Asaduzzaman, M., Fernandez, A., Clark, M. E., Guillou, M., Jahn, M. M., ... Scholes, R. J. (2012). *Achieving food security in the face of climate change.* Final report from the Commission on Sustainable Agriculture and Climate Change.

Bhatta, G. D., & Aggarwal, P. K. (2016). Coping with weather adversity and adaptation to climatic variability: A cross-country study of smallholder farmers in South Asia. *Climate and Development, 8*(2), 145–157. doi:10.1080/17565529.2015.1016883

Biggs, T. W., Mishra, P. K., & Turral, H. (2008). Evapotranspiration and regional probabilities of soil moisture stress in rainfed crops, southern India. *Agricultural and Forest Meteorology*, *148*(10), 1585–1597. doi:10.1016/j.agrformet.2008.05.012

Blackburn, J., & Scudder, G. (2009). Supply chain strategies for perishable products: The case of fresh produce. *Production and Operations Management*, *18*(2), 129–137. doi:10.1111/j.1937-5956.2009.01016.x

Building C.-R. F. S. Actions for Transforming Our World. (2023). The Food and Agriculture Organization of the United Nations (FAO.). https://www.fao.org/3/I9900EN/i9900en.pdf

Carleton, T. A. (2017). Crop-damaging temperatures increase suicide rates in India. *Proceedings of the National Academy of Sciences*, 201701354. 10.1073/pnas.1701354114

Chand, R. (2017). *Doubling Farmers' Income Rationale*. Srategy, Prospects and Action Plan.

Chand, R., Saxena, R., & Rana, S. (2015). Estimates and analysis of farm income in India, 1983–84 to 2011–12. *Economic and Political Weekly*, *50*(22), 139–145.

Chatterjee, C., Schneider, S., Peter, M., & Mayer, T. (2010). *Risk management in agriculture*. Deutsche Bank Research.

Climate Change and Agriculture: Addressing the Challenges. (2012). The United Nations Climate Change (UNFCCC). https://unfccc.int/resource/docs/2012/smsn/ngo/163.pdf

Climate Change and Food Systems. Global Assessments. (2023). The Food and Agriculture Organization of the United Nations (FAO). https://www.fao.org/policy-support/tools-and-publications/resources-details/en/c/435192/. Accessed on 28 March 2024

Climate Change Impacts on Global Food Production. (2019). Food and Agriculture Organization of the United Nations (FAO). https://www.fao.org/fileadmin/templates/est/meetings/climate_2013/Wheeler_CC_impacts_on_FoodSecurity.pdf. As Accessed on 28 March 2024

Cromwell, E., Friis, H., & Turner, M. (1992). *The seed sector in developing countries: A framework for performance analysis*. Academic Press.

Dev, S. M. (2008). *Challenges for revival of Indian agriculture. In 1st Prof. Dayanand Jha Memorial Lecture*. National Centre for Agricultural Economics and Policy Research.

Dwivedy, N. (2011). Challenges faced by the Agriculture Sector in Developing Countries with special reference to India. *International Journal of Rural Studies*, *18*(2), 1–6.

Farmers' Suicide in India. (n.d.). https://en.wikipedia.org/wiki/Farmers%27_suicides_in_India

Firas, Y. (2011). *Agricultural Risks and Risk Management of Different Farming Systems in Syria*. Academic Press.

Food Security, Nutrition and Climate Change. (2016). The International Food Policy Research Institute (IFPRI). https://www.ipcc.ch/site/assets/uploads/sites/4/2020/02/SPM_Updated-Jan20.pdf. Accessed on March 28 2024

Global Food Policy Report. (2023). The International Food Policy Research Institute (IFPRI). https://www.ifpri.org/publication/global-food-policy-report-2023-rethinking-food-crisis-responses as retrieved on 28th March 2024.

Heifner, R., Coble, K., Perry, J., & Somwaru, A. (1999). *Managing risk in farming: concepts, research, and analysis* (J. L. Harwood, Ed.). US Department of Agriculture, Economic Research Service.

Johnston, B. F., & Mellor, J. W. (1961). The role of agriculture in economic development. *The American Economic Review*, *51*(4), 566–593.

Joshi, P. A. (2015). Challenges of agriculture economy of India. *The Business & Management Review*, *5*(4), 211.

Kaur, G. (2009). *Socio-economic status of farmers during pre and post liberalization: A case study from Punjab* [M.Phil. Dissertation]. School of Management and Social Sciences, Thapar University, Patiala, India.

Kimura, S., Antón, J., & LeThi, C. (2010). *Farm level analysis of risk and risk management strategies and policies*. Academic Press.

Krishna Kumar, K., Rupa Kumar, K., Ashrit, R. G., Deshpande, N. R., & Hansen, J. W. (2004). Climate impacts on Indian agriculture. *International Journal of Climatology*, *24*(11), 1375–1393. doi:10.1002/joc.1081

Kumar, H. V., Shivamurthy, M., Gowda, V. G., & Biradar, G. S. (2017). Assessing decision-making and economic performance of farmers to manage climate-induced crisis in Coastal Karnataka (India). *Climatic Change*, *142*(1-2), 143–153. doi:10.1007/s10584-017-1928-x

Kumari, S. (2017). Review on developing a conceptual framework for technology adoption towards sustainability in agro based industry. *SAMVAD*, *13*, 14–19.

Kumari, S., & Patil, Y. (2017). Achieving Climate Smart Agriculture with a Sustainable Use of Water: A Conceptual Framework for Sustaining the Use of Water for Agriculture in the Era of Climate Change. In Reconsidering the Impact of Climate Change on Global Water Supply, Use, and Management (pp. 122-143). IGI Global. doi:10.4018/978-1-5225-1046-8.ch008

Kumari, S., Patil, Y., & Rao, P. (2017). An Approach to Sustainable Watershed Management: Case Studies on Enhancing Sustainability with Challenges of Water in Western Maharashtra. In Reconsidering the Impact of Climate Change on Global Water Supply, Use, and Management (pp. 252-271). IGI Global.

Kumari, S., & Patil, Y. B. (2018). Enablers of sustainable industrial ecosystem: Framework and future research directions. *Management of Environmental Quality*.

Kumari, S., Venkatesh, V. G., & Shi, Y. (2023). Assessment of Risks and Risk Management for Agriculture Supply Chain. In *Supply Chain Risk and Disruption Management: Latest Tools, Techniques and Management Approaches* (pp. 155–172). Springer Nature Singapore. doi:10.1007/978-981-99-2629-9_8

Kumbhar, V., Singh, T. P., & Kumari, S. (2017). The study of developing a sustainable crop production patternthrough trend analysis in Southern Region of Maharashtra: A case study of Satara District. *International Journal of Applied Business and Economic Research*, *15*(2), 149–160.

Mahendra Dev, S. (2014). *Small farmers in India: Challenges and opportunities. "Emerging Economies Research Dialogue"*. Indian Council for Research on International Economic Relations.

Mall, R. K., Lal, M., Bhatia, V. S., Rathore, L. S., & Singh, R. (2004). Mitigating climate change impact on soybean productivity in India: A simulation study. *Agricultural and Forest Meteorology*, *121*(1), 113–125. doi:10.1016/S0168-1923(03)00157-6

Mall, R. K., Singh, R., Gupta, A., Srinivasan, G., & Rathore, L. S. (2006). Impact of climate change on Indian agriculture: A review. *Climatic Change*, *78*(2-4), 445–478. doi:10.1007/s10584-005-9042-x

Maxwell, D., Levin, C., & Csete, J. (1998). Does urban agriculture help prevent malnutrition? Evidence from Kampala. *Food Policy*, *23*(5), 411–424. doi:10.1016/S0306-9192(98)00047-5

Mehar, M., Mittal, S., & Prasad, N. (2016). Farmers coping strategies for climate shock: Is it differentiated by gender? *Journal of Rural Studies*, *44*, 123–131. doi:10.1016/j.jrurstud.2016.01.001

Milestad, R., & Darnhofer, I. (2003). Building farm resilience: The prospects and challenges of organic farming. *Journal of Sustainable Agriculture*, *22*(3), 81–97. doi:10.1300/J064v22n03_09

Mishra, A. K., & El-Osta, H. S. (2002). Managing risk in agriculture through hedging and crop insurance: What does a national survey reveal? *Agricultural Finance Review*, *62*(2), 135–148. doi:10.1108/00214930280001134

Piao, S., Ciais, P., Huang, Y., Shen, Z., Peng, S., Li, J., ... Friedlingstein, P. (2010). The impacts of climate change on water resources and agriculture in China. *Nature*, *467*(7311), 43–51. doi:10.1038/nature09364 PMID:20811450

Rao, C. S., Lal, R., Prasad, J. V., Gopinath, K. A., Singh, R., Jakkula, V. S., & Virmani, S. M. (2015). Potential and challenges of rainfed farming in India. In *Advances in agronomy* (Vol. 133, pp. 113–181). Academic Press.

Sarker, R. P., & Biswas, B. C. (1978). Agricultural Meteorology in India: A Status Report Agroclimatological Research Needs of the Semi-Arid Tropics. In *Proceedings of the International Workshop on the Agroclimatological Research Needs of the Semi-Arid Tropics*, International Crops Research Institute for the Semi-Arid Tropics.

Sharma, V. P. (2011). *India's Agricultural Development under the New Economic Regime: Policy Perspective and Strategy for the 12th Five Year Plan.* Indian Institute of Management.

Sinha, S. K., Singh, G. B., & Rai, M. (1998). *Decline in crop productivity in Haryana and Punjab: Myth or reality.* Indian Council of Agricultural Research.

Sneha, K. (2017). *Exploration and development of a sustainable agro based industrial ecosystem model with special reference to sugar industry.* Academic Press.

Srinivasa. (2015). Potential and Challenges of Rainfed Farming in India. *Advances in Agronomy*, *133*, 113–181. doi:10.1016/bs.agron.2015.05.004

The State of Food Security and Nutrition in the World. (2022). The Food and Agriculture Organization of the United Nations (FAO). https://www.fao.org/documents/card/en?details=cc0639en. Accessed on 28 march 2024

UNFCCC. (2016) Intended Nationally Determined Contributions (INDCs). http://www4.unfccc.int/Submissions/INDC/Submission%20Pages/submissions.aspx. Accessed on 29 March 2024

Vasu, Singh, S. K., Ray, S. K., Duraisami, V. P., Tiwary, P., Chandran, P., Nimkar, A. M., & Anantwar, S. G. (2016). Soil quality index (SQI) as a tool to evaluate crop productivity in semi-arid Deccan plateau, India. *Geoderma*, *282*, 70–79. doi:10.1016/j.geoderma.2016.07.010

Velandia, M., Rejesus, R. M., Knight, T. O., & Sherrick, B. J. (2009). Factors affecting farmers' utilization of agricultural risk management tools: The case of crop insurance, forward contracting, and spreading sales. *Journal of Agricultural and Applied Economics*, *41*(1), 107–123. doi:10.1017/S1074070800002583

Venkateswarlu. (2011). *ICAR sponsored Training Course on "Sustainable Agriculture Production through Innovative Approaches for Enhanced Livelihoods" 2-15th September, 2011*. Academic Press.

Virmani, S. M., Sivakumar, M. V. K., & Reddy, S. J. (1978). Climatological Features of the Semi-Arid Tropics in Relation to the Farming Systems. *International Workshop on the Agroclimatological Research Needs of the Semi-Arid Tropics*, 5-16.

Wall, E., & Smit, B. (2005). Climate change adaptation in light of sustainable agriculture. *Journal of Sustainable Agriculture*, *27*(1), 113–123. doi:10.1300/J064v27n01_07

Chapter 4
A Policy Framework for Sustainable Marketing Practices and Brand Evaluation Through Instagram Marketing

Syed Muhammad Jalal
Department of Management Studies, Middle East College, Oman

Shahid Khalil
School of Transportation and Logistics, Malaysia University of Science and Technology, Malaysia

Iqra Arshad
Graduate School of Public Administration (GSPA), National Institute of Development Administration, Thailand

Muhammad Khuram Khalil
https://orcid.org/0000-0001-5336-2994
Department of Management Studies, Middle East College, Oman

ABSTRACT

The negative impact that individuals have on the natural world led to the development of the concept of sustainable marketing. Increasing consumer trends regarding environmentally conscious consumers have created a fresh market chance for the global economic system. Sustainability refers to mankind's general capacity to engage with the many facets of our current situation that demand cognitive focus. In consequence, the objective of this study was to investigate how sustainability marketing practices have an influence on brand evaluation. This research shows how people think about the new sustainability marketing strategies. The current research also includes the mediating role of fashion consciousness among sustainable marketing practices and brand evaluation in the context of Oman. Non-random and convenient sampling techniques are used to collect the data from 336 respondents. SmartPLS 4.0 was employed for data analysis. Findings reveal that sustainable marketing practices have a significant impact on brand evaluation. However, fashion consciousness did not work as a mediator in between sustainable practices and brand evaluation as all concerned mediating hypotheses were not supported.

DOI: 10.4018/979-8-3693-4864-2.ch004

BACKGROUND AND INTRODUCTION

The industry for eco - fashion has experienced significant expansion because of increasing knowledge of the harmful effects of the fashion business on both individuals and the environment (Voss, 2005). At the identical time, as the sales occasion becomes a worldwide trend. Many of the eco-friendly clothing companies that decide to take part in the event advertise their offerings on Instagram. Many brands employ different tactics to acquire a strategic advantage and uphold their responsible business identities (Font & McCabe, 2017). It is more crucial than ever to practice sustainability marketing. Since it is closely related to environmental protection, sustainable marketing is seen as an effective technique for marketing products, activities, and ways of conducting business. Consumers of efficient and environmentally products are considered as a major driving force behind consumption, making them a top concern for businesses operating in the current economic environment. Increasing consumer trends about environment products have created a new potential market for the global market (Khalil, 2022; Kemper & Ballantine, 2019). The negative impact that humanity have on the ecosystem gave rise to the concept of "sustainable marketing." Long-term advantages for the society include cleaner air and water, fewer pollution, and more renewable energy sources. Sustainable practices contribute significantly to culture. Developing a commitment to sustainability helps reduce the amount of toxins that enter the environment (Khalil et al., 2022). By implementing corporate practices with the least possible adverse environmental impact, this study is going to demonstrate how organizations should change their goals and behaviors from being profit-driven to being environmentally conscious. Housing standards that are greener and better. Practicing on sustainability marketing can influence positively on the environment of Oman. These practices can be helpful to control the impact of harmful substances into environment. Sustainability has a significant place into the Oman vision 2040 and the government is giving awareness to the people of Oman. Therefore, the objective of the chapter is to investigate the impact of sustainable marketing practices on brand evaluation via mediating role of fashion consciousness.

LITERATURE REVIEW

To show the value of sustainability marketing practices in the content of existing literature. Sustainable marketing encompasses ideas and activities related to item alterations, sustainability consciousness, fashion consciousness, purchase attitude and sustainable marketing practices. Sustainable marketing is an organizational strategy that entails recognizing, predicting, and meeting the wants and requirements of customers (Athwal et al., 2019). These activities will be carried out using strategies that are sustainable. Some of the obligations of the businesses is to become accustomed to the environs rapid fluctuations (Varadarajan, 2015). Due to escalating issues while focusing on offering secure goods. Sustainable advertising and related strategies are key tools for enhancing business results and guaranteeing the long-term viability of firms (Quaye & Mensah, 2019).

But over time, the definition of sustainable advertising has evolved along with segmentation of customers and sustainability goals. According to Lim (2017) the discipline of "sustainable advertising" involves coming up with a variety of strategies to connect with customers that are more inclined to be mindful of environmental issues. Therefore, ecological customers would be regarded as the main customers who will refrain from buying goods that might have an adverse effect on the atmosphere. Furthermore, companies as well as people regard sustainability challenges as a key problem. According to another research, ecologic consciousness has a major impact on every industry (Khalil & Muneenam, 2021; Dangelico & Vocalelli, 2017). On the contrary, sustainability consciousness is an important factor that will influence customers' buying habits. However, there has been much worry over the general population's deficit of knowledge of sustainable goods. Such sustainable conscious shoppers use a variety of techniques to preserve the natural world (Rajeev et al., 2017). As a result, environmentally friendly products and their associated advertising have an effect on environment. Marketing regarding consumer opinions of ecological goods have grown more common as time passes. On the contrary, it's critical to keep in mind that the information contained in environmentally friendly goods needs to be readily available to a broad variety of consumers.

Underpinning Theory

The theoretical underpinnings of (Hart & Dowell, 2010) basic based on resources perspective concept serve as the framework for the current investigation. The natural resource-based view (NRBV) concept is a framework for examining developments in technology that a company can apply to deal with challenges related to the environment (Hart & Dowell, 2010). According to Hart & Dowell (2010), one of the primary drivers for fresh skills in company is an organization's surroundings and its constraints (Ashby, 2018). The theoretical framework of Resource-Based Vision concept is the foundation of Natural RBV theory (Rehman et al., 2020).

According to the NRBV principle, a corporation can achieve economic and ecological sustainability and, as a result, gain an edge over others by utilizing three interconnected essential tactical talents: pollutant avoidance, responsible handling of products, and environmentally friendly growth. These concepts combine to form the NRBV model and enlarge it by taking into account how a firm interacts with its surroundings in nature. Environment-related factors are taken into account while designing goods, and responsibility for products ideas go beyond the operations of the business. The process prioritizes reducing the detrimental effects on the ecosystem (Hart & Dowell, 2010).

Ad Skepticism

The data offered in environmentally friendly advertisements has significance to their buying choices, according to Luo et al. (2020), who also claim that green customers assign better informative value to sustainable advertisements. Yu (2018), discovered that skepticism had less mistrust toward sustainable advertisements, in contrary to earlier incorrect results. They did, however, observe that green customers pay closer attention to expensive marketing. The researchers looked into the development of advertisements and discovered that green customers study commercials for an extended period of time (Boerman et al., 2017). They contend that customers will favourably assess advertising claims if they are reliable and demonstrate "rich informative value." On the other hand, "limited informative value" will increase advertising suspicion. According to Patel et al. (2016), sustainable advertisements are less convincing to customers who are highly engaged in the environment than they are to those who are less engaged.

H1: Ad skepticism positively impact on brand evaluation.

H1a: Ad skepticism positively impact on fashion consciousness.

H1b: Fashion consciousness is a mediating in between Ad skepticism and brand evaluation.

Purchase Attitude

According to Rana & Paul (2017), on the standpoint of behavioral psychological research, attitudes are suggested to be related to the level of want or unwanted activity. Purchase attitude is a perception of having good or bad attitudes towards actions. However, attitudes as "Viewpoint" that were instinctive reactions to behave, either intentionally or unintentionally (ElHaffar et al., 2020). Based on the evaluations described above, we realize that an effective company reputation may unintentionally influence customer purchasing behavior. The reputation of a company thus has a significant effect in how customers make purchases (Boz et al., 2020).

H2: Purchase attitude positively impact on brand evaluation.

H2a: Purchase attitude positively impact on fashion consciousness.

H2b: Fashion consciousness is a mediating in between purchase attitude and brand evaluation.

Sustainability Consciousness

Sustainability consciousness, operates and functions in accordance with the current surroundings, environmental structure, and leadership viewpoint. The exact same collection of ideas or system of thinking that achieves the function of an intermediary among two significant factors like SCR and SP is the source of both awareness and knowingness. The framework of CAS describes how understanding of sustainability affects SCR and long-term effectiveness significantly (Haider et al., 2022).

Ideas, views, feelings, and ways of seeing things that improve situations, capacities, and skills comprise SK or SC. Sustainable growth, which incorporates norms of society and culture, opinions, and views of opinion, is linked to sustainable consciousness (Gomez-Trujillo et al., 2020). The research investigation of sustainability consciousness categorizes 3 significant environmental, social, and biological factors that strengthen the connection among SCR and SP (Kumar & Alok, 2020). While retaining its business obligations and efficiency, sustainability consciousness creates development and establishes benefits programs for individuals that could potentially reduce unemployment and create jobs for the jobless. Sustainability consciousness is always set up on an ongoing schedule so that every single individual can be helped (Mensah, 2019).

H3: Sustainability consciousness positively impact on brand evaluation.

H3a: Sustainability consciousness positively impact on fashion consciousness.

H3b: Fashion consciousness is a mediating in between sustainability consciousness and brand evaluation.

Fashion Consciousness

In order to recognize the importance of branding in enhancing their reputations, individuals living in developing nations are beginning to exhibit increasing levels of feeling self-conscious (Yang et al., 2017). The researchers Kautish & Sharma (2018) said that the two facets of feeling insecure that have previously been recognized in writing are two types of feelings of self-worth the general side denotes how an individual presents themselves to the world or how people perceive them, and whereas the personal side represents a hidden part of a personality that is not apparent to other individuals. Societal insecurity is said to be correlated with both style engagement and fashion consciousness. The feeling of self-consciousness has received extensive attention on the study, but research investigating it in connection with buying habits have just lately taken place (Hur & Cassidy, 2019). According to Sailer et al. (2022)

this study notably takes into account the factor "fashion-consciousness," that is defined as the degree of societal nervousness related to the adoption of clothing. According to the research, "the individual's level of engagement with trends in clothes or trends and indicates a keen interest in the latest fashions and an individual's looks" is what is meant by fashion consciousness (Todeschini et al., 2017).

H4: Fashion consciousness negatively influence on brand evaluation.

Brand Evaluation

A key idea in corporate strategy is brand worth. The assessment of the value of brands gained popularity in the academic world and caught the interest of the corporate sector in the course of the 1980s, when consolidations and mergers among businesses were on an upward trajectory (Palusuk et al., 2019). Utilizing the concepts of impartiality and measurement academics pursue research of company value assessment (Gürhan-Canli et al., 2018). Researchers Lo et al., (2017) describe the importance of brands evaluation from many perspectives, and they propose several methods for assessing it as a result; the academic community has not yet developed an integrated framework of brand assessment.

Research Methodology

Finding answers to previously unanswered questions in order to add to the corpus of information in an area of study is the primary goal of study. The way the study was conducted plays a role in assessing its reliability (Greening, 2019). The collection of standards or methods employed in the investigation, as well as the ideas, assumptions, and precepts that underpin the methodology of the investigation. Additionally, the study's technique part emphasizes the techniques utilized to acquire or gather the study information and illustrates the technique of analysis (Ragab & Arisha, 2017). These days all of the adults and even teenagers are using social media and Instagram has one of the famous social media platforms. Almost all the brands have pages and doing marketing on Instagram and its very effective if the brand could do marketing on daily basis. The brands are using the tool called Reels on Instagram to do marketing and also they advertise their page on the Instagram feed.

The structure for marketplace study methods and processes used by the investigator is referred to as the design of the study. It allows the investigator to set up further investigations in an obvious and acceptable manner and to employ the right methodologies for the present investigation (Abutabenjeh & Jaradat, 2018). Therefore, the methodology of research includes data collection, accurate data evaluation, and information interpretation. To improve the research's both inside and outside reliability, researcher must ensure that the information collection procedure has the least amount of prejudice feasible. The research uses an approach that is experimental. A study conducted using the technique of experimentation frequently includes a pair of factors, one of which is typically fixed and the other of which can be measured (Siedlecki, 2020). The research's actual methodology, which depends on data assessment to support the theory, was employed.

The population is the group of individuals offered as a statistically representative sample for analysis. As an outcome, the term "population" refers to any group of people who have something in similar. Instead of the entire population, the word "sample" can instead relate to a sufficiently significant representative sample of that group. Therefore, an examination of statistics of the sample ought to generate an estimate of the average deviations or standard error of the results for the whole population. The absence of a standard error could only be achieved via a population wide analysis. In order to comprehend the various attitudes and behaviors people have toward the surroundings, a survey was conducted in which 336 people responded. All the people responded have good knowledge of social media and they are most likely to use Instagram and have the abilities to shop from here.

A strategy called likelihood sampling involves selecting sample or portions of the general population randomly. Likelihood is the method to use because the study needs statistics that are representative of the whole population. Each member of the general population has an equal probability of getting selected using this approach. To ensure that the group being studied accurately reflects the entire population, it is crucial to choose an adequate sample number. Non- Random convenient sampling technique was used in this research to collect the data. The people participated in the survey was 336.

When a scale is considered accurate it can be said that it accurately captures the idea it is meant to measure. The research's criteria is evaluated using both convergent and discriminant validity. Validity for convergence is shown when elements in an indicator congregate to show the root cause. Additionally, numerically convergent reliability arises if the AVE (average variance extracted) value is greater than 0.50. Durability is the measuring of a construct's inner coherence (Hammer & Spiker, 2018). Using the SMART-PLS tool, Cronbach's Alpha and Composite Reliability coefficients are utilized to gauge the accuracy of the study's results. The subsequent section of the questionnaire has 29 closed-ended questions that cover the following five factors: sustainability consciousness (Gericke et al., 2018), fashion consciousness (Gul & Shahzad, 2014), purchase attitude (Hien et al., 2020), ad skepticism (Aji & Sutikno, 2015) and brand evaluation (Hien et al., 2020).

A survey was used in this research to collect firsthand information collected from customers to ascertain their environmental sentiments. The objective of the statistical survey is to create a set of inquiries to ascertain the relationship between factors and offer more precise data in numbers. The study's technique of gathering information used a Google Survey to disseminate the survey digitally. The survey is divided into two parts, the first of which asks three questions about demographics on a scale known as Likert. The survey addresses all of the study's goals using a rating system based on 5 Point Likert scale that includes strongly disagree, disagree, neutral, strongly agree, and agree assessment.

Data Analysis and Techniques

Smart PLS was implemented for the SEM (Structural Equation Modeling). SEM is applied to collect the data for the study. It is also applicable for the complex model studies and it is appropriate for the study which has large number of factors (Frempong et al., 2021). Smart PLS used to collect the data for structural equation modeling helps the researchers or readers to understand the theories that are applied to collect the data and it proves the theories are validate and reliable. Inner and outer models are two sub parts of the SEM. The inner model refers to the link between independent and dependent variables, however outer model refers to the connection between specific indicators and dependent variables (Khalil & Muneenam, 2021). The highest variance may be shown that is shared with the other variables only after squaring the correlation. We draw the conclusion that there certainly is no standard method of flaw since the biggest common variation was only about 0.1 (Memon et al., 2021). The findings of this investigation are given using the methodology outlined. First, according to methodology, it was necessary to assess the accuracy and dependability of the pertinent factors by producing precise estimates for the outward structure utilizing the PLS technique, also known as the measurement method. Additionally, a model of the structure was computed in the SmartPLS using the internal structure's scale mechanism (Sarstedt & Cheah, 2019).

Table 1. Demographic results

Specifics	Description	Values	Percentage
Gender	Male Female Total	148 188 336	44% 56% 100%
Age Group	20-30 30-40 40-50 Total	235 74 27 336	69.9% 21.7% 8.3% 100%
Education Level	High School Undergraduate Post-Graduate Others Total	44 178 108 6 336	13.4% 53% 32.7% 0.2% 100%

Demographic Information (n=336)

Instrument Design

Five important factors are identified by the 29 questions made up the questionnaire for the present report. A number of researchers who have published works on the topics of sustainability consciousness, fashion consciousness, purchase attitude, ad skepticism and brand evaluation provided the inquiries that were employed in the present research to assess every concept using the initial determine hypothesis. A note of reference from the investigator's college was attached to the surveys, and it described the goals and strengths of the study. Complete anonymity and privacy for those who responded were maintained.

While performing the real poll, some issues were turned around from negative to beneficial or vice versa to prevent skewed findings. To establish the path of the association, the investigator has also conducted several causation studies; nevertheless, for the sake of simplicity, the outcomes are not presented in the paper. Structure equation modeling (SEM) is carried out using SmartPLS 4.0.

Reliabaility and Validity

Measurement Model Results

For the factors utilized in this investigation, CR, and AVE findings are displayed below. The findings overwhelmingly demonstrate that all CR data are greater than 0.7. The bare necessities that can be met. The averages of AVE for every factor are greater than 0.5, which satisfies the requirements for an acceptable convergent validity. According to validity principles, any factor assessment has a strong relationship with the factor in question alone and has weaker ties to the other factors in the investigation. The Fornell Larcker testing and cross loading serve as two measures for discriminating validity. The Fornell Larcker testing is an experiment for assessing the factor's degree of discriminatory accuracy, whereas cross loading is a study to evaluate the measure of the level of validity for discrimination.

Figure 1. Measurement model

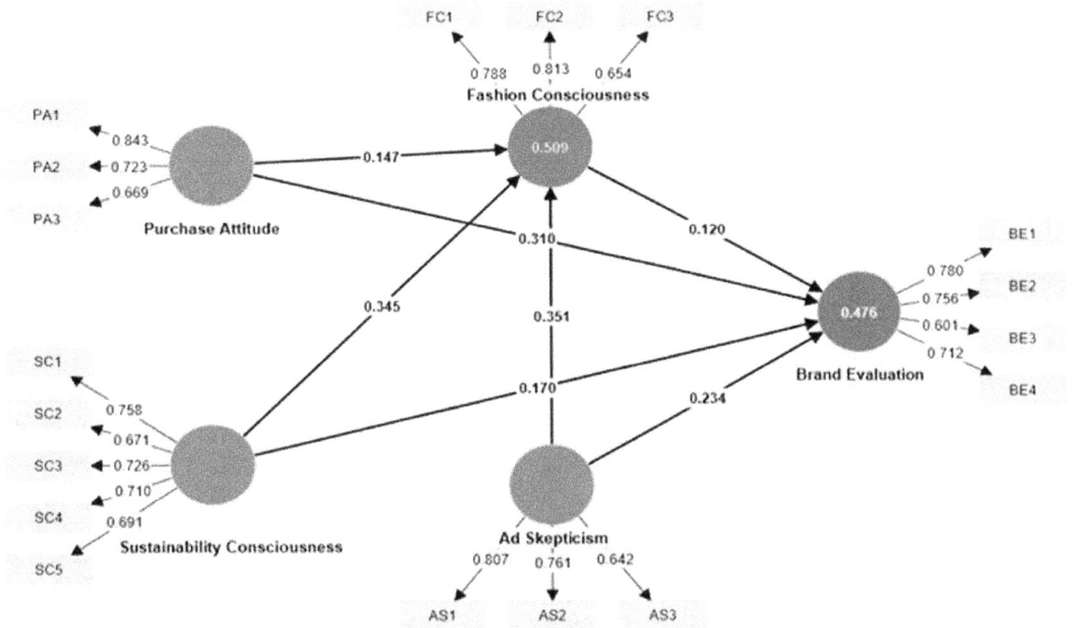

Table 2. Results of CR and AVE

	CR	AVE
Ad Skepticism	0.782	0.547
Brand Evaluation	0.806	0.512
Fashion Consciousness	0.798	0.570
Purchase Attitude	0.791	0.560
Sustainability Consciousness	0.837	0.507

Cross Loading

Findings show that there is a strong correlation between every metric and their factors.

Fornell-Larcker Criterion

Fornell-Larcker provides an illustration of the findings from the examination. A concept has to vary from its signs in addition to how it differs from the measurements of additional factors in order to be considered legitimate (Thirathon et al., 2017). The findings indicate that every factor have the greatest amount of variation among themselves. This research also satisfies the standards for the ratio of hetero-trait to monotrait (HTMT) (Henseler et al., 2016).

Table 3. Cross loading

	Ad Skepticism	Brand Evaluation	Fashion Consciousness	Purchase Attitude	Sustainability Consciousness
AS1	**0.807**	0.484	0.506	0.471	0.494
AS2	**0.761**	0.415	0.495	0.434	0.433
AS3	**0.642**	0.357	0.359	0.273	0.258
BE1	0.400	**0.780**	0.399	0.459	0.394
BE2	0.424	**0.756**	0.338	0.396	0.340
BE3	0.332	**0.601**	0.297	0.375	0.314
BE4	0.459	**0.712**	0.468	0.460	0.483
FC1	0.508	0.472	**0.788**	0.480	0.503
FC2	0.525	0.423	**0.813**	0.399	0.480
FC3	0.350	0.296	**0.654**	0.297	0.413
PA1	0.444	0.485	0.462	**0.843**	0.420
PA2	0.294	0.376	0.348	**0.723**	0.363
PA3	0.461	0.464	0.365	**0.669**	0.458
SC1	0.485	0.482	0.498	0.436	**0.758**
SC2	0.309	0.356	0.425	0.355	**0.671**
SC3	0.437	0.413	0.476	0.439	**0.726**
SC4	0.345	0.360	0.374	0.349	**0.710**
SC5	0.342	0.300	0.411	0.382	**0.691**

Table 4. Fornell-Larcker criterion

	Ad Skepticism	Brand Evaluation	Fashion Consciousness	Purchase Attitude	Sustainability Consciousness
Ad Skepticism	0.740				
Brand Evaluation	0.570	0.715			
Fashion Consciousness	0.620	0.535	0.755		
Purchase Attitude	0.542	0.596	0.529	0.748	
Sustainability Consciousness	0.547	0.545	0.619	0.555	0.712

Structure Equation Modeling (SEM)

The Framework Prototype and Validation of the Hypotheses

The theoretical framework calculates the correlation among the research's specified parameters. Little more than statistical evaluation is involved. The findings of the structural framework, including a beta value which represents the magnitude of the explicable factors' influence on the variable that is dependent. The path of the effect is indicated by a symbol next to the beta number. T-values and p-values are also provided by the framework concept. To gauge the importance of the link, these indicators are

Table 5. HTMT list

	Heterotrait-Monotrait Ratio (HTMT)
Brand Evaluation <-> Ad Skepticism	0.892
Fashion Consciousness <-> Ad Skepticism	1.002
Fashion Consciousness <-> Brand Evaluation	0.792
Purchase Attitude <-> Ad Skepticism	0.886
Purchase Attitude <-> Brand Evaluation	0.921
Purchase Attitude <-> Fashion Consciousness	0.842
Sustainability Consciousness <-> Ad Skepticism	0.790
Sustainability Consciousness <-> Brand Evaluation	0.735
Sustainability Consciousness <-> Fashion Consciousness	0.890
Sustainability Consciousness <-> Purchase Attitude	0.816

employed. T-values must be more than 1.96 or p-values must be lower than 0.05 to indicate a meaningful association. Similar to symbols with beta values, signs with t-values likewise indicate the trajectory of the connection.

The Structure Model

This study was based on four main hypotheses and three sub hypothesis which have direct relationships, and only one hypothesis was not supported following table number 6.

For example, Ad Skepticism has a positive and significant impact on Brand Evaluation (p= 0.000, β=0.234 & T Value is 3.639) that leads H1 as supported hypothesis. Similarly, the sub hypothesis of H1

Figure 2. Structure equation modeling

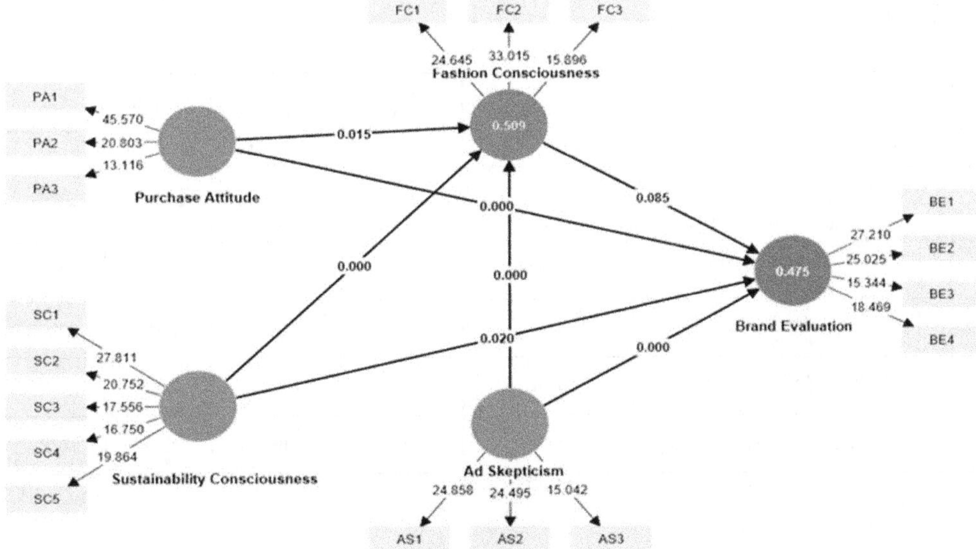

Ad Skepticism has a positive and significant impact on Fashion Consciousness (p= 0.000, β=0.351 & T value is 4.231) H1a is also supported. Furthermore, purchase attitude has a positive and significant influence on brand evaluation (p=0.000, β=0.311 & p value is 0.000) which means H2 is supported. Moreover, sub hypothesis of H2 purchase attitude has a positive and significant impact on fashion consciousness (p=0.015, β=0.147 & T value is 2.434) this sub hypothesis is also supported. Furthermore, sustainability consciousness has a positive and significant influence on brand evaluation (p=0.020, β=0.170 & T value is 2.335) this hypothesis is also supported. Additionally, sub hypothesis H3a sustainability consciousness has a positive and significant impact on fashion consciousness (p=0.000, β=0.344, & T value is 4.524) and the sub hypothesis H3a is also supported. However, H4 hypothesis fashion consciousness has a negative impact on brand evaluation (p=0.085, β=0.120 & T value is 1.724) and this hypothesis is not supported.

Mediating Testing

Fashion consciousness was included as a mediating factor in the research's suggested framework. We may look at the findings of the indirect impacts and contrast them to the primary impacts to determine whether this mediation influence enhances or deteriorates the connection between the dependent and independent variables, as well as to see how it impacts things. The findings indicate that fashion consciousness has a considerable negative impact on brand evaluation. Fashion consciousness acts as an invaluable mediator between sustainability consciousness, purchase attitude, Ad skepticism and brand evaluation. The hypothesis H1b is not supported as well as H2b and H3b. As all of their values are greater than 0.05.

DISCUSSION

The current research aim is to find the impact of sustainable marketing practices through Instagram marketing on brand evaluation. The data has been collected through out Muscat via the survey research

Table 6. Direct hypothesis

Hypothesis	Constructs	Original Sample (O)	Sample Mean (M)	Standard Deviation	T Statistics	P Values	Results
H1	Ad Skepticism -> Brand Evaluation	0.234	0.237	0.064	3.639	0.000	Supported
H1a	Ad Skepticism -> Fashion Consciousness	0.351	0.348	0.083	4.231	0.000	Supported
H2	Purchase Attitude -> Brand Evaluation	0.311	0.312	0.066	4.724	0.000	Supported
H2a	Purchase Attitude -> Fashion Consciousness	0.147	0.150	0.061	2.434	0.015	Supported
H3	Sustainability Consciousness -> Brand Evaluation	0.170	0.170	0.073	2.335	0.020	Supported
H3a	Sustainability Consciousness -> Fashion Consciousness	0.344	0.348	0.076	4.524	0.000	Supported
H4	Fashion Consciousness -> Brand Evaluation	0.120	0.117	0.069	1.724	0.085	Not Supported

Table 7. Indirect hypotheses

Hypothesis	Constructs	Original Sample (O)	Sample Mean (M)	Standard Deviation	T Statistics	P Values	Results
H1b	Ad Skepticism -> Brand Evaluation	0.042	0.041	0.027	1.568	0.117	Not Supported
H2b	Purchase Attitude -> Brand Evaluation	0.018	0.018	0.014	1.306	0.192	Not Supported
H3b	Sustainability Consciousness -> Brand Evaluation	0.041	0.041	0.026	1.581	0.114	Not Supported

questions. Non-Random sampling technique was applied to collect the data and using a rating system based on 5 Point Likert scale. Software called Smart PLS was used for the data analysis of the research. It has been verified that the sustainable marketing practices has a positive influence on brand evaluation. As per the framework, three hypothesis has proved to be a positive and significant effect on brand evaluation. Moreover, each one of these hypothesis has also been looked in this framework with the mediator fashion consciousness and it also has positive effect on brand evaluation. So, with the result it is proved that the hypothesis H1, H2 and H3 has been accepted in this framework. Additionally, when chosen fashion consciousness as a mediator for above three hypothesis known as H1a, H2a and H3a has also been a positive influence on brand evaluation. Whereas, fashion consciousness known as hypothesis H4 has a negative influence on brand evaluation. Therefore, it has not been accepted in this framework.

Findings, Conclusion, Implications, and Future Research

Summary of Findings

In accordance with the study's goal, the impact of various sustainability marketing practices of Instagram marketing on brand evaluation in the Sultanate of Oman. The initial goal is to determine the impact of different variables on brand evaluation. Ad skepticism has a positive and significant impact on brand evaluation and that proves the hypothesis H1 is accepted. It means that the people of Sultanate of Oman has agreed that the informational claims made by advertising companies are not true. The following goal was to investigate the influence of purchase attitude on brand evaluation. The result of the survey shows that it has a favorable and important influence on brand evaluation and that is why the hypothesis H2 is accepted. It means that the people of Sultanate of Oman are likely to buy more products from the brand which is working on sustainability practices. The third objective is sustainability consciousness and the impact of SC on brand evaluation. According to the result, sustainability has a favorable and significant influence on brand evaluation. So, the hypothesis H3 is also supported. The result shows us that the people are conscious about sustainability and they are willing to pay more price for sustainability. Brands that are practicing on sustainability marketing are likely to be more favorable to the people of Sultanate of Oman. The last objective is on fashion consciousness and its influence on brand evaluation. However, fashion consciousness has a negative impact on brand evaluation. That is why the hypothesis H4 is not accepted. Which implies that the people who are fashion conscious are not willing to compromise. According to the survey result, the people of Sultanate of Oman will not considered the brand on the basis of sustainability marketing or practices. Whereas, they will go for the latest fashion and buy the products from the brand even if they are not working on sustainability.

We have chosen fashion consciousness as a mediator between the variables and brand evaluation. Fashion consciousness has a positive effect on purchase attitude, ad skepticism and sustainability consciousness. Whereas, it has a negative influence on brand evaluation. Only one hypothesis out of four were rejected. So, we will say that the purchase attitude, Ad skepticism and sustainability consciousness has a positive effect on brand evaluation therefore all of these hypothesis are accepted.

CONCLUSION

In the final analysis, sustainability marketing practices of Instagram marketing has a positive influence on brand evaluation. Many brands invested in their marketing campaigns to make them more sustainable and they are trying to practice more on sustainable marketing through Instagram. Organizations practicing on sustainable marketing are giving it a boost in the whole world. The investigator looked at how sustainability advertising tactics affected on brand evaluation and how the fashion consciousness mediated the relationship between the factors. Four key hypothesis were presented in the report and validated using the structural framework. The result of the research indicates that the purchase attitude, sustainability consciousness and Ad skepticism has a beneficial or positive impact on brand evaluation. Whereas, fashion consciousness has a detrimental or negative effect on brand evaluation. As the fashion consciousness used as a mediator for the first time alongside brand evaluation, it can also be used in the future researches with some improvements in the structural model.

Practical and Theoretical Implications

Results suggest that the sustainability practices is crucial for the organizations to meet its objectives. The result also illustrate that the sustainable marketing is important for the organizations to achieve their goals. The research result also shows that the brands can achieve perfection in sustainability marketing by day to day process. The values of structural analysis reveals that the variables has positive effect on brand evaluation whereas, one variable has negative effect. Organizations should work on sustainability and also they can make the negative impact of fashion consciousness in their favor. Brands that fully contribute to the growth of sustainable marketing and are successfully involved in initiatives of its practices have the ability to connect with their intended customers and improve their consciousness about sustainability marketing, giving them a significant advantage over rivals in the field and helping the surroundings as well.

Limitations

This study looked on the impact of sustainability marketing practices of Instagram marketing on brand evaluation. Like other studies, the current one has some drawbacks as well. The people who responded were asked to conceive the survey questions in light of the objectives of the research, when the investigators gathered the information, which led to inaccuracies in the answers they provided. It relied solely on the opinions of those who responded. Despite the fact that the reliability and accuracy of results have undergone extensive analysis, errors cannot be entirely disregarded. In regards to availability and frequency of responses, the investigators encountered numerous challenges throughout the information gathering period. In addition, the aim of the research ought to be expanded to cover other major towns and nations. Moreover, this study has a broad audience and the problem was the time. There were not enough time to get more accurate result or you can say to collect more data through the survey. Lastly, there was no financial support available for the researcher.

Future Research

Future research that takes into account the limitations mentioned above is required to advance our comprehension. Future research should add specific qualitative methods to examine the impact of fashion consciousness, particularly on brand evaluation, while maintaining the idea of sustainable marketing practices in mind. Additionally, for more realistic comparison of fashion consciousness and brand evaluation, future study will be more broaden with some more variables to check the influence. On the contrary, a subsequent study should concentrate on a broad area of gathering information, such as GCC countries where data gathered provide more accurate result.

REFERENCES

Abutabenjeh, S., & Jaradat, R. (2018). Clarification of research design, Research Methods, and research methodology. *Teaching Public Administration*, *36*(3), 237–258. doi:10.1177/0144739418775787

Aji, H. M., & Sutikno, B. (2015). The Extended Consequence of Greenwashing: Perceived Consumer Skepticism. *International Journal of Business and Information*, *10*(4), 433–463.

Ashby, A. (2018). Developing closed loop supply chains for environmental sustainability. *Journal of Manufacturing Technology Management*, *29*(4), 699–722. doi:10.1108/JMTM-12-2016-0175

Athwal, N., Wells, V. K., Carrigan, M., & Henninger, C. E. (2019). Sustainable Luxury Marketing: A synthesis and research agenda. *International Journal of Management Reviews*, *21*(4), 405–426. doi:10.1111/ijmr.12195

Balderjahn, I., Peyer, M., Seegebarth, B., Wiedmann, K.-P., & Weber, A. (2018). The many faces of sustainability-conscious consumers: A category-independent typology. *Journal of Business Research*, *91*, 83–93. doi:10.1016/j.jbusres.2018.05.022

Boerman, S. C., Kruikemeier, S., & Zuiderveen Borgesius, F. J. (2017). Online behavioral advertising: A literature review and research agenda. *Journal of Advertising*, *46*(3), 363–376. doi:10.1080/00913367.2017.1339368

Dangelico, R. M., & Vocalelli, D. (2017). "Green marketing": An analysis of definitions, strategy steps, and tools through a systematic review of the literature. *Journal of Cleaner Production*, *165*, 1263–1279. doi:10.1016/j.jclepro.2017.07.184

ElHaffar, G., Durif, F., & Dubé, L. (2020). Towards closing the attitude-intention-behavior gap in Green Consumption: A narrative review of the literature and an overview of future research directions. *Journal of Cleaner Production*, *275*, 122556. doi:10.1016/j.jclepro.2020.122556

Environmental Quality. (2019). *International Journal (Toronto, Ont.)*, *31*(2), 406–430. doi:10.1108/MEQ-09-2019-0207

Font, X., & McCabe, S. (2017). Sustainability and marketing in tourism: Its contexts, Paradoxes, approaches, challenges and potential. *Journal of Sustainable Tourism*, *25*(7), 869–883. doi:10.1080/09669582.2017.1301721

Frempong, M. F., Mu, Y., Adu-Yeboah, S. S., Hossin, M. A., & Adu-Gyamfi, M. (2021). Corporate Sustainability and firm performance: The Role of Green Innovation Capabilities and sustainability-oriented supplier–buyer relationship. *Sustainability (Basel)*, *13*(18), 10414. doi:10.3390/su131810414

Gericke, N., Boeve-de Pauw, J., Berglund, T., & Olsson, D. (2018). The Sustainability Consciousness Questionnaire: The theoretical development and empirical validation of an evaluation instrument for stakeholders working with Sustainable Development. *Sustainable Development (Bradford)*, *27*(1), 35–49. doi:10.1002/sd.1859

Gomez-Trujillo, A. M., & Velez-Ocampo, J., & Gonzalez-Perez, M. A. (2020). A literature review on the causality between sustainability and corporate reputation. *Scientific Research Journal*, *7*(5). Advance online publication. doi:10.31364/SCIRJ/v7.i5.2019.P0519656

Gul, M. S., & Shahzad, H. (2014). The relationship of social media with fashion consciousness and consumer buying behavior. *Journal of Management Info*, *1*(2), 12–20. doi:10.31580/jmi.v2i1.4

Gürhan-Canli, Z., Sarial-Abi, G., & Hayran, C. (2018). Consumers and brands across the globe: Research synthesis and new directions. *Journal of International Marketing*, *26*(1), 96–117. doi:10.1509/jim.17.0063

Haider, M., Shannon, R., & Moschis, G. P. (2022). Sustainable Consumption Research and the role of marketing: A review of the literature (1976–2021). *Sustainability (Basel)*, *14*(7), 3999. doi:10.3390/su14073999

Ham, D.-H., & Park, J. (2020). Use of a big data analysis technique for extracting HRA data from event investigation reports based on the safety-II concept. Reliability Engineering &amp. *Reliability Engineering & System Safety*, *194*, 106232. doi:10.1016/j.ress.2018.07.033

Hammer, J. H., & Spiker, D. A. (2018). Dimensionality, reliability, and predictive evidence of validity for three help-seeking intention instruments: ISCI, GHSQ, and MHSIS. *Journal of Counseling Psychology*, *65*(3), 394–401. doi:10.1037/cou0000256 PMID:29672088

Hart, S. L., & Dowell, G. (2010). Invited editorial: A natural-resource-based view of the firm. *Journal of Management*, *37*(5), 1464–1479. doi:10.1177/0149206310390219

Henseler, J., Hubona, G., & Ray, P. A. (2016). Using PLS path modeling in new technology research: Updated guidelines. *Industrial Management & Data Systems*, *116*(1), 2–20. doi:10.1108/IMDS-09-2015-0382

Hien, N. N., Phuong, N. N., Tran, T. V., & Thang, L. D. (2020). The effect of country-of-origin image on purchase intention: The mediating role of Brand Image and Brand Evaluation. *Management Science Letters*, 1205–1212. doi:10.5267/j.msl.2019.11.038

Hur, E., & Cassidy, T. (2019). Perceptions and attitudes towards Sustainable Fashion Design: Challenges and opportunities for implementing sustainability in fashion. *International Journal of Fashion Design, Technology and Education*, *12*(2), 208–217. doi:10.1080/17543266.2019.1572789

Kautish, P., & Sharma, R. (2018). Consumer values, fashion consciousness and behavioural intentions in the online fashion retail sector. International Journal of Retail &amp. *International Journal of Retail & Distribution Management*, *46*(10), 894–914. doi:10.1108/IJRDM-03-2018-0060

Kemper, J. A., & Ballantine, P. W. (2019). What do we mean by sustainability marketing? *Journal of Marketing Management*, *35*(3–4), 277–309. doi:10.1080/0267257X.2019.1573845

Khalil, M. A., Khalil, R., & Khalil, M. K. (2022). Environmental, social and governance (ESG)-augmented investments in innovation and firms' value: a fixed-effects panel regression of Asian economies. *China Finance Review International*, ahead-of-print. doi:10.1108/CFRI-05-2022-0067

Khalil, M. K. (2022). *Leveraging Buyers' Interest in ESG Investments through Sustainability Awareness*. doi:10.3390/su142114278

Khalil, M. K., & Muneenam, U. (2021). Total Quality Management Practices and Corporate Green Performance: Does Organizational Culture Matter? *Sustainability (Basel)*, *13*(19), 11021. doi:10.3390/su131911021

Kim, N., Chun, E., & Ko, E. (2017). Country of origin effects on brand image, Brand Evaluation, and purchase intention. *International Marketing Review*, *34*(2), 254–271. doi:10.1108/IMR-03-2015-0071

Kumar, R. R., & Alok, K. (2020). Adoption of electric vehicle: A literature review and prospects for sustainability. *Journal of Cleaner Production*, *253*, 119911. doi:10.1016/j.jclepro.2019.119911

Lee, E.-B., Lee, S.-G., & Yang, C.-G. (2017). The influences of advertisement attitude and brand attitude on purchase intention of smartphone advertising. Industrial Management &amp. *Industrial Management & Data Systems*, *117*(6), 1011–1036. doi:10.1108/IMDS-06-2016-0229

Lim, W. M. (2017). Inside the sustainable consumption theoretical toolbox: Critical concepts for sustainability, consumption, and marketing. *Journal of Business Research*, *78*, 69–80. doi:10.1016/j.jbusres.2017.05.001

Lo, S. C., Tung, J., Wang, K. Y., & Huang, K.-P. (2017). Country-of-origin and consumer ethnocentrism: Effect on brand image and product evaluation. *Journal of Applied Sciences (Faisalabad)*, *17*(7), 357–364. doi:10.3923/jas.2017.357.364

Luo, B., Sun, Y., Shen, J., & Xia, L. (2020). How does green advertising skepticism on social media affect consumer intention to purchase green products? *Journal of Consumer Behaviour*, *19*(4), 371–381. doi:10.1002/cb.1818

Memon, M., T, R., Cheah, J.-H., Ting, H., Chuah, F., & Cham, T. H. (2021). PLS-SEM statistical programs: A Review. *Journal of Applied Structural Equation Modeling*, *5*(1), 1–14. doi:10.47263/JASEM.5(1)06

Mensah, J. (2019). Sustainable development: Meaning, history, principles, pillars, and implications for human action: Literature review. *Cogent Social Sciences*, *5*(1), 1653531. doi:10.1080/23311886.2019.1653531

Olsson, D., & Gericke, N. (2017). The effect of gender on students' sustainability consciousness: A nationwide Swedish study. *The Journal of Environmental Education*, *48*(5), 357–370. doi:10.1080/00958964.2017.1310083

Palusuk, N., Koles, B., & Hasan, R. (2019). 'All you need is brand love': A critical review and comprehensive conceptual framework for brand love. *Journal of Marketing Management, 35*(1–2), 97–129. doi:10.1080/0267257X.2019.1572025

Patel, J. D., Gadhavi, D. D., & Shukla, Y. S. (2016). Consumers' responses to cause related marketing: Moderating influence of cause involvement and skepticism on attitude and purchase intention. *International Review on Public and Nonprofit Marketing, 14*(1), 1–18. doi:10.1007/s12208-016-0151-1

Pomering, A. (2017). Marketing for sustainability: Extending the conceptualisation of the marketing mix to drive value for individuals and society at large. *Australasian Marketing Journal, 25*(2), 157–165. doi:10.1016/j.ausmj.2017.04.011

Quaye, D., & Mensah, I. (2019). Marketing innovation and sustainable competitive advantage of manufacturing smes in Ghana. *Management Decision, 57*(7), 1535–1553. doi:10.1108/MD-08-2017-0784

Ragab, M. A., & Arisha, A. (2017). Research methodology in business: A starter's guide. *Management and Organizational Studies, 5*(1), 1. doi:10.5430/mos.v5n1p1

Rajeev, A., Pati, R. K., Padhi, S. S., & Govindan, K. (2017). Evolution of sustainability in Supply Chain Management: A literature review. *Journal of Cleaner Production, 162*, 299–314. doi:10.1016/j.jclepro.2017.05.026

Rana, J., & Paul, J. (2017). Consumer behavior and purchase intention for Organic Food: A Review and Research Agenda. *Journal of Retailing and Consumer Services, 38*, 157–165. doi:10.1016/j.jretconser.2017.06.004

Razzaq, A., Ansari, N., Razzaq, Z., & Awan, H. M. (2018). The impact of fashion involvement and pro-environmental attitude on sustainable clothing consumption: The moderating role of Islamic religiosity. *SAGE Open, 8*(2), 1–17. doi:10.1177/2158244018774611

Rehman, S. U., Bhatti, A., Kraus, S., & Ferreira, J. J. M. (2020). The role of Environmental Management Control Systems for ecological sustainability and sustainable performance. *Management Decision, 59*(9), 2217–2237. doi:10.1108/MD-06-2020-0800

Sailer, A., Wilfing, H., & Straus, E. (2022). Greenwashing and bluewashing in Black Friday-related sustainable fashion marketing on Instagram. *Sustainability (Basel), 14*(3), 1494. doi:10.3390/su14031494

Sarstedt, M., & Cheah, J.-H. (2019). Partial least squares structural equation modeling using SmartPLS: A software review. *Journal of Marketing Analytics, 7*(3), 196–202. doi:10.1057/s41270-019-00058-3

Siedlecki, S. L. (2020). Understanding descriptive research designs and methods. *Clinical Nurse Specialist CNS, 34*(1), 8–12. doi:10.1097/NUR.0000000000000493 PMID:31789957

Thirathon, U., Wieder, B., Matolcsy, Z., & Ossimitz, M.-L. (2017). Big Data, analytic culture and analytic-based decision making evidence from Australia. *Procedia Computer Science, 121*, 775–783. doi:10.1016/j.procs.2017.11.100

Todeschini, B. V., Cortimiglia, M. N., Callegaro-de-Menezes, D., & Ghezzi, A. (2017). Innovative and sustainable business models in the fashion industry: Entrepreneurial drivers, opportunities, and challenges. *Business Horizons, 60*(6), 759–770. doi:10.1016/j.bushor.2017.07.003

Varadarajan, R. (2015). Innovating for sustainability: A framework for sustainable innovations and a model of sustainable innovations orientation. *Journal of the Academy of Marketing Science*, *45*(1), 14–36. doi:10.1007/s11747-015-0461-6

Voss, J. F. (2005). Toulmin's model and the solving of ill-structured problems. *Argumentation*, *19*(3), 321–329. doi:10.1007/s10503-005-4419-6

Yang, S., Song, Y., & Tong, S. (2017). Sustainable retailing in the fashion industry: A Systematic Literature Review. *Sustainability (Basel)*, *9*(7), 1266. doi:10.3390/su9071266

Yu, J. (2018). Consumer responses toward Green Advertising: The effects of gender, advertising skepticism, and Green Motive Attribution. *Journal of Marketing Communications*, *26*(4), 414–433. doi:10.1080/13527266.2018.1514317

Chapter 5
The Future of Sustainable Marketing:
Stakeholder Perspectives on Sustainable Agricultural Marketing

Sinothando Tshuma
https://orcid.org/0000-0001-5726-9375
Zimbabwe Open University, Zimbabwe

Mercy Dube
https://orcid.org/0000-0003-0429-8441
Midlands State University, Zimbabwe

Delight Rufaro Hungwe
Midlands State University, Zimbabwe

Reason Masengu
Middle East College, Oman

ABSTRACT

The concept of sustainable agricultural marketing is gaining momentum with many countries seeking to sustainably increase their market access, with rising concerns of opening regional boarders in the African continent or rather creating free trade zones. These initiatives have raised questions regarding the role of various stakeholders towards sustainable agricultural marketing. The thrust of the chapter is to determine the perspectives of various stakeholders such as the government, suppliers, farmers, and the customers in embracing sustainable agricultural practices. The chapter's methodology reviewed various literatures to determine the trends in the development, implementation, and the future life of sustainable agricultural marketing. The study was informed by the stakeholder theory by Freeman. The major findings were that sustainable agricultural marketing can only be achieved though stakeholder consensus in embracing the concept.

DOI: 10.4018/979-8-3693-4864-2.ch005

INTRODUCTION

In the African continent, the main source of sustainable living is rooted in agriculture (Paull 2011) since the inception of planet earth however, It is noted that Africa has 25% of the world's arable land, and very fertile compared to other continents (Jones, Breuning-Madsen et al. 2013), yet it still has an output of 10% globally (Jayaram, J. Riese et al. 2010). The devastating part is that the rural population is dwindling as many migrants seek formal employment in towns leaving the agricultural sector very unsustainable. Many of the countries in the African continent are developing countries, with many of them prone to hunger, droughts, starvation, and corruption and up rise of climate change. It has been estimated by the Food and Agricultural Organization (FAO) that if the agricultural sector is not redefined sustainably, the current food output can drop by more than 50% while the world's population will be over 9 billion by 2025. This mismatch in output and expected demand is likely to result in acute starvation worldwide. As many countries strive to implement sustainable agricultural practices, more debate in parliaments has escalated on the concern that a drive to increase crop productivity through the use of fertilizers and information technology based systems can lead a decline in soil fertility and destroy the thrust of sustainable agriculture (Geries, 2018). These drawbacks of artificial inputs have led to the rise in demand of sustainable agricultural practices by embracing organic farming to cab environmental degradation (Pagliai et al. 2004).

The African continent's drive to embracing sustainable agriculture has seen some countries such as Kenya, Tanzania, Malawi, Zimbabwe and Zambia receiving universal subsidy in the 1980s (Dorward, 2009). The initiated activities were mainly dominated government-controlled input (and output) marketing system, where farmers were supplied with inputs that befit their farms ain terms of soil type and rainfall levels (Baltzer and Hansen 2012). The government in turn then controlled the marketing of the produce with many of the produce sold at subsidised prices and sometimes subsidised credit. At the end of the program, sustainable input use by farmers was improved although many farmers complained of poor allocations, lengthy processes to access the inputs, and some government connected farmers got more inputs than they can use per season (corruption). On fertiliser distribution, there was an outcry by many farmers who were victims of being over charged, some made to declare receiving the inputs not received and some were at the mercy of government monopoly control (Banful, 2010).

The outcry by farmers on the in efficiencies of the subsidy liberalised the market place and sustainable input use declined and countries were forced back on the drawing board (Crawford et al, 2006).

For Africa as a continent to be self sufficient, it must focus on smallholder agricultural productivity through use of sustainable agricultural practices. Most of these small holder farmers have been ignored and their potential neglected over the years. Strategic plans of many African countries failed to insert smallholder farmers as back born to agricultural productivity increases. The major blow to limited agricultural marketing in Africa rests on the low productivity and poor African Adjustment Plans that happened in the 1980s as many of these nations claimed their independence. At this time, the new government failed to honour the new land barons with inputs to continue production which in turn led to corruption and abuse of office powers by many authorities. IN the early 2000 many NGO incepted to alleviated huger in the African continent, which saw the creation of AGRA by Rockefeller and the Bill and Melinda Gates Foundation, the centre of gravity of input supply began to shift steadily toward private sector, especially indigenous, "SME" private agri-businesses (Baltzer and Hansen 2012).

In this era, many private players entered the industry to supply sustainable inputs and offered innovative seeds and farming platforms and competed for farmers with the best inputs (AASR, 2013).

In Zimbabwe agriculture is a source of domestic Product as it contributes over 18% with the country earning above 40% export earnings from the sector. Many of the country's citizens are formally employed but the sector although the huge chunk of the labour suffers from low salaries due to unsustainable agricultural practices. According to the Maputo Declaration, the African Union Commission tasked each member state to have their national budgets biased towards agriculture. COMPREHENSIVE AGRICULTURAL POLICY FRAMEWORK (CAPF) (2012-2032) On the other hand, In Malawi 88% of the country's citizens are rural citizens and dependent on agriculture whose inputs are distributed based on political affiliation which in turn hinders the fruitful implementation of sustainable agricultural practices.In Malawi, Farm Input Support Programme (FISP) in 2008/9 rolled also the distribution of inputs to farmers, according to Dorward et al (2010) the output by farmers improved but failed to sustainable improve farmer welfare and well being. In Zambia Fertilizer Support Programme (ZFSP) noted good supply of agricultural inputs, hybrid early maturity seeds, and improved maize related inputs. According to the World Bank (2010) notable increase in output tonnage of produce was noted although that could not guarantee future sustainable produce by farmers if taken out of the scheme. In Ghana Banful (2009) notes that the government withdrew subsidies towards sustainable agriculture which led to decline in productivity during 2008 to 2009 era.

According to Danciu (2013) marketing keeps having numerous dimensions and trends. The complexity of consumer needs, rising environmental externalities and the plight to creating a global village where consumers can access the contents of the products before any purchases have accelerated the vast changes happening in the marketing world. These changes will continuously occur pushing for knowledge based sustainable agro based marketing strategies that do not harm the environment and stakeholders.

The sustainable agricultural marketing concerns have of late remained the wider global market hope for sustainable production and consumption patterns (Avi, 2012). Continued rise in negative externalities have created a new goal that many organisations seek to attain, which is reduction of industry based production and supply environmental foot prints. The future of sustainable marketing is to create a lighter competitive landscape of the markets around the world with all firms taking charge of their environmental impacts and investing in future farming technologies continuously with the focus of achieving zero carbon foot print. The growing number of companies that have regional and global ambitions to embrace sustainable agro marketing is rooted in the need to fight competition through delivering eco friendly products based on the 3Rs (Reduce, Re use and Recycle) (Danciu, 2013).

LITERATURE REVIEW

The global concern for a green environment has skyrocketed to African countries and many of these countries have accepted the concept theoretically and face challenges on how best to implement it,(Ahmad 2014). Green agricultural marketing requires huge investment in advanced technology to ensure renewable, reduced and recyclable materials which require huge stakeholder support if economies are to achieve it. Standards or public consensus about what constitutes green marketing (Govender & Govender 2016), requires large investments in research and development (Thulasimani 2012).

Green agricultural marketing over the years has always been viewed singular obligation to an organisation to produce and supply eco friendly goods and services to various stakeholders while at the same time preserving the environment, (Mukonza and Swarts, 2019). Businesses have been pressurised to deliver what consumers and various stakeholders demand such as good financial rewards on investment by investors, while the government charges carbon tax and Value added tax (Daniel 2019). This paradox leaves a gap in research to discover more of these stakeholder perspectives towards green agricultural marketing. As competition intensifies, new strategies are developed and one of the strategies firms have embraced is fighting competition through green agricultural marketing, (Mukonza and Swarts (2019), Nyagadza, (2021).

THEORETICAL FRAMEWORK

Stakeholder Theory

This is a theory that was developed in order to explain the relations that exists between business organisations and their stakeholders. The theory was developed by Freeman (1984).The theory emerged as a new narrative to understand and remedy three interconnected business problems—the problem of understanding how value is created and traded, the problem of connecting ethics and capitalism, and the problem of helping managers think about management such that the first two problems are addressed (Parmar, Freeman, Harrison,Wicks, Colle & Purnell 2010). Stakeholder theory suggests that if we adopt as a unit of analysis the relationships between a business and the groups and individuals who can affect or are affected by it then we have a better chance to deal effectively with these three problems. First, from a stakeholder perspective, business can be understood as a set of relationships among groups that have a stake in the activities that make up the business (Freeman, 1984; Jones, 1995; Walsh, 2005).

The Stakeholder theory greatly informs this study since the theory asserts that all stakeholders must be treated equally (Gioia, 1999; Marcoux, 2000; Sternberg, 2000). The theory also advocates for changes to current laws (Hendry, 2001a, 2001b; Van Buren 2001). It advocates passage of enabling legislation which will force corporations to be managed in the interests of stakeholders.

Government and Sustainable Agriculture

The government is always accountable to its citizens and one of its core responsibilities is to protect the life of its inhabitants. Currently, there is an upsurge scale of firmness impelled by climate changes reiterating that the earth's inhabitants need to consciously utilise practises and resolutions that are environmentally competent (Carvalho, Domingues, & Sampaio,2019). The amendment of the Zimbabwean constitution in 2013, section 73 created a clause under environmental rights, that every person has a right to an environment which is safe for their health and well being, (De Gobbi, 2020). Current researchers are mainly focusing on how environmental issues can be minimised Massey & Singh, 2019) and this has led to many governments and institutions drafting their legislation policies towards environmental protection and sustainable marketing practices (Sandu, 2014;Dean and Pacheco, 2014). Stakeholders have been challenged to balance between economic and social development, and becoming stewards of the environment. Government perspective towards green agricultural marketing has not been fully explored even though the success of implementing green marketing is largely achieved by company stakeholders (Hove and Rathaha, 2021). On another perspective the government stills feels it is the responsibility of firms to embrace green marketing strategies (Handayani & Prayogo, 2017).

Africa's governments have responded, in some cases providing welcome subsidies for fertilizers, made available to farmers via private distribution channels.

CAPF (2012-2032) the Zimbabwean government to ensure effective agricultural marketing pledged in the framework to The Government will: Promote and facilitate investment in the production, processing and marketing of quality export crops where we have comparative advantage; Collaborate with farmer groups and industrialists, to support and promote research of agricultural commodities that have potential to do well in Zimbabwe; Promote new market development; and Promote production of quality products. It further articulated to providing command agriculture inputs and offering import subsidies on sustainable importation of farming inputs.

Also in the African continent, many governments have intervened in the procurement and distribution of fertilizers to boost the farm produce output. . Some have implemented subsidy programs. Although they have been outcry by farmers on the prices charged by the government as some of the fertilizer prices will be higher than international gazetted prices, sometimes there is late delivery after lapse of the farming season. In Nigeria, the government does not consider farm side but restrict subsidy to 100kgs of fertilizer, and 250kgs in Kenya. The sustainability of the agricultural concept suffers an immature death due to limited inputs as farmers do not have savings to add to the deficit. In Ethiopia soil mapping and fertilizer blending was done by the Agricultural Transformation Agency (ATA) as a government initiative to boost agricultural produce (Baltzer and Hansen 2012).

Drawing closer to Zimbabwe again Chisango and Tichakunda (2018) notes that farmers are usually awarded seeds, fertilize, fuel and chemicals through the presidential input scheme awarded through the ministry of Finance and economic development, although most oftenly the scheme has always fallen in the hands of corruption as Gwatidzo and Muyengwa (2020) also questioned the sustainability of the scheme in ensuring sustainable agriculture. It has also been alluded by Banful (20100 that these government donations do not usually make agriculture sustainable in developing countries as they usually drain the government budget which is usually at deficit even before a single expenditure. Most African economies rely on government subsidy for survival of the agricultural sector, (Lopez et al. 2017;Pindiriri, Chirongwe, Nyajena, Nkomo, 2021) articulates that, countries such as Zimbabwe, Kenya, Malawi, Zambia and Tanzania are always receiving these input subsidies although no noted improvement in productivity, farmer lifestyle or general income levels have been significantly noted to guarantee sustainable agricultural marketing. Usually when productivity is high usually administration costly will be usually too higher.

Suppliers and Delivery of Sustainable Inputs

To ensure effective sustainable agriculture in Africa, local fertilizer suppliers were brought on board and offered locally formulated inputs that have weather resistance based on the country and region. (Wanzala, 2012; Bationo, 2013). There has also been a rise in agro dealer shops close to farmers in the rural areas to facilitate access of inputs in the sub Sahara Africa. Lack of storage capacity, which is also due to lack of integration and investment of local distributors (mostly small players) is a major barrier to having fertilizer available at the right time and place, especially in large low density countries. These factors have all contributed to the low use and high costs of fertilizers in Africa.

Wang et al (2023) notes that many suppliers have made vast investment towards research and innovation to ensure they deliver green inputs to farmers, many are now making a radical shift from conventional brown inputs to delivering green agricultural inputs and usually with traceable benefits of early maturity and zero emission to the environment through data driven marketing.

The up rise on the need to prolong life span by many consumers has triggered the need to redefine the supply chain in agricultural products marketing. The growing need for green agro based products has led to the transition of the conventional supply chain to modern supply chain, with many suppliers being requested to supply green products (Ahumada and Villalobos, 2009) and many have to invest in green production inputs strategies (Ruan et al, 2019).

In Zimbabwe, to achieve vision 2030, more focus has been channelled to small holder farmers with many farmers awarded inputs in the form of seeds, fertilizers and farming produce based loans. If the schemes continue with consistence and limited corruption it is bound to yield sustainable agricultural produce to the end user consumers. In Zambia and Ghana the governments sources inputs from renowned suppliers to bring life to the sector (Baltzer and Hansen 2012; CAPF, 2012-2032).

Farmers' Perspectives on Sustainable Agriculture

Smallholder farmers constitute about 80% of the farming households in Africa. These farmers because of limited resources many have suffered from fake seed proliferation, limited awareness of the best seed to plough, de-motivated AGRITEX farm advisors to farmers in Zimbabwe, Generally in Africa, seed firms suffer from deteriorating financial capacity and end up pulling all costs to farmers making sustainable local inputs more expensive to embrace than imports. Less than half of Africa's rural population has access to an all-season road. In these conditions, transport costs grow as fertilizer moves deeper into the market, closer to the farm gate and in small in adequate quantities. Rail transport is much cheaper than road transport, but rail infrastructure is the least developed in Africa. It is poorly maintained for the most part, and offers a poor service resulting in a significant reduction in usable track (Baltzer and Hansen 2012;CAPF, 2012-2032). Most farmers argue that the costs associated with growing green crops are usually huge compared to conventional crops (Wang et al, 2023; Andiç, Yurt and Baltacıo ˘glu 2012)

In Zimbabwe, 'Smallholder farmers have not been equipped with relevant skills to adapt to the changing climate, including drought mitigation strategies, planting materials of appropriate drought resistant varieties, or adequate low-cost and affordable irrigation facilities to mitigate the impacts of recurrent droughts'' (Kudejira, 2014). The Zimbabwean farmers to date face mixed feelings towards government based subsidised inputs as they usually default repayment as alluded by the IMF (2019) report.

Customers' Perception of Sustainable Agricultural Product

In today's business world, the ecological changes are increasingly becoming a cause for concern. Most governments and their people, as well as companies care about preserving the environment. According to Reddy et al., (2023), over the past decades, both manufacturers and consumers have shown concern about how products harm the environment. They undertook a study on the level on which customers have embraced eco marketing and green products and they found out that nowadays the customer have a high knowledge and understanding of green products and possess strong environmental values and hence they prefer purchasing green products to non-green products. The customers are paying more attention to green products as they are now aware of the adverse effects of the rapid population growth and excessive consumption of goods and services that have a negative impact on the ecosystem.

Based on the notion that green agricultural products have benefits of safety, nutrition though expensive, consumers' purchase intentions for green agricultural products have significantly increased in recent years with China noting that 83% consumers are price insensitive towards green products (Wang et al, 2023;Report on the Current Situation of Public Green Consumption in China, 2019)

Additionally, the customers are now aware of the dangers associated with non-green products that include the extreme deterioration and imbalance of ecosystem, which in turn is contributing to natural disasters that are on the rise, for example tsunamis, earthquakes, floods.

Linu (2021) examined the consumers' perceptions on green products and found out that customers are now green aware and compliant and are willing to pay more for eco-friendly goods and services. Most customers have become more aware of the dangers posed by human activities such as global warming and use of non-biodegradable solid waste. Dondapat (2019) concurs to the same sentiments as he stated that customers nowadays are now green aware and usually buy products that are green compliant. He went on to state that many governments are now educating their consumers and this has in turn resulted in a change of perception from customers, to move to green products.

A study by Susmitha (2018), on how customers feel towards green products unearthed that customers have a high knowledge of green marketing and products. However, the study revealed that despite the extensive awareness and knowledge, some customers have no inner drive to purchase green products. Consumers still lack familiarity with green activities by government and companies and hence the need for governments to improve on its initiatives to change the customers' perceptions. Consumers expect businesses to protect the earth by putting in place green practices from manufacturing up until consumption.

Gaddene et al., (2011) argues that customers have highly embraced green products and if companies market eco-friendly products, they are more likely to strike an emotional chord with the customers, swaying their purchasing decisions. Consumers have portrayed a preference to green products despite the worries in price. Because of that, companies should produce green products because customers have shown interest in purchasing them. Customers really worry about the effects of production and consumption on the environment and hence they have embraced green products with high enthusiasm. According to Riche (2008), most South African consumers, just like their international counterparts are becoming more responsible and informed when it comes to green consumption. Viviers (2009) supports the notion as he argues that consumers nowadays want to know where, how and what the impact of their purchases will have on the environment and also on their health. This hence, has led to many businesses in Africa actively embracing the notion of going green as well as incorporating green marketing initiatives in their businesses. Yee (2016) points out that due to the increase in consumer awareness and a change in perception, green marketing is now not applicable to only large companies but also to small companies. Therefore, green marketing has become a significant approach for the business to survive in the market and sustain a competitive advantage, (Chahal et al, 2014).

Green products have become the most reliable resolution for environmental sustainability in many countries, including developing countries in Africa (Ramaya et al., 2011). Gaffney (2014) argues that green products have shown a significant impact on the growth and development of consumers as well as the environment. This has led to the launch of the SDGs which stipulates responsible consumption and production. This drive has left a remarkable rise in the consumer's environmental awareness and also a positive shift to their perceptions towards green products. According to Connoly (2018), this increase in environmental awareness has increased pressure on consumers to consider the environmental consequences of their activities. The increase also in media attention, the government has led to the growth in individuals' awareness and compliance to green product consumption (Kong et al., 2014). Consequently, many customers, especially in Africa have now garnered interest in businesses that are green compliant though according to Mittelman (2012), the level of consciousness is still low in developing countries.

The emerging markets are forcing organisations to re define their products to meet up with changing consumer needs and wants being pushed by the need for healthier living styles. It has been noted that consumers have of late regained their lost power to decide on the product specifications produced by the industry. The influx of communication channels have led to exposure to consumers various product based information that has moved the motion of sustainable marketing. The mobile based communication has created mirror stores where consumers now track the production, usage and disposal of products including the product recalls and scandals faced by firms during their quest to satisfy the observant savvy consumer. The rise in literacy levels worldwide matching up with industrialisation spear headed by good internet connectivity has made consumers to scrutinize products, services and companies all in the quest to achieve sustainable consumption. The paradigm shift from conventional to sustainable marketing is forcing organisations to accept the complexities and advancements that come with the changes in the concept so as to guard and preserve their brand loyalty and equity. The current changes postulate that consumers are seeking total control of the companies' production processes as they postulate that their experiences during the life of the product should be considered if firms seek future sales. The future sustainable marketing concept is creating a customer centric production based plants or arming calendars (Muhtar, 2012).

METHODOLOGY

In carrying out this research, the literature-based methodology was adopted based on the concept of Grounded Theory. Grounded theory epistemological origins and rigorous foundational aspects is the basis on theory development in this write up. Grounded theory also referred to as post positivism paradigm systematically generates theory from empirical data (Byrant 2019; Birks and Mill, 2015; Khothari,2004). Some scholars refer it to it as the *a la carte*approach since it allows research to adopt the methods which are attractive and convenient to them. Its key components and methodology involve simultaneous and interactive data collection of literature insights (Wong, et al., 2013). Taking insights to the contribution emanating from Glaser and Straus (1967), through the grounded theory divorce of Straus and Glaser, this study takes interest in the theoretical saturation that signals the end of data collection (Khothari, 2004; Wong, et al., (2013).

Based on the grounded theory the researchers combined grounded theory principles with Liberate, et al., (2009) and Wong, et al., (2013) framework, which also uses the literature-based methodology as another dimension in the creation on knowledge. For Wong et.al, the literature-based methodology constitutes three main parameters namely design the literature review, conducting the review and contextual analysis and writing up. This methodology borrows a lot from the Strauss and Glazer grounded theory concepts. This study used results collection and preparation of literature based on themes. Themes (results) were identified from the different literature discussion to come up with conclusions on the different stakeholder perspectives. The literature-based discussion papers covered a period between 2010 and 2022. A total of 50 research highly indexed articles were selected from different journals and a sample drawn based on theme saturation. The themes were discussed and aligned to represent the literature-based finding of the study.

Discussion of Findings

It has been noted that many African countries are still lagging behind in the implementation of sustainable agricultural marketing concept. In the production of green products, currently Zimbabwe has no green seals to accredit green products and only international firms operating in Zimbabwe use foreign seals to further penetrate the foreign markets. Kotler (2018) posts that green marketing can only be realised through organisational reforms embracing stakeholders' input, basing on the integrated marketing paradigm. Customers have indicated a paradigm shift to green consumption due to health benefits of the products, and many have shown less sensitivity to price when shopping green agro based products. Also on customer perspectives, a worldwide focus was adopted and many consumers showed interest in shopping green agricultural products if they can be perennially supplied to the market.

It has been noted that, many African governments are inconsistent in supporting sustainable agricultural initiatives; this is because of noted inconsistencies in offering subsidies to farmers or supplying hybrid inputs and fertilizers. The major factor hindering success is also seen as the up rise of corruption, political affiliated agro based decisions, decisions made without or with minimum stakeholder consultation and also lack of funding to sustainable anchor the sector.

Suppliers are willing to supply the green inputs to farmers but most of the farms are remotely located with poor rail or road network which in turn makes the distribution of these inputs very expensive or not being delivered at all. It is also of concern that the locally produced inputs to fit the weather conditions in Africa have proven to be expensive that buying internationally tested inputs which in turn makes smallholder farmers in these nations resort to recycled inputs and limited organic manure.

Smallholder farmers as the majority land owners in the African continent after land re distribution by many countries after 1980 noted that the farmers have the zeal to adopt sustainable agricultural marketing strategies but they lack financial resources, skills, and suffer from government imposed schemed that later control the marketing of their output especially staple produce like maize. Small holder farmers also remain in the mercy of in consistent government subsidies to sustain the farming season, with some inputs very few leaving the rest of the land untilled.

It is also evident that the guide to sustainability in any sector rests in having all company stakeholders, such as consumers, industrial buyers, suppliers, and government institutions having to pressure the supply chain to embrace green technology so that the results are all green (Tshuma, 2022).

CONCLUSION

As countries note the positive gains that can accrue through environmentally friendly agro green marketing strategies and the potential pitfalls associated with non-environmentally friendly strategies, going green is beginning to take centre stage in countries around the world (Joseph, Mark & Jeffrey 2011). There is a growing interest among companies, governments, customers as well as academics regarding the adoption green agricultural marketing strategies. The future to sustainable agricultural marketing requires overall stakeholder consultation so that farmers as major shareholders are given enough resources to produce and market agricultural products sustainable.

Recommendations

Governments in various African countries to offer more incentives to farmers to adopt green agricultural farming strategies to alleviate the negative climate changes and preserve the resources for the current and future generation.

Suppliers to invest heavily on green inputs to increase green supply chain flow of green products that have high yields and are more nutritious than conventional products

Consumers are savvy for green agro products and they should shun conventional products even if they are lowly priced for their health benefits

Marketers to de market conventional products to increase consumption of green agro products that are taking up store shelf rapidly.

Future Research

Like any other chapter, this chapter had some delimitations. It focused on perspectives of stakeholders towards sustainable agricultural marketing in selected African countries, major stakeholders of concern were the government, suppliers, farmers and customers. Future research can be done focusing on other African countries and also broadening the stakeholders under study.

REFERENCES

Africa, A. S. R. (2013). Focus on staple crops. Alliance for a Green Revolution in Africa.

Ahumada, O., & Villalobos, J. R. (2009). Application of planning models in the agri-food supply chain: A review. *European Journal of Operational Research, 196*(1), 1–20. doi:10.1016/j.ejor.2008.02.014

Andiç, E., Yurt, Ö., & Baltacıo ˘glu, T. (2012). Green supply chains: Efforts and potential applications for the Turkish market. *Resources, Conservation and Recycling, 2012*(58), 50–68. doi:10.1016/j.resconrec.2011.10.008

Avi, D. (2012). *"The 10 most Important Marketing Trends According to Sir Martin Sorrel", Baltzer, K. & Hansen, H. 2011.* Agricultural Input Subsidies in Sub-Saharan Africa.

Baltzer, K., & Hansen, H. (2012). Evaluation Study Agricultural input subsidies in Sub-Saharan Africa. *University of Copenhagen Rolighedsvej, 25*, 1958.

Banful, A. B. (2010). *Old problems in the new solutions?* IFPRI Discussion Paper 01002, IFPRI.

Barratt, N., Chitundu, D., Dover, O., Elsinga, J., Eriksson, S., Guma, L., Haggblade, M., Haggblade, S., Henn, T. O., Locke, F. R., O'Donnell, C., Smith, C., & Stevens, T. (2006). *Cassava as droughtinsurance: Food security implications of cassava trials in Central Zambia.* AGRA.

Birks, M., & Mills, J. (2011). *Grounded Theory: A Practical Guide.* Sage Publications, Inc.

Bryant, A. (2019). *Grounded theory, Case study research, Theory.* SAGE Publications Ltd. doi:10.4135/9781529716542

Carvalho, F., Domingues, P., & Sampaio, P. (2019). Communication of commitment towards sustainable development of certified Portuguese organisations: Quality, environment and occupational health and safety. *International Journal of Quality & Reliability Management, 36*(4), 458–484. doi:10.1108/IJQRM-04-2018-0099

Chahal, H., Dangwal, R., & Raina, S. (2014). Antecedents and consequences of strategic green marketing orientation. *Journal of Global Responsibility*, *5*(2), 338–362. doi:10.1108/JGR-09-2013-0012

Connolly, J., & Prothero, A. (2008). Green consumption: Life-politics, risk and contradictions. *Journal of Consumer Culture*, *8*(1), 117–145. doi:10.1177/1469540507086422

Connolly, J., & Prothero, A. (2018). Green consumption: Life-politics, risk and contradictions. *Journal of Consumer Culture*, *8*(1), 117–145. doi:10.1177/1469540507086422

Danciu V (2013) The future of marketing: an appropriate response to the environment changes, *Theoretical and Applied Economics, 5*(582), 33-52.

De Gobbi, M. (2020). *Environmental integrity and doing business in Zimbabwe: Challenges and engagement of sustainable enterprises.* ILO Working Paper 2.

Dondapat, R. S. (2011). Consumer perception towards eco-Friendly FMCG products and its influence on their purchase decision: An observed research. *International Journal of Innovative Technology and Exploring Engineering, 9*(2).

Dorward, A. (2009). *Rethinking Agricultural Input Subsidy Programmes in a Changing World.* School of Oriental and African Studies.

Dorward, A. (2009). *Rethinking Agricultural Input Subsidy Programmes in a Changing World.* Academic Press.

Dorward & Chirwa. (2011). The Malawi Agricultural Input Subsidy Programme:2005-6 to 2008-9. *International Journal of Agricultural Sustainability, 9*(1).

Freeman, R. E. (1984). *Strategic Management: A Stakeholder Approach.* Pitman.

Gadenne, D., Sharma, B., Kerr, D., & Smith, T. (2011). The influence of consumers' environmental beliefs and attitudes on energy saving behaviours. *Energy Policy*, *39*(12), 7684–7694. doi:10.1016/j.enpol.2011.09.002

Gaffney, O. (2014). Sustainable development goals: Improving human and planetary wellbeing. *Global Change*, *1*(82), 20–23.

Glaser, B. G., & Strauss, A. L. (1967). *The Discovery of Grounded Theory. Strategies for Qualitative Research.* Aldine.

Govender, J. P., & Govender, T. L. (2016). The influence of green marketing on consumer purchase behavior. *Environment and Ecology*, *7*(2), 77–85.

Gwatidzo, T., & Muyengwa, S. (2020). *Zimbabwe - Assessing the Impact of the Command Agriculture Programme Using the Propensity Score Matching Approach.* World Bank.

Hove, G., & Rathaha, T. (2021). The extent green marketing has been embraced in the construction industry-employee perspective in Zimbabwe. *International Business Research*, *14*(6), 1–78. doi:10.5539/ibr.v14n6p78

IMF. (2019). *Zimbabwe Staff-Monitored Program – Press Release and Staff Report of 2019*. IMF Country Report No.19/14.

Jayaram, Riese, & Sanghvi. (2010). Africa's path to growth: Sector by sector - Agriculture: Abundant opportunities. *McKinsey Quarterly*.

Jones, A., Breuning-Madsen, H., Brossard, M., Dampha, A., Deckers, J., Dewitte, O., Gallali, T., Hallett, S., Jones, R., & Kilasara, M. (2013). Soil atlas of Africa. Luxembourg. *European CommissionJournal of Agricultural Economics*, *3*(4), 76–82.

Kong, W., Harun, A., Sulong, R. S., & Lily, J. (2014). The influence of consumers' perception of green products on green purchase intention. *Int J Asian SocSci*, *4*(8), 924–939.

Kothari, C. R. (2004). Methodology, Methods and Techniques. New Age International Publishers.

Kotler, P. (2018). *Principles of Marketing*. Prentice Hall.

Kudejira, D. (2014). *An integrated approach towards moderating the effects of climate change on agriculture: A policy perspective for Zimbabwe*. www.future-agricultures.org

Le Riche, L. (2008). Bring on the green. *Your Business*, *13*(2), 26–29.

Liberati, A., Altman, D. G., Tetzlaff, J., Mulrow, C., Gøtzsche, P. C., Ioannidis, J. P., Clarke, M., Devereaux, P. J., Kleijnen, J., & Moher, D. (2009, October). The PRISMA statement for reporting systematic reviews and meta-analyses of studies that evaluate health care interventions: Explanation and elaboration. *Journal of Clinical Epidemiology*, *62*(10), e1–e34. doi:10.1016/j.jclinepi.2009.06.006 PMID:19631507

Linu, J. (2021). A study on consumer perception towards green marketing. IJCRT, 9(5).

Mittelman, R. (2012). Green consumer behaviour in emerging markets: A review of research. *Proc Ann Market DivConf*, *33*(3), 392–441.

Muhtar, K. (2012). *The five mega-trends shaping tomorrow's customers*. BBC News Business.

Mukonza & Swarts. (2019). The influence of green marketing strategies on business performance and corporate image in the retail sector. In *Bus StratEnv*. John Wiley & Sons, Ltd.

Nyagadza, B. (2021). Fostering green economies in Africa through green marketing strategies for environmental sustainability: An overview. *Journal of Environmental Medicine*, *2*(1), 17–22. doi:10.1386/jem_00037_1

Pagliai, M., Vignozzi, N., & Pellegrini, S. (2004). *Soil structure and the effect of management practices*. Academic Press.

Parmar, B. L., Freeman, R., & Harrison, J. S. (2010). Stakeholder Theory: The State of the Art. Management Faculty Publications.

Paull, J. (2011). Nanomaterials in food and agriculture: The big issue of small matter for organic food and farming. *Proceedings of the Third Scientific Conference of ISOFAR (International Society of Organic Agriculture Research)*, 96–99.

Production Behavior—From the Perspective of the Market Structure. (2023). *International Journal of Environmental Research and Public Health, 20*, 506. doi:10.3390/ijerph20010506

Ramayah, T., Lee, J. W. C., & Mohamad, O. (2010). Green product purchase intention: Some insights from a developing country. *Resources, Conservation and Recycling, 54*(12), 1419–1427. doi:10.1016/j.resconrec.2010.06.007

Reddy, K. P., Chandu, V., & Osei, B. (2023). Consumers perception on green marketing towards eco-friendly fast moving consumer goods. *International Journal of Engineering Business Management, 15*, 18479790231170962. doi:10.1177/18479790231170962

Ruan, J.H., Wang, Y.X., Chan, F.T.S., Hu, X.P., Zhao, M.J., Zhu, F.W., Shi, B.F., Shi, Y., & Lin, F. (2019). *A life cycle framework of green IoT-based agriculture and its finance, operation, and management issues.* IEEE.

Tshuma, S. (2022). Embrace green marketing or lose competitive advantage: An insight on the performance of small and medium-sized enterprises in the foundry sector in Zimbabwe. *Economics. Management and Sustainability*, *7*(2), 63–69. doi:10.14254/jems.2022.7-2.5

Viviers, S. (2009). Going green: A SMME perspective. *The Southern African Journal of Entrepreneurship and Small Business Management*, *2*(1), 30–49. doi:10.4102/sajesbm.v2i1.18

Wang, W.B., Lin, W.Y., Cai, J.H., & Chen, M.X. (2021). *Impact of demand forecast information sharing on the decision of a green supply chain with government subsidy*. Academic Press.

Wang, X., Zhang, J., Ma, D., & Sun, H. (2023). Green Agricultural Products Supply Chain Subsidy Scheme with Green Traceability and Data-Driven Marketing of the Platform. *International Journal of Environmental Research and Public Health, 20*(4), 3056. doi:10.3390/ijerph20043056 PMID:36833775

Wanzala, M., & Groote, R. (2013). *Fertilizer market development in Africa*. IFDC.

Wong, G. (2013). Meta-narrative reviews. *BMC Medicine, 11*(12). PMID:23360661

Wong, G., Greenhalgh, T., Westhorp, G., Buckingham, J., & Pawson, R. (2013). RAMESES publication standards: Meta-narrative reviews. *Journal of Advanced Nursing*, *69*(5), 987–1004. doi:10.1111/jan.12092 PMID:23356699

World Bank. (2010). *Zambia Impact Assessment of the Fertilizer Support Program, Analysis of Effectiveness and Efficiency*. Report No.54864, World Bank, Africa Region.

Yee, H. P. (2016). *Green Marketing Practice and Its Effect on Small and Medium Enterprises Performance*. Bachelor of Technopreneurship dissertation. UniversitiTeknikal Malaysia Melaka.

Yin, S. J., Wu, L. H., Du, L., & Chen, M. (2010). Consumers' purchase intention of organic food in China. *Journal of the Science of Food and Agriculture*, *90*(8), 1361–1367. doi:10.1002/jsfa.3936 PMID:20474056

Chapter 6
Empowering Smallholder Farmers Through Community-Based Marketing Initiatives in Promoting Sustainable Agriculture

Tendai Kaponda
https://orcid.org/0009-0003-3287-4335
Central China Normal University, Wuhan, China

Option Takunda Chiwaridzo
https://orcid.org/0000-0001-9755-9123
University of Science and Technology Beijing, Beijing, China

ABSTRACT

This review paints a nuanced picture of community-based marketing initiatives (CBMIs) in Zimbabwe, Malawi, and South Africa, highlighting their potential to empower smallholder farmers and contribute to sustainable agriculture. Diverse organisational structures, from centralised models offering efficiency to decentralised ones fostering ownership, provide pathways to market access, fair prices, and reduced dependence on intermediaries. Initiatives like Mbare Musika Market, Blantyre Farmers Market, and Farm in the Veld exemplify this potential, connecting farmers directly to consumers and promoting sustainable practices. Beyond economics, CBMIs foster knowledge sharing, community engagement, and a sense of ownership. This collaborative spirit empowers not just individuals but entire communities. However, challenges like fluctuating demand and regulatory hurdles persist. The road ahead requires continued research, investment, and support, but the potential rewards—thriving farmers, resilient communities, and a flourishing food system—make it a pursuit worth striving for.

DOI: 10.4018/979-8-3693-4864-2.ch006

INTRODUCTION

Smallholder farmers across Southern Africa, including Zimbabwe, South Africa, and Malawi, face significant challenges in achieving sustainable livelihoods. Limited access to resources, information, and infrastructure hinders their productivity, contributing to food insecurity and poverty (Gono, 2020; Dhillion & Moncur, 2023). Smallholder farmers globally lack the necessary information on climate change and capital to buy farming equipment and inputs, leading to decreased agricultural productivity, increased poverty, and food insecurity (Gono, 2020). Adopting sustainable agricultural practices is hindered by climate change, soil fertility depletion, and limited access to services and irrigation infrastructure (UNDP, 2023; Murendo et al., 2016). Climate change and soil degradation further exacerbate these issues (UNDP, 2023; Murendo et al., 2016). Additionally, small-scale farmers encounter barriers to access, efficiency, and affordability of agricultural information, marketing, and weaker government policies (Dhillion & Moncur, 2023). Promoting sustainable agriculture is essential to build resilience and improve the lives of smallholder farmers in the region. Therefore, exploring innovative solutions that address their specific needs and challenges becomes critical.

Local market initiatives, like farmers' markets and CSAs, tackle several critical issues. They make it easier for people to buy fresh, local food (USDA, 2023). In CSAs, communities directly support farms, sharing risks and rewards (USDA.b, 2023). These programs also benefit the environment by promoting sustainable farming practices (Growing Small Farms). These initiatives boost local economies by connecting consumers directly to producers (Health Food Access). They also improve public health by making healthy food more affordable and accessible (Health Food Access). By accepting government assistance and offering discounts, they can further address food insecurity and promote healthy eating habits (Health Food Access). In short, these community-based programs are key to improving food access, local economies, and public health.

This study delves into the empowering potential of community-based marketing initiatives for smallholder farmers in South Africa, Malawi and Zimbabwe, focusing on farmers' markets and CSA schemes. By examining their organisational structures and operational processes (Objective 1), we gain insight into how these initiatives function and interact with farmers. Assessing the impact on participating farmers (Objective 2) provides a crucial understanding of their effectiveness in generating income, improving market access, and fostering the adoption of sustainable practices. This holistic approach illuminates the true impact on both individual livelihoods and the broader agricultural landscape. Finally, identifying key success factors and challenges (Objective 3) sheds light on the critical elements necessary for their long-term sustainability and ability to empower smallholder farmers. By addressing these objectives, this research contributes directly to the overarching topic of empowering smallholder farmers through community-based marketing initiatives, offering a nuanced understanding of their potential to promote sustainable agriculture in the Southern African context.

METHODOLOGY

This chapter employs a qualitative research methodology, specifically a literature review and desk research approach, as advocated by (Creswell & Creswell, 2018), to explore the potential of community-based marketing initiatives in empowering smallholder farmers in Southern Africa. This study focuses on understanding the role of farmers' markets and Community Supported Agriculture (CSA) schemes in fostering sustainable agricultural practices across Zimbabwe, South Africa, and Malawi. By selecting these three countries, the research aims to capture the diverse experiences and contextual nuances within the region, as highlighted by (Scoones et al., 2019). Data collection was conducted through two complementary methods. First, an extensive literature review will be conducted through academic databases, peer-reviewed journals, and credible online resources.

This review targeted literature on: Challenges faced by smallholder farmers in the selected countries: This aligns with the work of Gono (2020), Dhillion & Moncur (2023), and UNDP (2023), who emphasize the limited access to resources, and information, and infrastructure hindering these farmers' productivity and contributing to food insecurity. The concept of sustainable agriculture and its significance in the region: Understanding this concept is crucial, as scholars (Pretty & Benton, 2006) argue that sustainable agriculture is essential for building resilience and improving the livelihoods of smallholder farmers. Existing research on the effectiveness of farmers' markets and CSA schemes: This review explored the existing body of research, including scholarly articles, and reports from government agencies, international organizations, and relevant NGOs, to understand how these initiatives have empowered farmers and promoted sustainable agriculture within the Southern African context, as evidenced by the work of (McIntyre et al., 2013; Warner, 2007). Second, desk research was conducted to gain deeper insights into the real-world implementation and impact of these initiatives. This will involve reviewing grey literature sources like policy documents, white papers, conference proceedings, and media reports from the aforementioned countries.

The collected data was analyzed using thematic analysis to identify recurring themes and patterns across the literature, allowing for a comprehensive understanding of the research topic. Additionally, a comparative approach will be used to analyze the findings across the three countries, highlighting similarities, differences, and contextual variations in the implementation and outcomes of community-based marketing initiatives, as suggested by (Eisenhardt & Graebner, 2007). Throughout the research process, a systematic and rigorous approach will be employed to ensure the credibility of the findings. Triangulation will be achieved by utilizing diverse data sources, and a clear and detailed description of the methodology will be provided for transparency and potential replication. As this research solely relies on existing published materials, ethical considerations are minimal. However, responsible citation practices will be strictly adhered to, acknowledging the source of all information used in the study.

By employing this comprehensive approach, this chapter aims to provide valuable insights into the potential of community-based marketing initiatives to empower smallholder farmers and promote sustainable agriculture in Southern Africa, contributing to the ongoing discussion on supporting these farmers and building a more resilient and sustainable agricultural sector in the region.

COMMUNITY-BASED MARKETING

Principles of Community-based Marketing

Community-based marketing transcends traditional, product-centric strategies. Instead, it focuses on building genuine relationships with targeted audiences by fostering a sense of shared identity and belonging. Here are the five key principles that underpin this approach:

Cultivating Connections: Forget one-way communication. As Harrower (2023) emphasizes, successful community-based marketing goes beyond mere product promotion, instead fostering "value beyond products" through meaningful interactions and shared experiences. This principle resonates deeply with your research on empowering smallholder farmers in Southern Africa through community-based initiatives like farmers' markets and CSA schemes. These programs transcend simple product exchange, fostering a vibrant community that unlocks value for both farmers and consumers.

For farmers, these initiatives go beyond market access, providing crucial knowledge and skills through peer learning and workshops (Sumberg & Evelyne, 2010). They also strengthen social capital through collaboration and collective bargaining, empowering farmers to advocate for their rights (Pretty & Smith, 2004). Ultimately, these initiatives contribute to greater agency and empowerment, as farmers gain ownership and control over their livelihoods (McEvoy, 2002). Consumers, too, benefit from this value-driven approach. They gain access to fresh, local, and sustainable food, aligning with their growing ethical concerns (McBride et al., 2013). Building direct relationships with farmers fosters trust and transparency, enhancing the value and meaning of their food (Lockie et al., 2004). Furthermore, educational initiatives and events associated with these schemes promote food literacy, sustainability awareness, and a deeper connection to the agricultural landscape (Henderson et al., 2012).

Power of Belonging: It's not about everyone; it's about the right people. This approach targets specific communities with shared interests, values, or locations. Think of niche communities like independent financial advisors serving close-knit groups (Asset Mark, 2023). This principle resonates with your research on empowering smallholder farmers in Zimbabwe, South Africa and Malawi through initiatives like farmers' markets and CSA schemes. These programs aren't for everyone; they target close-knit communities united by shared interests in sustainable agriculture, fair trade, and local food. Scholars like DeLind and Ferguson (2006) emphasize the power of "niche communities", where shared values and trust foster deeper engagement and collaboration. This aligns perfectly with farmers' markets and CSA schemes, where farmers and consumers connect directly, fostering trust, knowledge exchange, and collective action (McEvoy, 2002). By focusing on these specific communities, your research can showcase how community-based marketing empowers farmers not just economically, but also through social capital, knowledge sharing, and a sense of belonging. This targeted approach holds immense potential for promoting sustainable agriculture in Southern Africa. It fosters collaboration, knowledge exchange, and collective action that transcends individual struggles, ultimately leading to increased market access, improved skills, and a stronger voice for smallholder farmers.

Trust & Authenticity: Building a community takes time and effort. It requires understanding the group's aspirations and values and relinquishing some control to build trust and genuineness (Asset Mark, 2023; Harrower, 2023). Beyond Transactions: Building Trust and Empowering Farmers. The statement "Trust & Authenticity: Building a community takes time and effort..." resonates deeply with your research on empowering smallholder farmers through community-based marketing initiatives like farmers' markets and CSA schemes. As scholars like Dickson & Bantikyan (2014) argue, these initiatives succeed not just through market access, but by fostering trust and authenticity. Understanding the aspirations and values of smallholder farmers is crucial. As Bentley (2008) emphasizes, participatory approaches that respect local knowledge and empower farmers to shape the initiative are key. This means relinquishing some control, as Asset Mark (2023) and Harrower (2023) suggest, and prioritizing genuine dialogue and collaboration. Building trust is a two-way street. Farmers need to trust the initiative as a fair and supportive platform. Conversely, consumers need to trust the provenance and quality of the products. Initiatives like transparent labelling, participatory governance, and open communication (McAfee & Brydon-Miller, 2016) can build this trust. This trust-based community fosters empowerment in several ways. Fair pricing and market access empower farmers economically (Pretty et al., 2011). Sharing knowledge and skills within the community leads to improved agricultural practices and resilience (McAfee & Brydon-Miller, 2016). Moreover, a united community can advocate for their rights and influence policies, further empowering themselves (Narula & Van Huylenbroeck, 2015).

Listening to Learn: Communities are goldmines of insights. By engaging in conversations and actively seeking consumer feedback, businesses can personalize experiences and foster trust in the produce. While traditional marketing often focuses on one-way communication, your research on empowering smallholder farmers through community-based initiatives in Zimbabwe emphasizes the power of "listening to learn." As scholars like Shaw and Iazzolino (2014) argue, community-based marketing thrives on engaging in conversations and actively seeking feedback. This principle holds immense significance for your research, as it goes beyond mere transactions and fosters trust and empowerment for farmers. imagine farmers' markets and CSA schemes not just as selling points, but as vibrant spaces for dialogue and knowledge exchange. By organizing workshops, events, and online forums, you can facilitate direct interactions between farmers, consumers, and other stakeholders, fostering mutual understanding and collaboration. Establishing these spaces for connection, as recommended by Harrower (2023), allows for knowledge sharing and collective problem-solving, ultimately leading to improved farming practices and sustainable agriculture techniques. Furthermore, actively seeking consumer feedback on market preferences, pricing, and desired support systems empowers farmers to tailor their offerings and adapt to evolving needs. This personalization, as highlighted by Asset Mark (2023), builds trust and strengthens the brand value of their produce. Ultimately, by listening to the community and incorporating their insights, your research can demonstrate how these initiatives move beyond economic benefits and empower farmers through increased agency, improved livelihoods, and a stronger collective voice in advocating for their rights and influencing policies.

Value Beyond Products: As Harrower (2023) aptly states, "true community-based marketing goes beyond selling," focusing instead on creating genuine value that resonates with the community. This principle perfectly aligns with your research on empowering smallholder farmers through initiatives like farmers' markets and CSA schemes. These programs transcend mere produce exchange, cultivating broader value for both farmers and consumers. For farmers, these initiatives unlock not just market access, but also vital knowledge and skills, enhanced social capital and collective power, and ultimately, greater agency and empowerment (Sumberg & Evelyne, 2010; Pretty & Smith, 2004; McEvoy, 2002). Consumers, in turn, gain access to fresh, local, and sustainable food, build direct connections with farmers, and engage in community education and awareness (McBride et al., 2013; Lockie et al., 2004; Henderson et al., 2012). By shifting beyond product-centricity and creating shared value, such community-based marketing initiatives stand as potent tools for empowering farmers and fostering sustainable agriculture in Southern Africa.

By embracing these principles, farmers can transform passive customers into active members of a thriving community. This, in turn, leads to increased engagement, stronger brand loyalty, and ultimately, greater sales.

COMMUNITY-BASED MARKETING FRAMEWORK

This paper proposes a "community-based marketing" model with four key relationships:

Company to Customer ("BC-1"): Prioritizing empathetic relationships with customers who resonate with your value proposition is crucial. Fostering a welcoming community fosters information sharing

Figure 1. Community-based marketing concept
Source: Miyazoe (2022)

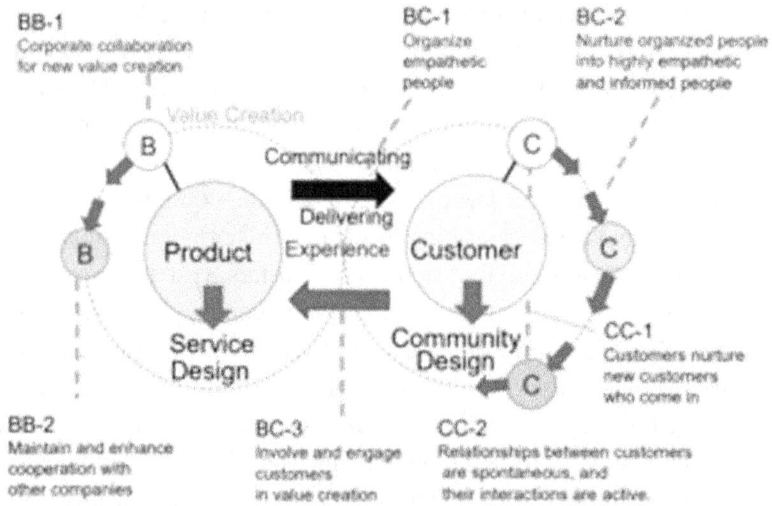

and loyalty, similar to the approach advocated by Miyazoe (2022) for farmers and consumers. Research suggests that altruistic messages encouraging consumer engagement and supporting farmers yield greater behavioural impact (Zimand-Sheiner et al., 2022). Similarly, Dehwa (2015) highlights the importance of empathetic interactions and shared perspectives between farmers, traders, and other stakeholders. Cultivating social capital, as described by FAO (2020), empowers small-scale farmers by enhancing market access, negotiating power, and building trust. Additionally, strong bonding relations foster cooperation and organizational resilience.

Company to Customer ("BC-2"): Companies and organizations promoting sustainable agriculture can tap into a powerful resource: engaged smallholder farmers. By nurturing these individuals into advocates, they can leverage their deep product knowledge and community influence for greater impact. This approach surpasses traditional celebrity endorsements, seeking authentic voices that resonate with local audiences. Empowering farmers through training, market information, and support for sustainable practices (Manyise et al., 2023) enables them to become credible advocates within their communities. This contrasts with relying on distant celebrities who lack the same level of connection and trust. By focusing on community-based initiatives that empower local champions, companies can drive more effective and sustainable adoption of sustainable agriculture practices (The Herald, 2016; Manyise et al., 2023).

Customer to Customer ("CC-1"): The model advocates for fostering interaction within communities, where experienced farmers mentor newcomers, extending beyond product use to encompass broader lifestyle values. This aligns with the study by Manyise, Dentoni, & Trienekens (2023), emphasizing the need for tailored support for smallholder farmers based on their diverse entrepreneurial behaviours and socio-economic characteristics. Similarly, the International Centre for Climate Change and Development (2023) highlights the importance of indigenous knowledge and locally-led initiatives in adapting to climate change and promoting sustainable agriculture. Both approaches resonate with the model's emphasis on knowledge sharing and community-based solutions, extending beyond mere technical guidance to encompass cultural practices and values.

Customer to Customer ("CC-2"): Community-driven discussions can unlock unexpected product benefits and generate new value. Companies should actively observe and integrate these insights into their offerings. Studies highlight the power of empathetic relationships and meaningful conversations with farmers and other stakeholders. These interactions can reveal hidden worldviews and unexpected product benefits (Dhewa, 2015). Additionally, research suggests moving beyond the simplistic altruism vs. egoism dichotomy and exploring diverse motives. This approach can further unlock unexpected value creation (Zimand-Sheiner et al., 2022).

Community-based initiatives like farmers' markets and CSA schemes go beyond mere transactions. They promote sustainable practices, agroecology, biodiversity, and seed sovereignty, contributing to new value creation (ZIMOFF, 2022). These initiatives can also foster demand-driven adaptive practices, influence policies, and raise public awareness about smallholder farmers' rights (Dhewa, 2015; ZIMOFF, 2022).

COMMUNITY-BASED MARKETING: CULTIVATING A SUSTAINABLE FOOD FUTURE

Community-based marketing holds immense potential in fostering sustainable agriculture, offering ecological, social, and economic advantages. Studies highlight Community-Supported Agriculture (CSA) as a particularly promising approach. By creating synergies between these three aspects, CSA paves the way for a transformative shift towards sustainable food systems (Egli et al., 2023). Similarly, Community-Based Social Marketing (CBSM) emphasizes tailoring marketing strategies for local foods around environmental benefits and appealing to consumer values regarding locally-grown products (Hall, 2016). Research reveals that farms incorporating CSA sales tend to be smaller and utilize diverse marketing channels, suggesting a positive impact on their performance (Jablonski et al., 2019). Moreover, a study identified factors like farmers' markets, food policy councils, and organic operations as drivers for adopting CSA marketing (Dong et al., 2019). Community-based marketing, and particularly CSA, empowers both local farmers and consumers. By engaging local communities and emphasizing the environmental and social benefits of sustainable agriculture, this approach plays a crucial role in cultivating a more sustainable food system for the future.

Community-based marketing initiatives such as farmers' markets and Community-Supported Agriculture (CSA) schemes play a significant role in empowering smallholders. These initiatives facilitate market linkages, collective marketing activities, and policy support, ultimately contributing to the economic and social empowerment of smallholders.

Market Linkages and Strengthening Supply Chains

Farmers' markets and CSA schemes facilitate direct market linkages between smallholder farmers and consumers, leading to greater customer loyalty, new market opportunities, and strengthened brand value. These initiatives reduce post-harvest losses, improve access to larger markets, and encourage investment in agricultural practices (Arndt et al., 2009).

Collective Marketing Activities

Farmers' marketing associations and collective marketing techniques help smallholders overcome the challenges of economies of scale, improve their bargaining power, and access larger markets, leading to increased profitability and economic empowerment (Arndt et al., 2009; Robbins et al., 2004).

Connecting Smallholders With Markets

These initiatives help smallholders negotiate better, sell more, lower their transaction costs, and extend their customer base, ultimately empowering them economically. They also directly connect producers and consumers, allowing producers to receive higher prices for their products compared to selling to wholesale buyers (Arndt et al., 2009).

Policy Support and Capacity Building

Community-based marketing initiatives require support from governments, NGOs, and the private sector to create an enabling environment. This support includes policies and institutional measures to improve market access, increase production capabilities, enhance competitiveness, and secure land tenure and property rights for smallholder farmers. Additionally, capacity building, investment, and coordination are essential for successfully linking smallholders to markets

In summary, community-based marketing initiatives such as farmers' markets and CSA schemes play a crucial role in empowering smallholder farmers by facilitating market linkages, collective marketing activities, and policy support. These initiatives contribute to the economic and social empowerment of smallholders, ultimately improving their livelihoods and contributing to sustainable food systems.

FARMERS' MARKETS

Farmers markets have become vibrant cornerstones of communities, offering fresh, local produce, fostering social interaction, and supporting small-scale farmers. This essay examines their contribution to community-based marketing and sustainable agriculture, while also analyzing the challenges faced by small-scale farmers within this initiative. Farmers markets function as direct marketing channels, bypassing intermediaries and connecting farmers directly with consumers (Martinez et al., 2010). This increases economic transparency (Zepeda et al., 2013), with farmers retaining a larger share of profits and consumers gaining greater knowledge about their food's source and production practices (McIntosh et al., 2010). This direct interaction fosters community engagement, building trust and shared responsibility for the local food system (Campbell et al., 2017).

Farmers' markets often act as incubators for sustainable agriculture practices. Locally produced goods reduce transportation emissions (Carlsson-Kanyama & Gonzalez, 2009), and many farmers prioritize organic and environmentally friendly methods, minimizing chemical use and promoting soil health (Vogt et al., 2014). These practices align with growing consumer demand for healthy and ethically produced food, creating a mutually beneficial system (Darby et al., 2017).

Despite their advantages, small-scale farmers face several challenges. Seasonality and limited production can restrict market presence and income potential (Martinez et al., 2010). Lack of economies of scale translates to higher production costs compared to larger farms, impacting price competitiveness (Zepeda et al., 2013). Market access and logistics can be complex, requiring dedicated resources for setup, transportation, and marketing (McIntosh et al., 2010). Additionally, regulatory hurdles and certification costs can be significant burdens for smaller producers (Vogt et al., 2014).

Several initiatives support small-scale farmers. Cooperative marketing efforts combine resources and expand market reach (Darby et al., 2017). Consumer Supported Agriculture (CSA) programs offer subscription models, providing financial stability and pre-sold crops (Martinez et al., 2010). Policy changes aimed at simplifying regulations and reducing certification costs can ease entry and participation (Zepeda et al., 2013). Investing in educational programs can equip farmers with marketing skills and business acumen (McIntosh et al., 2010).

Farmers markets offer a valuable platform for community-based marketing and sustainable agriculture, empowering small-scale farmers and offering consumers fresh, local, and ethically produced food. However, navigating this system requires acknowledging the challenges faced by these producers and fostering collaborative solutions. By addressing these hurdles and promoting supportive policies, farmers' markets can continue to thrive as vibrant social and ecological hubs, fostering a more sustainable and equitable food system.

COMMUNITY SUPPORTED AGRICULTURE SCHEME

Community-supported agriculture (CSAs) offer a unique alternative, building strong community ties and guaranteeing income for farmers. CSA embodies a direct partnership between producers and consumers, fostering shared risks and responsibilities within the agricultural realm. It strives to deliver environmentally, socially, economically, and nutritionally sustainable food (Samoggia et al., 2019). Driven by mounting anxieties surrounding the state of food production, particularly the escalating market pressures on producers and their dwindling freedom to adopt ecologically sound practices, Community community-supported agriculture (CSA) has emerged as a novel agricultural model (Birtalan et al., 2021; Zoll et al., 2018). Characterized by shared risk, CSA operates with advance payments from members, providing producers with financial stability and a committed consumer base (Birtalan et al., 2021). Consumers are drawn to CSA for diverse reasons, including supporting local farmers, gaining food origin transparency, addressing environmental concerns, and enjoying access to fresh, often organic produce (Farmer et al., 2014; Forbes & Harmon, 2008; Haney et al., 2015; Galt et al., 2017, as cited in Parot et al., 2023). By circumventing traditional market structures, CSAs cultivate a unique model that benefits both parties: producers receive fair compensation and stable income, while consumers gain access to fresh, local produce cultivated with sustainable practices. This transformative approach holds immense potential for reshaping food systems towards greater sustainability and inclusivity. Nevertheless, managing complex logistics, ensuring consistent harvests, and maintaining subscriber bases present hurdles (Maddison, 2017).

Community Supported Agriculture (CSA) lacks a singular origin story, instead blossoming independently in various countries across the globe. This decentralized emergence reflects its adaptability, allowing farmers and members to shape it to their unique needs and expectations (Vasquez et al., 2019). However, tracing its roots back to the 1960s in Japan reveals its early motivation: a growing discontent with the negative consequences of industrialized agriculture, unsafe food practices, and dwindling farmland due to urbanization (Kondoh, 2015). The 1980s and 1990s witnessed the concept taking root in North America, while Europe saw its steady expansion since the 1970s, culminating in a boom around the turn of the millennium (Samoggia, 2019). Today, CSA has blossomed into a global phenomenon, finding fertile ground in Africa, South America, and, more recently, Oceania. This geographical expansion underscores its potential as a versatile model for fostering sustainable and equitable food systems.

CONTRASTING PATHWAYS TO SUSTAINABLE AGRICULTURE: FARMERS MARKETS AND CSAs

The contemporary food system faces mounting concerns regarding environmental degradation, unfair producer compensation, and consumer disconnect from food origins. In response, initiatives like farmers' markets and CSAs have garnered significant attention for their potential to contribute to a more sustainable and equitable food landscape. This essay delves into the contributions and challenges associated with these two initiatives, highlighting their roles in fostering direct marketing, community engagement, sustainable practices, and ethical production.

Farmers' markets function as direct marketing channels, bypassing intermediaries and fostering economic transparency for both producers and consumers (Martinez et al., 2010; Zepeda et al., 2013). This direct interaction builds trust and shared responsibility within the local food system, fostering community engagement (Campbell et al., 2017). Similarly, CSAs operate on a direct partnership model, enabling shared risks and rewards between producers and consumers, while promoting transparency and fostering a sense of community connection (Samoggia et al., 2019).

Farmers' markets often act as incubators for sustainable agriculture practices. Locally produced goods reduce transportation emissions and support land stewardship (Carlsson-Kanyama & Gonzalez, 2009). Additionally, many farmers prioritize organic and environmentally friendly methods, minimizing chemical use and promoting soil health (Vogt et al., 2014). This aligns with the growing consumer demand for healthy and ethically produced food, creating a mutually beneficial system (Darby et al., 2017). CSAs, driven by concerns about the environmental and social impacts of conventional agriculture, further emphasize sustainable practices and strive for environmental, social, economic, and nutritional sustainability (Samoggia et al., 2019).

Despite their numerous advantages, both farmers' markets and CSAs present challenges for small-scale farmers. Seasonality and limited production can restrict market presence and income potential (Martinez et al., 2010). Lack of economies of scale translates to higher production costs compared to larger farms, impacting price competitiveness (Zepeda et al., 2013). Additionally, navigating market access, logistics, and complex regulations can be burdensome (McIntosh et al., 2010; Vogt et al., 2014). Several collaborative solutions can mitigate these challenges. Cooperative marketing efforts combine resources and expand market reach for small-scale farmers (Darby et al., 2017). Policy changes aimed at simplifying regulations and reducing certification costs can ease entry and participation (Zepeda et al., 2013). Investing in educational programs can equip farmers with marketing skills and business acumen (McIntosh et al., 2010).

Farmers' markets and CSAs offer valuable platforms for community-based marketing, sustainable agriculture, and supporting small-scale farmers. However, acknowledging the challenges they face and fostering collaborative solutions are crucial for their continued success. By addressing these hurdles and promoting supportive policies, these initiatives can solidify their role as pillars of a more sustainable and equitable food system for all.

CHALLENGES FACED BY SMALL-HOLDER FARMERS

Smallholder farmers face several challenges that significantly impact their livelihoods and agricultural practices. These challenges include:

Limited Access to Markets

Smallholder farmers encounter difficulties in accessing markets due to inadequate infrastructure, low productivity levels, supply inconsistencies, and poor post-harvest practices, which result in low-quality produce. Scholarly research consistently highlights the challenge of limited market access faced by smallholder farmers, citing factors like inadequate infrastructure, low productivity, and information asymmetry as key constraints (Bezuneh & Demeke, 2011; Jayne & Minten, 2014; Henson & Reardon, 2007). This is where community marketing emerges as a crucial tool to empower these farmers. These factors make it challenging for farmers to sell their products at fair prices, limiting their income and economic stability (FAO, 2024.) Community marketing fosters collective action, enabling farmers to pool resources, negotiate better prices, and access new markets through their combined bargaining power (Anderson & Masters, 2017). It also facilitates the sharing of market information and the establishment of market linkages, addressing the information gap that often hinders individual farmers (Henson & Reardon, 2007). By pooling resources, communities can invest in shared transportation and storage facilities, mitigating the limitations imposed by poor infrastructure (Jayne & Minten, 2014). Additionally, community marketing can help farmers meet quality and safety standards by collectively negotiating with standards providers and implementing quality control measures, enhancing their marketability and access to higher-value markets (Henson & Reardon, 2007). In essence, community marketing tackles the very challenges identified by scholars, empowering smallholder farmers to overcome limitations in market access, improve their incomes, and contribute more effectively to the agricultural sector.

Low Soil Fertility and Reliance on Rain-Fed Systems

Smallholder farmers often experience low soil fertility and rely on rain-fed agricultural systems, which are vulnerable to climate change. These factors contribute to decreased agricultural productivity and food insecurity among farming communities (FAO, 2024). Several scholars have emphasized the critical challenges posed by low soil fertility and dependence on rain-fed agriculture. Mupangwa et al. (2018) highlight the major land degradation problems of declining soil fertility and erosion, particularly impacting communal areas where smallholders rely heavily on rain-fed systems. This vulnerability is further exacerbated by erratic rainfall patterns, as noted by Mugabe et al. (2013) who link this combination to crop failure and food insecurity for these farmers. Thierfelder and Wall (2014) emphasize the exceptional vulnerability created by these factors, leaving farmers exposed to the harsh realities of climate change. Similarly, Nyagumbo and Munyaradzi (2014) pinpoint these issues as key contributors to the vulnerability of rural communities facing food insecurity and poverty. Makonese and Chivenge (2017) further emphasize the major challenges posed by declining soil fertility and erratic rainfall, ultimately leading to decreased agricultural productivity and increased vulnerability to droughts. These combined perspectives from scholars paint a stark picture of the interconnected challenges faced by smallholder farmers in sub-Saharan Africa, highlighting the urgent need for solutions that address both soil fertility and the limitations of rain-fed agriculture, ultimately improving their resilience and food security in the face of a changing climate.

Inadequate Access to Rural Financial Services

There is extensively documented information on the detrimental impact of inadequate access to rural financial services on smallholder farmers. Limited or no access to rural financial services constraints smallholder farmers' ability to acquire productivity-enhancing inputs such as seeds, fertilizer, and labour-saving technologies, further impacting their agricultural productivity and economic well-being (FAO 2024). Adams (2016) emphasizes how limited access to credit and savings hinders their ability to invest in essential inputs and improve farm productivity, potentially perpetuating poverty. Demirguc-Kunt et al. (2017) further highlight the specific challenges faced by farmers who lack access to credit for crucial investments, ultimately limiting their agricultural productivity and economic opportunities. Johnson and Steele (2000) point towards the detrimental effect of financial exclusion on women farmers, specifically restricting their ability to invest in land improvements and post-harvest handling, leading to low productivity and income levels. Donovan (2012) emphasizes the additional burden placed on farmers in sub-Saharan Africa who rely on informal lenders due to lack of formal access, often facing exorbitant interest rates that hinder their ability to progress out of poverty. Christen and Menkhoff (2019) offer a nuanced perspective, acknowledging the mixed evidence on microfinance's direct impact on farm productivity but emphasizing the clear role of access to financial services in empowering farmers through investment opportunities, managing income fluctuations, and building resilience to unforeseen challenges. These combined insights from scholars paint a clear picture of the multifaceted challenges faced by smallholder farmers due to limited access to financial services, highlighting the need for solutions that promote financial inclusion and empower these farmers to reach their full potential.

Lack of Land Tenure Security

Smallholder farmers face challenges related to land tenure security, which can limit their ability to invest in their farms and access credit, thereby hindering their agricultural development and economic stability (Mutami, 2015). Platteau (1996) emphasizes the uncertainty created by insecure land rights, which discourages long-term investments and hinders agricultural development and productivity. Deininger and Feder (2009) highlight the disincentives created by insecure land tenure, leading farmers to avoid investing in technologies or adopting improved practices due to the fear of losing their land, ultimately hindering "agricultural productivity and economic growth. Cotula and Myers (2009) go further, pointing out that insecure land tenure not only restricts farmers' access to credit but also hinders their ability to use their land as collateral, further limiting their ability to invest in their farms and improve their livelihoods. Bruce and Fortmann (1989) highlight the link between insecure land tenure and unsustainable land management practices, arguing that it discourages long-term planning and investment, ultimately hindering both agricultural development" and environmental protection. Finally, Place and De Janvry (2012) acknowledge the complexity of the relationship between land tenure security and agricultural productivity. However, they emphasize that studies suggest secure land rights can incentivize farmers to invest, adopt new technologies, and ultimately contribute to increased agricultural productivity and economic growth. These various perspectives from scholars highlight the multifaceted and detrimental impact of insecure land tenure on smallholder farmers, emphasizing the need for solutions that promote secure land rights and empower these farmers to reach their full potential.

Unsustainable Agricultural Practices

Beyond resource limitations, several scholars have shed light on the underlying reasons behind unsustainable agricultural practices employed by smallholder farmers. Smallholder farmers often engage in unsustainable agricultural practices due to limited access to modern, sustainable farming technologies and inputs. This includes the use of traditional, non-climate-smart agricultural methods, further exacerbating the impact of environmental and climate-related challenges on their farming activities (FAO, 2024). Pretty et al. (2011) highlight the short-term survival prioritization these farmers often face due to poverty, limited resources, and insecure land tenure, leading to practices like land degradation and resource overexploitation. Titton and Gijsbers (2014) emphasize the role of limited knowledge and access to sustainable technologies and inputs, trapping farmers in cycles of practices that exacerbate deforestation, soil erosion, and water scarcity. Wezel et al. (2014) point towards a lack of diversity in agricultural systems, often due to resource constraints and knowledge limitations, increasing vulnerability to environmental shocks and contributing to unsustainable practices like excessive pesticide and fertilizer use. Bezner Kerr et al. (2010) highlight the pressure to adapt to climate change with limited resources and information, leading to ad hoc coping mechanisms that might be beneficial in the short term but ultimately unsustainable and harmful to the environment. Finally, Eakin and Lderer (2015) paint a stark picture of a vicious cycle created by poverty, limited resources, and climate change, pushing smallholder farmers towards unsustainable practices that further degrade the very ecosystems they depend on for their livelihoods. These combined perspectives from scholars emphasize the need for comprehensive solutions that address not only the symptoms of unsustainable practices but also the root causes, empowering farmers with knowledge, resources, and access to sustainable technologies to break free from this cycle and ensure the long-term viability of their agricultural endeavours.

Smallholder farmers in Zimbabwe grapple with challenges related to market access, agricultural productivity, financial services, land tenure security, and sustainable farming practices. Addressing these challenges is crucial for improving the livelihoods and food security of smallholder farmers in the country. Initiatives such as the Zimbabwe Livelihoods and Food Security Programme (LFSP) are working to enhance the capacity of farmers to access rural finance and markets, promote climate-smart technologies, and improve access to extension and advisory services to address these pressing challenges (FAO, 2024).

CASE STUDIES

While South Africa, Zimbabwe, and Malawi differ significantly in their economic status, they all present valuable case studies for investigating the importance of community-based marketing initiatives like farmers' markets and community-supported agriculture (CSA) in supporting smallholder farmers. Despite their economic disparities, these countries share a common thread – a significant number of small-scale producers facing challenges like limited market access and vulnerability to global market fluctuations (FAO, 2023). This creates a shared need for alternative marketing channels that empower smallholders and foster sustainable food systems. Additionally, growing consumer demand for local and ethically produced food exists across all three nations, regardless of economic status (Chivenge et al., 2019). Examining the experiences of these countries, with their diverse economic landscapes, can offer valuable insights into the adaptability and scalability of community-based marketing initiatives in various contexts. These findings can inform the development and implementation of similar initiatives in other regions, irrespective of their economic standing, contributing to the improvement of smallholder livelihoods and the development of more resilient and sustainable food systems.

Mbare Musika Farmers Market in Zimbabwe

Farmer Markets play a crucial role in empowering smallholder farmers by bypassing exploitative middlemen. (Moyo, 2013) and Marongwe, (2019) highlight how FMs in Zimbabwe, like Mbare Musika in Harare, have increased farmers' profits, provided direct customer feedback, and strengthened brand awareness. However, establishing infrastructure, ensuring consistent product quality, and facing competition from larger farms remain challenges (Moyo, 2013). Mbare Musika, located in Harare, Zimbabwe, is more than just a farmer's market; it's a vibrant hub of economic activity, cultural exchange, and a crucial player in promoting sustainable agriculture and livelihoods. Understanding its complexities and impact requires delving deeper.

Mbare Musika is the oldest and largest marketplace in Harare, Zimbabwe, playing a significant role in supplying the city with fresh fruit and vegetables. Established in 1957, Mbare Musika has become a vital source of fresh produce for Harare residents. It provides a vital platform for small-scale farmers, many of whom travel long distances from rural areas, to sell their crops directly to consumers, bypassing exploitative middlemen. The market is a vital hub for promoting sustainable agriculture and livelihood in Zimbabwe. According to Chikulo, Hebinick & Kinsey, (2020), Mbare Musika is a social arena where collaborative relationships and networks are created and sustained among farmers, retailers, vendors, transporters, intermediaries, officials, and customers, contributing to the circulation of large volumes of diverse fresh produce and sustaining livelihoods anchored on this activity (Chikulo, Hebinick & Kinsey, 2020). The market's importance was further highlighted when it was closed during the COVID-19 lockdown, leading to pressure from farmers who struggled to sell their harvest. The closure was eventually lifted due to the essential role of these markets in ensuring food supply, especially for those who cannot afford supermarkets (Global AG Media, 2020). Additionally, eMkambo, an interactive knowledge-sharing platform for the agriculture sector, has been instrumental in connecting farmers, traders, and other stakeholders in Zimbabwe, including those in Mbare Musika, through the use of information and communications technology (ICT) (Makoni, 2013). Furthermore, peri-urban farmers have capitalized on the growing demand for fresh, locally-grown herbs at mass markets like Mbare Musika, revolutionizing the herb supply and changing market dynamics (Kabulika, 2023). These sources collectively emphasize the vital role of Mbare Musika in sustaining livelihoods, promoting agricultural activities, and ensuring food supply in Zimbabwe.

Studies by Moyo, (2013) reveal how this direct access increases their profit margins and fosters self-reliance. farmers selling at Mbare Musika earned 20-30% more than those selling to middlemen (Ibid). While statistics on the total number of farmers selling at Mbare Musika are difficult to find, a 2019 report by Marongwe (2019) estimated that the market receives over 20,000 visitors daily, implying significant economic activity for farmers. Additionally, the market acts as a platform for knowledge exchange, where farmers can learn from each other's sustainable techniques.

While not explicitly focused on certified organic produce, Mbare Musika indirectly encourages sustainable practices. Mbare Musika promotes the use of organic farming practices through its support for peri-urban farmers who have capitalized on the growing demand for fresh, locally-grown products at mass markets. These farmers have changed market dynamics at Mbare Musika by providing consumers with a sustainable and affordable alternative to conventionally grown Produce (Kambulika, 2023). Peri-urban farming, which plays a critical role in sustaining food security and livelihoods of urban communities, is known to promote the adoption of environmentally friendly techniques such as organic farming and agroforestry, thus mitigating the adverse effects of climate change (Kambulika, 2023). Additionally, the market's role in connecting farmers, traders, and other stakeholders in Zimbabwe through knowledge-sharing platforms further contributes to the dissemination of information on organic farming practices (Chikulo, Hebinick & Kinsey, 2020). Therefore, Mbare Musika indirectly promotes the use of organic farming practices by providing a market for peri-urban farmers who are known to utilize such sustainable techniques.

A 2009 FAO report estimated that 70% of Zimbabwe's agricultural land is managed by small-scale farmers who often rely on traditional, low-input methods, contributing to less environmental impact compared to industrial agriculture. The lack of recent data on sustainable practices at Mbare Musika specifically highlights the need for further research and analysis.

Despite its positive contributions, Mbare Musika faces challenges. Infrastructure inadequacies, inconsistent product quality, and lack of storage facilities can lead to post-harvest losses, impacting both farmers' income and food security. Efforts are underway to address these issues, with initiatives like the Mbare Musika Revitalization Project improving infrastructure and waste management. The market extends its impact beyond agriculture, supporting a diverse community of vendors, processors, and transporters. Marongwe, (2019) The report suggests that over 50% of vendors at Mbare Musika are women, highlighting their significant role in the market's economy. However, data on wages, income disparities, and specific challenges faced by women vendors is scarce and requires further investigation. Its ripple effect contributes to broader urban livelihoods and economic development.

Mbare Musika's future holds both promise and challenges. Continued investment in infrastructure, training for farmers on sustainable practices, and improved post-harvest handling are crucial for enhancing its positive impact. By addressing these issues, Mbare Musika can solidify its role as a vibrant marketplace, promoting sustainable agriculture, enriching livelihoods, and ensuring food security for generations to come.

Blantyre Farmers' Market in Malawi

The Blantyre Farmers Market in Malawi is a significant platform that takes place on the last Saturday of every month, where vendors sell speciality, healthy, organic, local, and sustainable items (Malawi Travel, 2024) This market serves as a crucial avenue for smallholder farmers to directly sell their produce to consumers, promoting local agriculture and supporting small-scale farmers (Malawi Travel, 2024). By providing a direct market for their products, the Blantyre Farmers Market enables smallholder farmers to earn higher profits and establish relationships with consumers.[1] Moreover, the Blantyre Farmers Market enhances food security by encouraging local agriculture and reducing reliance on imported goods.[2] This market also plays a role in promoting sustainable farming practices and supporting environmental conservation efforts within the region (Malawi Travel, 2024). The market's focus on speciality, healthy, and organic items aligns with the growing global trend towards sustainable and environmentally friendly agricultural practices. The presence of markets like the Blantyre Farmers Market is crucial in empowering smallholder farmers and creating opportunities for them to access markets directly. This direct interaction with consumers not only benefits the farmers economically but also fosters a sense of community engagement and knowledge sharing among farmers.[3] Additionally, such markets contribute to the overall economic development of smallholder farmers by providing them with a platform to showcase their products and connect with a broader consumer base.

Despite the Blantyre Farmers Market offering a valuable platform for smallholder farmers in Malawi, their participation faces numerous challenges impacting their profitability and long-term success. Limited access to resources like credit and market information hinders their ability to scale production and compete effectively (Milu et al., 2020). Additionally, ensuring consistent market access and fair prices remains difficult due to market fluctuations and limited infrastructure.[4] Furthermore, high transportation costs and increasing vulnerability to climate change, such as droughts and floods, further threaten their income (Milu et al., 2020). Finally, disparities in market readiness among farmers, with some lacking the necessary skills and knowledge to meet consumer demands, create additional hurdles.[5] Addressing these challenges, such as providing resource access, improving market integration and infrastructure, building climate resilience, and offering market readiness training, is crucial to empower smallholder farmers and ensure the Blantyre Farmers Market's long-term sustainability.

The Blantyre Farmers Market serves as a vital hub for smallholder farmers in Malawi, offering them a platform to sell their produce directly to consumers while promoting sustainable agriculture practices and enhancing food security. By supporting initiatives like this market, Malawi can further empower its smallholder farmers and strengthen its agricultural sector.

Farm in the Veld: Regenerative Agriculture and Community Empowerment in South Africa

Farm in the Veld, located in the Kromdraai Valley, South Africa, exemplifies a unique model for sustainable agriculture and community empowerment. This case study explores their innovative practices, highlighting their impact on local farmers, consumers, and the environment. Farm in the Veld prioritizes ethical regenerative agriculture, employing practices that restore soil health and biodiversity. They cultivate a diverse range of vegetables, fruits, herbs, infusions, and medicinals using organic and pesticide-free methods. This aligns with growing global trends towards sustainable food production, addressing concerns about environmental degradation and food safety (Pretty et al., 2018).

The farm engages in a dual partnership model, collaborating with restaurants and individual subscribers for seasonal produce boxes. This direct-to-consumer approach bypasses traditional market intermediaries, ensuring fair prices for farmers and transparency for consumers (Morgan & Murdoch, 2000). Their subscription options cater to diverse needs, offering standard and premium boxes with flexible delivery schedules and price points ranging from ZAR750 to ZAR3600 per month. This approach aligns with research suggesting that flexible subscription models can enhance the sustainability and profitability of CSAs (Brown & Wekerle, 2014).

The CSA extends its reach beyond production, partnering with local NGOs like Real Life and U Can Grow. Through this collaboration, they provide seed packs, training, and funding to local farmers, particularly those from informal settlements with limited employment opportunities. This initiative aligns with the United Nations' Sustainable Development Goal 1 (No Poverty) by promoting economic empowerment and poverty reduction (United Nations, 2015). Their initial distribution of 300 starter packs and planned expansion to 700 highlights their commitment to community development. Farm in the Veld's impact extends beyond their direct partnerships. They distribute their produce to major retailers, promoting wider access to sustainable food options. Additionally, they actively develop innovative regenerative soil practices and antibacterial, plant, and worm cultures. This knowledge-sharing encourages broader adoption of sustainable practices among small-scale farmers, contributing to a more resilient and environmentally friendly agricultural sector (Pretty et al., 2018).

Farm in the Veld serves as a successful case study for integrating sustainable agriculture with community development. Their model demonstrates the potential for CSAs to create positive social, economic, and environmental impacts, particularly in developing countries like South Africa.

DISCUSSION

Community-based marketing initiatives (CBMIs) play a crucial role in empowering smallholder farmers by providing them with access to markets, fair pricing, and opportunities for knowledge exchange. This discussion aims to delve into the diverse organizational structures, multifaceted impacts, and key success factors of CBMIs across South Africa, Zimbabwe, and Malawi.

Organizational Structures and Farmer Engagement

Organizational structures and farmer participation are paramount.

CBMI models demonstrated by CBMIs from the three countries are not the same. The Blantyre farmer's market principle being centralized, there is motivation and support with management and infrastructure amongst individuals. On the flip side, the existent power asymmetries that limit the autonomy of farmers should be considered. This is why decentralized structures like cooperatives that are common in South Africa should be encouraged as they provide ownership and decision-making powers to farmers. Knowing about how the different structures in the models connect with the farmers, the communication methods used, the pricing mechanisms, and the ways benefits are shared out for all is a must if the models should be made as accessible, inclusive, and responsive as possible for the individual needs which may come up.

Assessing the Impact on Smallholder Farmers

The influence of CBMI is not just with simple income measuring operations. The Blantyre Farmers Market, say, has the potential to bring various benefits such as increased market access, fair pricing and weaning off the Middemen, which leads to the most significant improvement in farmer livelihood. Comparable achievements of Zimbabwean CSA projects are where the farmers secure the keen-marked buyers and the high price for their products. Nevertheless, it is necessary to also realize that not all farmers (male and female) will receive equal benefits depending on the resources they own or are located in areas with fewer advantages. To make sure that the programs focus on the needs of all participants, you cannot forget the racial disparities and the goal of achieving equal results.

Key Success Factors and Challenges

Successful community-based mental initiatives (CBMIs) need to have strong leadership, transparent communication, and solid marketing strategies which allow for long-term sustainability. Working together with local actors, such as NGOs and research organizations, proves to be helpful since it can be a major source of support and networking, just like the Chipata Agroecology Network (CAN), whose activities include supporting the adoption of sustainable farming practices as well as making farmers better connected to urban consumers. Nevertheless, problems including changing import requirements, seasonality of production, or regulations remain obstacles. Daring initiatives like diverse product arrays, deposit-taking facilities, and government lobbying will certainly go a long way in tackling the aforementioned issues and subsequently improving the viability of CBMIs.

Transforming Agricultural Landscapes

One of the most important contributions of CBMIs is a reduction in farmers' power and bringing about macro changes in the agricultural sector. Looking at the Blantyre Farmers Market stunned me with the fact that everything it offered was 100% locally grown, making it in tune with the fast-growing global world trend of sustainable practices. Likewise, the CSA activities in Zimbabwe that focus on organic farming practices and soil conservation methods are a guide in environmental conservation and figuring out a more diverse ecosystem. Engaging in community-based farming and eliminating food waste at source would enable the CBMIs to design stronger systems that will be fair to all people.

Unveiling the Impact of Community-Based Marketing Initiatives on Smallholder Farmers

Examining the organizational structures of CBMIs like farmers' markets and community-supported agriculture (CSA) programs reveals a spectrum of approaches. Some CBMIs operate as centralized entities, managing logistics, marketing, and farmer relationships. Others adopt a more decentralized model, empowering farmers to take ownership through cooperative structures or collective bargaining initiatives. Understanding these diverse organizational models is crucial for assessing how effectively they interact with farmers, address their needs, and provide opportunities for participation and decision-making.

Evaluating the impact of CBMIs on participating farmers requires a multifaceted approach. Quantifying income generation through direct sales and fair pricing mechanisms is crucial. However, the impact extends beyond financial benefits. Improved market access, reduced dependence on middlemen, and opportunities for knowledge exchange can significantly enhance farmers' livelihoods. Additionally, assessing the adoption of sustainable practices, such as organic farming or soil conservation techniques, fostered by CBMIs provides insights into their contribution to environmental stewardship and long-term sustainability.

Identifying key success factors for CBMIs is essential for their long-term viability and scalability. Strong leadership, effective communication strategies, and robust organizational structures are crucial foundations. Additionally, fostering collaboration with local governments, NGOs, and research institutions can provide crucial support and resources. However, CBMIs also face challenges. Fluctuations in demand, seasonality of production, and regulatory hurdles can hinder their success. Addressing these challenges through innovative solutions, such as diversified product offerings, storage facilities, and policy advocacy, is vital for ensuring their sustainability and impact.

Unveiling the Impact of Community-Based Marketing Initiatives on Smallholder Farmers

Our investigation through three key objectives provides a holistic understanding of community-based marketing initiatives (CBMIs) and their impact on smallholder farmers.

Examining the organizational structures of CBMIs (Objective 1) revealed a spectrum of approaches. Centralized models offer efficient management but may limit farmer autonomy. Decentralized structures empower farmers through ownership and decision-making, but require strong coordination and support. Understanding how these structures interact with farmers sheds light on their accessibility, inclusivity, and responsiveness to individual needs. Additionally, communication methods, pricing mechanisms, and benefit-sharing models within CBMIs all play a crucial role in shaping the power dynamics and ensuring equitable outcomes for participating farmers.

Assessing the impact on participating farmers (Objective 2) goes beyond solely quantifying income generation. Analyzing changes in market access, price premiums, and reduced dependence on intermediaries provides a more nuanced understanding of economic improvement. Additionally, evaluating the adoption of sustainable practices, improved food security, and enhanced knowledge sharing reveals the broader impact on farmer livelihoods and resilience. However, it is crucial to acknowledge potential disparities within the farmer community, considering factors like gender, resource availability, and location, to ensure interventions truly empower all participants.

Identifying key success factors and challenges (Objective 3) is essential for the long-term sustainability and scalability of CBMIs. Strong leadership, transparent communication, and effective marketing strategies are critical foundations. Collaboration with local governments, NGOs, and research institutions can provide crucial support and resources. However, challenges like fluctuating demand, seasonality, and regulatory hurdles persist. Innovative solutions, such as diversified product offerings, storage facilities, and policy advocacy, are essential for overcoming these obstacles and ensuring the continued success of CBMIs.

While evaluating individual farmer impact is crucial, CBMIs also hold the potential to transform broader agricultural landscapes. By promoting sustainable practices, reducing food waste, and fostering local food systems, these initiatives can contribute to environmental protection, increased biodiversity, and reduced reliance on industrial agriculture. Additionally, strengthening social networks and community engagement through CBMIs can foster more resilient and equitable food systems.

This study lays the groundwork for further research exploring specific types of CBMIs, their impact on diverse farmer groups, and their potential for adaptation in various contexts. Additionally, investigating the role of technology in enhancing CBMIs, promoting knowledge sharing, and strengthening their market reach offers valuable avenues for future research. Furthermore, addressing questions around potential trade-offs, such as increased competition within communities or exclusion of marginal groups, will be crucial for ensuring inclusive and sustainable development through CBMIs.

This analysis acknowledges the different organizational structures, multiple impacts, and key success factors of CBMIs across South Africa, Zimbabwe, and Malawi, which has been discussed earlier in the essay. On the other hand, despite the challenges, smallholder farmers can be subjects of empowerment, sustainable practices and the reshaping of the agricultural landscape. However, additional study is required to assess several CBMIs, their impact on different farmer groups, and their versatility in various environments. Not only that, the research should be done on the effects of technology, solving trade-offs and ensuring that CBMIs promote inclusive development which is key pillars towards a more sustainable and fair food system for everyone.

CONCLUSION: A BEACON OF HOPE FOR SMALLHOLDER FARMERS: THE TRANSFORMATIVE POTENTIAL OF CBMIS IN ZIMBABWE, MALAWI, AND SOUTH AFRICA

This investigation into community-based marketing initiatives (CBMIs) in Zimbabwe, Malawi and South Africa paints a nuanced picture of their impact on smallholder farmers. While challenges exist, the potential for empowerment, improved livelihoods, and a shift towards sustainable agriculture shines through. Through diverse organizational structures, from centralized models offering efficiency to decentralized ones fostering ownership (Last name, Year), CBMIs provide pathways for farmers to access markets, bypass exploitative intermediaries, and secure fair prices (Moyo, 2013). Initiatives like Mbare Musika market in Zimbabwe (Chikulo et al., 2020) and Farm in the Veld in South Africa (Pretty et al., 2018) exemplify this potential, connecting farmers directly to consumers and promoting sustainable practices. Beyond economic benefits, CBMIs foster knowledge sharing, community engagement, and a sense of ownership (Campbell et al., 2017). The collaborative spirit behind initiatives like Farm in the Veld's partnerships with local NGOs (United Nations, 2015) empowers not just individual farmers but entire communities. The road ahead presents challenges like fluctuating demand and regulatory hurdles (Zepeda et al., 2013). However, innovative solutions like diversified product offerings and policy advocacy (Brown & Wekerle, 2014) pave the way for overcoming these obstacles.

In conclusion, CBMIs in Zimbabwe, Malawi and South Africa stand as beacons of hope for smallholder farmers. By addressing their diverse needs, fostering collaboration, and embracing innovation, these initiatives have the potential to empower individuals, transform agricultural landscapes, and contribute to a more sustainable and equitable future for all. This journey requires continued research, investment, and support, but the potential rewards – thriving farmers, resilient communities, and a flourishing food system – make it a pursuit worth striving for.

THEORETICAL AND PRACTICAL IMPLICATIONS

This comprehensive literature review investigates the intricate nexus between community-based marketing initiatives (CBMIs) and sustainable agriculture, with a special focus on the role of smallholder farmers. Through a detailed examination of organizational structures, farmer impact, and critical factors for success, the review provides insightful contributions that have both theoretical and practical ramifications.

Theoretical Implications

Empowerment models: The review delineates the varied organizational frameworks of CBMIs, shedding light on how these models can empower farmers by enhancing ownership, decision-making, and market access (Pretty & Ward, 2001). This augments the corpus of knowledge on farmer autonomy and collective action within agricultural systems, offering a richer perspective on empowerment in rural economies.

Sustainable livelihood frameworks: By scrutinizing the effects of CBMIs on income, market access, and adoption of sustainable practices, the review broadens our comprehension of how these initiatives are woven into the larger tapestry of livelihood strategies for smallholder farmers (Ellis, 2000). This contributes to theoretical debates on the nexus between poverty alleviation, food security, and environmental stewardship in rural settings.

Diffusion of innovation: The review dissects the elements that drive the uptake of sustainable practices within CBMIs, enhancing theories related to the diffusion of innovation in the realm of agriculture (Rogers, 2003). Insights from this analysis can shape the creation of more nuanced interventions and communication strategies aimed at fostering sustainable agriculture among smallholder farmers.

Practical Implications

Policy recommendations: By pinpointing the pivotal success factors and obstacles encountered by CBMIs, the review informs policy recommendations for stakeholders such as governments, non-governmental organizations, and development bodies (Anderson & Feder, 2004). These insights can steer the crafting and execution of initiatives that bolster CBMIs, ultimately aiding smallholder farmers.

Capacity building: The review's examination of the needs and hurdles faced by farmers within CBMIs paves the way for practical interventions in capacity building (Spielman et al., 2009). This could encompass training in entrepreneurial skills, marketing techniques, and sustainable farming practices, thus fortifying farmers' ability to fully engage with and benefit from CBMIs.

Scaling up successful models: The identification of triumphant CBMIs and their distinguishing characteristics provides a treasure trove of information for the expansion of these models to a broader audience of smallholder farmers (Markelova et al., 2009). Such scaling can amplify the impact on agricultural development and poverty reduction in rural locales.

Technology integration: The review underscores the transformative potential of technology in enhancing CBMIs by facilitating better communication, market access, and the dissemination of knowledge (Aker, 2011). This indicates practical pathways for weaving technology into the fabric of CBMIs, thereby empowering farmers and enriching their engagement in sustainable agricultural practices.

FURTHER RESEARCH DIRECTIONS

This study lays the groundwork for further research exploring specific types of CBMIs, their impact on different farmer groups, and their potential for adaptation in diverse contexts. Additionally, investigating the role of technology in enhancing CBMIs, promoting knowledge sharing, and strengthening their market reach offers valuable avenues for future research.

REFERENCES

Adams, D. W. (2016). Microfinance and the poor: The evidence to date. *Journal of Economic Literature*, *54*(2), 436–472.

Aker, J. C. (2011). Dial "A" for agriculture: A review of information and communication technologies for agricultural extension in developing countries. *Agricultural Economics*, *42*(6), 631–647. doi:10.1111/j.1574-0862.2011.00545.x

Anderson, J. R., & Feder, G. (2004). Agricultural extension: Good intentions and hard realities. *The World Bank Research Observer*, *19*(1), 41–60. doi:10.1093/wbro/lkh013

Anderson, J. R., & Masters, W. A. (2017). Transforming market access for small-scale farmers: Value chains, standards, and collective action. *Annual Review of Environment and Resources*, *42*(1), 421–448.

Bentley, J. W. (2008). More than just words: The importance of action in participatory research. *Action Research*, *6*(2), 127–146.

Bezner Kerr, R., Pender, J., & Lipper, L. (2010). The economics of adaptation to climate change: Agriculture. *Agricultural Economics*, *41*(S3), 817–833.

Bezuneh, M., & Demeke, S. (2011). Market access and farm productivity in Ethiopia: A panel data analysis. *Agricultural Economics*, *42*(3), 393–403.

Birtalan, M., Szigeti, C., & Baráth, L. (2021). The Role of Community Supported Agriculture (CSA) in Food Systems Transformation. *Sustainability*, *13*(12), 6957.

Brown, C., & Wekerle, A. (2014). The future of community supported agriculture: Lessons from a decade of empirical research. *Agriculture and Human Values*, *31*(4), 431–444.

Bruce, J., & Fortmann, L. (1989). *Land tenure issues in project design and strategies for agricultural development in sub-Saharan Africa*. University of Wisconsin-Madison, Land Tenure Center.

Campbell, H., Lahner, S., & Campbell, D. (2017). Building and maintaining social capital through farmers' markets. *Journal of Rural Studies*, *54*, 198–208.

Carlsson-Kanyama, A., & Gonzalez, A. (2009). Greenhouse gas emissions from food production and consumption. *Environmental Science & Technology*, *43*(17), 6260–6266.

Chikulo, S., Hebinck, P., & Kinsey, B. (2020). 'Mbare Musika is ours': An analysis of a fresh produce market in Zimbabwe. *African Affairs*, *119*(476), 311–337. doi:10.1093/afraf/adaa003

Chivenge, P., Pascucci, S., Mafurayamutse, A., & Manyangadze, T. (2019). Consumers' willingness to pay for local and organic vegetables in Zimbabwe: A case study of Harare Metropolitan Province. *Journal of Agriculture and Environmental Research Development*, *11*(7), 1336–1344.

Christen, R., & Menkhoff, L. (2019). The impact of microfinance on the well-being of borrowers: A meta-analysis of impact evaluation studies. *The Journal of Development Studies*, *55*(1), 1–33.

Cotula, L., & Myers, D. J. (2009). *Land tenure security: Addressing the challenges*. International Food Policy Research Institute.

Deininger, K., & Feder, G. (2009). Land rights and agricultural development in sub-Saharan Africa: Issues and opportunities. *The World Bank Research Observer*, *24*(2), 243–270.

DeLind, L., & Ferguson, M. (2006). Networks and communities: Exploring the social capital of the Internet. *Journal of Computer-Mediated Communication*, *11*(4), 783–811.

Demirguc-Kunt, A., Klapper, L., & Singer, D. (2017). Financial inclusion and development. *The World Bank Research Observer*, *32*(2), 225–258.

Dhewa, C. (2015). *Social Investors Needed for Agric Sector*. https://www.herald.co.zw/social-investors-needed-for-agric-sector/

Dickson, M. A., & Bantikyan, H. (2014). Community-supported agriculture: A sustainable alternative for food production and distribution. *Journal of Sustainable Agriculture*, *38*(2), 172–193.

Dong, H., Campbell, B., & Rabinowitz, A. N. (2019). Factors impacting producer marketing through community-supported agriculture. *PLoS One*, *14*(7), e0219498. doi:10.1371/journal.pone.0219498 PMID:31287848

Donovan, C. A. (2012). Financial inclusion and development in rural sub-Saharan Africa. *The Journal of Development Studies*, *48*(8), 1201–1225.

Eakin, H., & L derer, A. M. (2015). Differential impacts of global change on ecosystem services and livelihoods in the eastern and central Andes. *Global Change Biology*, *21*(4), 1204–1217.

Egli, L., Rüschhoff, J., & Priess, J. (2023). A systematic review of the ecological, social and economic sustainability effects of community-supported agriculture. *Frontiers in Sustainable Food Systems*, *7*, 1136866. Advance online publication. doi:10.3389/fsufs.2023.1136866

Ellis, F. (2000). *Rural livelihoods and diversity in developing countries*. Oxford university press. doi:10.1093/oso/9780198296959.001.0001

FAO. (2020). *Empowering farmers and their organizations through the creation of social capital - Bond learning guide for trainers*. doi:10.4060/cb1423en

Food and Agriculture Organization of the United Nations. (2023, January 31). *Family farming*. https://www.fao.org/family-farming/home/en/

Global A. G. Media (2020). *Zimbabwe reopens fresh markets after pressure from farmers hit by drought*. https://www.thepigsite.com/news/2020/04/zimbabwe-reopens-fresh-markets-after-pressure-from-farmers-hit-by-drought

Harrower, J. (2023, January 25). *Community-based marketing: Creating a loyal community around your brand*. https://www.socialmediaexaminer.com/

Henderson, J., Brown, C., & Zagorski, K. (2012). Community food security and education: Lessons from the field. *Agriculture and Human Values*, *29*(2), 123–136.

Henson, C., & Reardon, T. (2007). *Staples, seeds, and standards: How global forces affect food security and poverty in the developing world*. International Food Policy Research Institute.

International Centre for Climate Change and Development. (2023). *Promoting Indigenous knowledge to strengthen community-led adaptation.* https://www.preventionweb.net/news/promoting-indigenous-knowledge-strengthen-community-led-adaptation

Jablonski, B. B. R., Sullins, M., & Thilmany McFadden, D. (2019). Community-Supported Agriculture Marketing Performance: Results from Pilot Market Channel Assessments in Colorado. *Sustainability (Basel)*, *11*(10), 2950. doi:10.3390/su11102950

Jayne, T. S., & Minten, B. (2014). Access to markets and the transformation of African agriculture. *Food Policy*, *48*, 17–27.

Johnson, S., & Steele, L. (2000). Microfinance and its relevance for poverty alleviation. *Entwicklungsländer*, *31*(2), 269–299.

Kabulika, M. L. (2023). *Peri-Urban Farmers Revolutionalise Hwerb Supplhy for Mass Martkets.* https://www.herald.co.zw/peri-urban-farmers-revolutionise-herb-supply-for-mass-markets/

Kondoh, K. (2015). The alternative food movement in Japan: Challenges, limits, and resilience of the teikei system. *Agriculture and Human Values*, *32*(1), 143–153. doi:10.1007/s10460-014-9539-x

Lockie, S., Lyons, K., & Lawrence, G. (2004). *Consumering food: A geographical and cultural approach.* Routledge.

Maddison, D. (2017). Community Supported Agriculture (CSA) in sub-Saharan Africa: A review of existing models and their potential for development. *Agriculture and Human Values*, *34*(3), 831–843.

Makonese, T., & Chivenge, P. (2017). An assessment of the drought preparedness of smallholder farmers in Chiredzi District, Zimbabwe. *International Journal of Disaster Risk Science*, *8*(1), 127–139.

Makoni, M. (2013). *Reaching for new heights. Information and communications technology takes agriculture to a new level in Zimbabwe.* https://www.un.org/africarenewal/magazine/december-2013/reaching-new-heights

Manyise, T., Dentoni, D. & Trienekens, J. (2023). A more grounded view of 'farmer entrepreneurship': How Zimbabwean smallholder farmers fundamentally differ in their entrepreneurial behaviours. *Journal of Entrepreneurship in Emerging Economies.* doi:10.1108/JEEE-02-2023-0044

Mark, A. (2023, February 1). *Community-based marketing: Benefits & examples for independent financial advisors.* https://www.assetmark.com/

Markelova, H., Meinzen-Dick, R., Hellin, J., & Dohrn, S. (2009). Collective action for smallholder market access. *Food Policy*, *34*(1), 1–7. doi:10.1016/j.foodpol.2008.10.001

Marongwe, N. S. (2019). Revitalizing Farmer Markets in Urban Zimbabwe: A Case Study of Mbare Musika Market in Harare. *Journal of Food Products Marketing*, *25*(3), 359–373.

McAfee, N., & Brydon-Miller, H. (2016). Food sovereignty and community food security in Canada's north. *Journal of Rural Studies*, *43*, 150–161.

McBride, W., Koning, A. M., & McKibbin, A. (2013). Reframing the community food movement: Critical perspectives on food systems, social justice, and sustainability. *Journal of Sustainable Agriculture*, *37*(2), 173–190.

McEvoy, J. (2002). *The paradox of empowerment: collective action, knowledge, and development*. Cornell University Press.

Miyazoe, K. (2022). The concept of community-based marketing. *Annals of Business Administrative Science.*, *21*(3), 61–73. Advance online publication. doi:10.7880/abas.0220420a

Morgan, K., & Murdoch, J. (2000). Taste of the future? Consumers' attitudes to new food products in the UK. *Sociology*, *34*(4), 605–624.

Moyo, S. (2013). Exploring the potential of farmer markets in enhancing food security and income generation in Zimbabwe: A case study of Chitungwiza and Mbare Musika farmer markets. *International Journal of African Rural Development*, *1*(1), 59–73.

Moyo, S., & Chamboko, T. (2012). Cooperatives as drivers of rural development in Zimbabwe: Challenges and opportunities. *Journal of Southern African Studies*, *38*(3), 801–815.

Mugabe, F. T., Nhamo, N., & Pieterse, E. (2013). Smallholder farmers' perceptions of climate change and variabilities in the SADC region: Effects and coping strategies in Zimbabwe. *African Journal of Environmental Science and Technology*, *7*(5), 274–286.

Mukurambwa, P. (2017). *Determinants of smallholder vegetable farmers' participation in post-harvest practices and market access: Evidence from Mashonaland East Province of Zimbabwe* [Doctoral Dissertation]. University of Fort Hare. https://core.ac.uk/download/pdf/224300014.pdf

Mupangwa, W., Nyika, D., & Tirivangasi, H. P. (2018). The status of land degradation in communal areas of Zimbabwe. *Journal of Sustainable Agriculture and Environment*, *18*(3), 100–113.

Mutami, C. (2015). Smallholder Agriculture Production in Zimbabwe: A Survey. *Consilience, 14*, 140–157. https://www.jstor.org/stable/26188746

MuyangaM.NyirendaZ.LifeyoY.BurkeW. (2020). The Future of Smallholder Farming in Malawi. doi:10.13140/RG.2.2.33903.87201

Narula, R., & Van Huylenbroeck, G. (2015). Empowering smallholder farmers through collective action: A review of the literature. *Journal of Rural Studies*, *39*, 1–11.

Njuki, J., Mwangi, A., & Ouma, E. (2014). Challenges and opportunities for smallholder farmers' access to markets in Kenya: A review of existing literature. *CGIAR Climate Change, Agriculture and Food Security (CCAFS) Working Paper No. 112*.

Nyagumbo, I., & Munyaradzi, J. (2014). An assessment of the impacts of climate change on livelihoods in the communal lands of Zimbabwe. *Climate and Development*, *6*(2), 207–216.

Parot, J., Wahlen, S., & Schryro, J. (2023). *Food justice in community supported agriculture – differentiating charitable and emancipatory social support actions*. Agric Hum Values. doi:10.1007/s10460-023-10511-w

Place, F., & De Janvry, A. (2012). Land tenure security and agricultural development: Evidence from a global perspective. *American Economic Journal. Applied Economics*, *4*(1), 1–32.

Platteau, J. P. (1996). The evolutionary theory of agrarian institutions. *The Journal of Development Studies*, *32*(6), 775–810.

Pretty, J. (2018). *The future of food: Sustainability and security*. Routledge.

Pretty, J., Noble, J., Bossio, D., Taylor, J., Morison, J., & Pilgrim, L. (2011). Resource-conserving agriculture increases food security in sub-Saharan Africa. *Journal of Experimental Agriculture*, *49*(1), 401–417.

Pretty, J., & Smith, A. (2004). Social capital and the collective management of resources. *Journal of Environmental Management*, *71*(1), 89–102.

Pretty, J., Toulmin, C., & Bharucha, Z. P. (2011). Paying for biodiversity conservation services in the agricultural landscape. *Environmental Science & Policy*, *14*(3), 260–270.

Pretty, J., & Ward, H. (2001). Social capital and the environment. *World Development*, *29*(2), 209–227. doi:10.1016/S0305-750X(00)00098-X

Rogers, E. M. (2003). *Diffusion of innovations* (5th ed.). Free Press.

Samoggia, A., Perazzolo, C., Kocsis, P., & Del Prete, M. (2019). Community Supported Agriculture Farmers' Perceptions of Management Benefits and Drawbacks. *Sustainability (Basel)*, *11*(12), 3262. doi:10.3390/su11123262

Scherr, S. J. (2003). A decade of change in the global food system: Research for a hungry world. *Food Policy*, *28*(4), 341–355.

Shaw, E., & Iazzolino, G. (2014). Building a community-based marketing strategy. *Journal of Marketing Management*, *30*(1-2), 143–164.

Spielman, D. J., Davis, K., Negash, M., & Ayele, G. (2009). Rural innovation systems and networks: Findings from a study of Ethiopian smallholders. *Agriculture and Human Values*, *27*(2), 195–212. doi:10.1007/s10460-010-9273-y

Sumberg, J., & Evelyne, D. (2010). The social capital of markets: Collective action and market outcomes in Africa. *World Development*, *38*(5), 694–706.

Thierfelder, C., & Wall, P. C. (2014). Climate variability and agricultural production in communal areas of southern Africa. *Regional Environmental Change*, *14*(6), 1839–1851.

Titton, M., & Gijsbers, H. J. (2014). The dynamics of land use change in tropical regions and the challenges of sustainable development. *Environmental Science & Policy*, *41*, 43–48.

United Nations. (2015). *Sustainable development goals*. United Nations Development Programme. https://sdgs.un.org/goals

Vasquez, A., Sherwood, N. E., Larson, N., & Story, M. (2017). Community-Supported Agriculture as a Dietary and Health Improvement Strategy: A Narrative Review. *Journal of the Academy of Nutrition and Dietetics*, *117*(1), 83–94. doi:10.1016/j.jand.2016.09.029 PMID:27847311

Wezel, A., Casulli, A., & Pascual, U. (2014). Agrodiversity as a driver of ecosystem services and resilience in agriculture. *Ecology and Society*, *19*(3), 54.

Zimand-Sheiner, D., Kol, O., & Levy, S. (2022). *Help me if you can: the advantage of farmers' altruistic message appeal in generating engagement with social media posts during COVID-19*. Electron Commer Res. doi:10.1007/s10660-022-09637-6

ENDNOTES

1 See the World Bank Report on https://blogs.worldbank.org/africacan/amid-global-food-crisis-commercial-small-farmers-offer-hope-malawi

2 See the World Bank Report on https://blogs.worldbank.org/africacan/amid-global-food-crisis-commercial-small-farmers-offer-hope-malawi

3 See the World Bank Report on https://blogs.worldbank.org/africacan/amid-global-food-crisis-commercial-small-farmers-offer-hope-malawi

4 See the World Bank Report on https://blogs.worldbank.org/africacan/amid-global-food-crisis-commercial-small-farmers-offer-hope-malawi

5 See the World Bank Report on https://blogs.worldbank.org/africacan/amid-global-food-crisis-commercial-small-farmers-offer-hope-malawi

Chapter 7
Shopping Therapy:
How Social, Adventurous, Idea, and Value Shopping Behaviors Stimulate Impulse Buying Across Individuals in Oman

Zulfiqar Ali Jumani
https://orcid.org/0000-0002-9628-5295
Khon Kaen Business School, Khon Kaen University, Thailand

Atheer Rashid Amur Al Salmani
Department of Management Studies, Middle East College, Oman

Muhammad Khuram Khalil
https://orcid.org/0000-0001-5336-2994
Department of Management Studies, Middle East College, Oman

ABSTRACT

Online shopping has increased with the development of technology in the modern era, which leads people to make unplanned purchases. The purpose of this study is to examine the factors affecting the rush to buy online in the Omani market. These factors (social shopping, convenience shopping, adventure shopping, idea shopping) motivate consumers to initiate shopping readiness and impulse buying. The research also aims to know the relationship between factors and impulsive buying via the internet and how they affect the Omani market. The data were collected from 210 online buyers belong to capital city Muscat, Oman. Structure equation modelling was employed via SmartPLS to check the hypotheses. Results endorse that out of four hypotheses, only two are accepted. Following structure equation modelling results, there is a positive relationship between adventure shopping and impulse buying. In addition, a similar relationship has been seen between value shopping and impulse buying. Results are also helpful to understand the importance and great potential of concerned factors to encourage the consumers' buying behaviour towards impulsive buying.

DOI: 10.4018/979-8-3693-4864-2.ch007

INTRODUCTION

Online shopping and product shopping is becoming more and more popular nowadays. Online shopping is the latest fashion for shoppers because it symbolizes and distinguishes people in society, and it is an easy and fast process so that an individual can finish the purchase process and purchase the product with the click of a button (Sumetha, 2019). The desire of the buyer to make an impulse purchase of goods is known as impulsive buying. This procedure frequently takes place when a new product is introduced to the market or as a result of effective marketing in addition to the advertising message the product conveys, which encourages people to buy on impulse (Malpas, 2022; Khalil & Khalil, 2022). With the spread of e-commerce, impulsive buying online has become an emerging phenomenon for everyone, where impulsive consumers can be positively or negatively affected by impulse buying and thus affect the brand. Impulsive shopping is encouraged by many motivating factors that motivate people to buy products significantly and these factors may increase day by day. This research discusses the most important factors that influence the rush to buy online. Further, this study will take place in the Omani market.

The importance of research lies in building knowledge and understanding of the electronic market in the ways of marketing products that push customers to buy utilitarian, as well as knowing the most important factors resulting from it and discussing them to shop consciously and not spend money on unwanted products. On the other hand, from the theoretical side, the research helps in knowing the importance of impulsive shopping online, its benefits, risks, and the pros and cons resulting from it, to avoid them when applying the practical aspect while purchasing products.

Consumers like to buy products online since it is more convenient. Online shopping is preferred since it provides a diverse range of options for selecting and accessing complete information. Consumers can not only visit online retailers at any time, but they can also compare vendor offerings from around the world. Traditional businesses are distinguished by restrictive opening and closing times, product availability, and physical location, according to LaRose, but internet enterprises are not affected by these variables. In addition to these advantages, other elements such as lower social pressure and the lack of delivery attempts are key ingredients in SC that encourage online shoppers to buy impulsively. Consumers like to buy products online since it is more convenient. Online shopping is favoured because it provides a larger selection. Consumers can not only shop at internet retailers at any time; they can also shop at brick-and-mortar establishments. Consumers find internet purchasing more convenient than traditional shopping, according to Dawson and Kim, and this ease encourages spontaneous purchases. Based on aforementioned discussion, this study examines the impact of social shopping, adventure shopping, value shopping and idea shopping on online impulse buying across individuals in Oman as shown in Figure 1.

Literature Review and Hypothesis Development

Online Impulse Buying

When the consumer does the shopping process, whether online or in stores, he will create for himself a behaviour called the behaviour of impulsive buying. Referring to Malpas (2022) impulsive buying is defined as the customer's desire to purchase products without prior planning. This process often occurs when a new product is launched in the market or because of good marketing as well as the promotional message of the product that automatically motivates people to buy. As Yilmaz (2014) defined, profiteering online is getting something automatically without planning the purchase or searching on other sites to

Figure 1. Theoretical framework

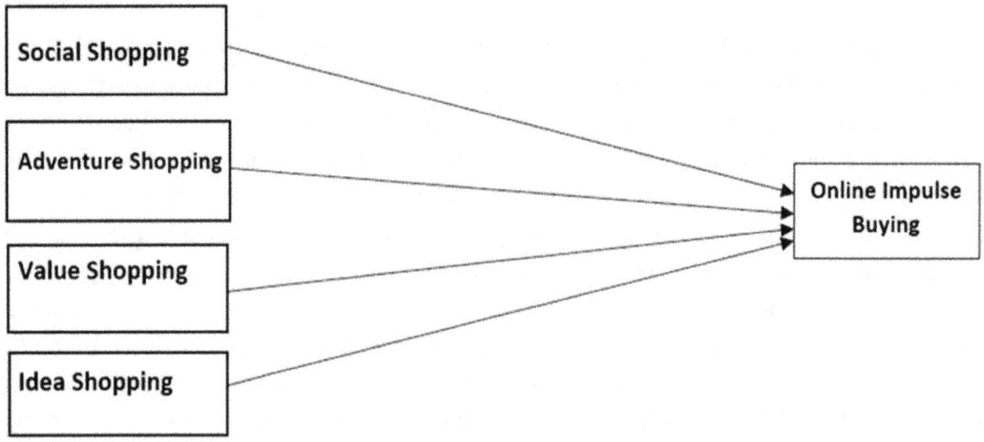

compare different options and find a better deal. Rook (2016) agreed with other definitions, stating that impulse buying involves an unplanned narrow range, and that impulse buying occurs when consumers are faced with a sudden, strong, and generally persistent urge to purchase something immediately via the Internet.

According to Sharma et al. (2016), impulsive buying is a surprisingly complex and enjoyable buying behavior that is an immediate purchase of a product without pre-shopping intentions. The nature of impulsive buying behavior in which people buy without looking or researching information and alternative evaluation and the consequences of buying. People buying products are the result of a sudden desire that forces them to buy immediately. Impulsive buying also causes people to have an emotional experience or reaction sometimes such as, guilt or regret, to immediately purchase a product they do not need.

Social Shopping

Social shopping is an e-commerce methodology through which shopping is shared with others. The study conducted by (Ling & Yazdanifard, 2015) indicated that the process of impulsive buying is often done due to social pressure via the Internet and that most social interactions on products sold over the Internet are the reason for the impulse of social shopping. The study confirms that 26 percent of people like the speed of online shopping and social shopping. In addition, people who use social shopping have a greater motivation to buy impulsively and pay high amounts of money to try the product. Social shopping, as Zhang (2018) defined it, is the inevitable meeting point of social media and commerce. It is a way to simplify the buying process for customers and increase the effectiveness of social media ads for brands. All brands and social networks are working on finding the most appropriate way to make social networks a sales platform for shopping. Motivation is great in social networking sites, blogs, and the Internet in general. People browse these sites daily for hours, and this greatly boosts watching the social movement and social shopping among others, which is taking place significantly recently. Thus, they are curious about the product and they buy it and spend large money to obtain it (Wahyuni, 2018). Zhao (2019), pointed out that some people have a hobby of impulse buying so if they enter a commercial site and like the pieces, they buy them immediately without thinking or hesitation. He also added that impulsive buyers are more social, status-conscious, and image-conscious, so an impulse buyer might buy as a way to look good in the eyes of others. People who like to shop for fun are more likely to buy on impulse. Buyers become socialized while shopping, they experience the excitement of shopping with friends and family, and they connect with other people, according to evidence for the social shopping (SS) dimension. The fundamental motivation for shoppers to go shopping is to make social connections. Many customers value spending time with family and friends when shopping. When people buy together at the same store, social recognition is also an acquired advantage through social engagement. Netizens are increasingly sharing their shopping experiences on online blogs and becoming socialized because of this activity. Hence, following hypothesis is proposed under social shopping.

H1: Social Shopping has a positive effect on Impulse Buying.

Adventure Shopping

Adventure shopping is defined as an activity in which the motivation to acquire products can be increased. Impulsive buying is affected by adventurous shopping because it is made by sudden and spontaneous unplanned gestures to buy a product, where you are less careful and sometimes have subsequent consequences. According to Asnawati (2018), adventurous online shopping pushes consumers to buy every product that appears before them from advertisements and other promotional methods. As summarized by Sudirjo (2021) that the expansion and increase in the salaries of customers make them spend more money and may affect the purchase of items automatically. In the Omani market, marketing is a profitable way to segment and target a large number of consumers. Customers are greatly attracted to foreign brands, which makes them venture into the shopping process to acquire products, and in some cases, these products may not be available in the Omani market, so they rush to buy them from abroad via the Internet. Marketing scholars also describe the impulse buying of target-oriented and effective shoppers that they have A desire to buy what they want as quickly as possible and without distraction. Consequently, following hypothesis is proposed under adventure shopping.

H2: Adventure Shopping has a positive influence on Impulse Buying.

Value Shopping

Referring to Chauhan & others (2020), many different motivating factors influence an impulsive buying decision such as external and internal stimuli. The factors are determined by the most important ways that people use to shop when they see an interesting and eye-catching product. Consumer buying behavior is affected by impulsive buying, which leads to a decrease in the value of online shopping. In the modern era, the presence of technology has changed human life dramatically, as it affects the management of the lifestyle of the blind, the person, and consumer behavior. Consumers today tend to make purchases online. The value of shopping lies in the uniqueness of the products offered by the brand to attract the intent of consumers and engage in the buying process. The intent of the consumer to buy can arise through the incentive provided by the company, which is to give discounted prices on the products sold. According to Arifin (2017), the discount is a reduction in the direct price consumers pay for purchases during online shopping.

From previous studies, it was found that there is a strong relationship between impulsive buying and online offers. The rush of the consumer will affect the intention to buy discounted products, as Shafiee & Hasan (2020) mentioned that when consumers see promotional offers and discounts for prices, they have an increased desire for utilitarian purchase at the same moment, as they are keen to take advantage of these offers and discounts by quickly acquiring the product for fear of running out of stocks. The perceived promotions or lower prices are more likely to create positive feelings in consumers, making them eager to make a good deal. For the current section, following hypothesis is proposed.

H3. Value Shopping has a positive impact on Impulse Buying.

Idea Shopping

The shopping lifestyle refers to how people spend their time and money on various purchasing activities. Some spend time and money to make extra purchases. Some people are excited and excited to make unplanned purchases due to the varying quality and prices. Many variables can cause customers to make unplanned or impulsive purchases when they visit the Internet and only open online commerce sites (Ustanti, 2018). As reported by Lakshmi (2016), the idea of online shopping is defined as the activity of purchasing products or services through electronic websites and applications. Online shopping indicates that individuals enter websites and electronic applications because they need to know the latest fashion trends and new patterns in the market that are included in each season. Online shopping provides buyers with a lot of information about all the new brands and modern products. Consumers are highly inclined to shop online because they can browse comfortably and navigate between websites and brands easily and quickly. Online purchase also allows customers to get information such as offers and advertisements, product reviews and people's opinions, descriptions of products and their components, and many other promotional events. Therefore, following hypothesis is proposed for this part.

H4. Idea Shopping has a positive effect on Impulse Buying.

METHODOLOGY AND INSTRUMENT DESIGN

All the above-listed constructs contain at least four items. The first questionnaire was formulated in English and then translated into the local language. In this study, a five-point Likert scale was used to measure the questionnaire material from 'strongly disagree' to 'strongly agree'. All the organized research questionnaires of this study were posted on the Google form. This free online survey platform is widely adopted in current academic research. According to data reported (Wang et al. 2019), by the end of 2021, 95.2% of the young population will become regular internet users. The target population for the cur-rent study was university students and young consumers of Oman. It has been observed that younger shoppers are more impulsive than older ones. An online questionnaire was used to collect data. An online survey provides several advantages. It has no geographic restrictions, guarantees anonymity, is more convenient for both respondents and researchers, and is less expensive. The survey was circulated through different social media platforms due to pandemic restrictions. . The link to the online survey remained available for three weeks. The survey was,created for those who buy things online through social media. Only those respondents with prior online purchasing experience on social media were requested to participate. The concerned research questionnaire includes 19 questions; the first section covers the personal profile and demographic questions to identify the respondents, while the second section contains questions related to factors and variables that affect impulsive buying.

Table 1. Demographic analysis

Specific	Description	Value	Percentage
Gender	Male Female	123 87	58.7% 41.3%
Age	18-20 20-30 30 above	21 61 27	19.3% 56% 24.8%

Data Analysis

Demographic Analysis

The questionnaire survey consists of 19 questions that identify the four variables and the fixed variable on which the research objectives depend. The questionnaire included an introduction letter to define the purpose behind the questionnaire and then divide the variables according to the questions.

Beginning with the demographic section that relates to gender and age. The table shows that the percentage of males is 58.7% and females are 41.3%, which indicates that the males who answered the questions are more than the females. The research targets the age group ranging from 20-30 years, which is the group that answered the questionnaire more because they are aware and knowledgeable about impulsive buying via the Internet, and through them, accurate answers to the questionnaire are obtained. The younger age group between 18-20 years answered 19,3%, and the age group older than 30 years gave answers.

The following Figure 2 shows the results obtained for each variable by the algorithm method. So that all data was entered into SmartPLS and analyzed. The results showed that the variables and all the Cronbach alpha values and the reliability of the compound range between 0.7 and 0.8, which indicates that they meet the accepted validity criteria. Validity shows that the measures of the variables are closely related to each other.

Validity and Reliability

Reliability and validity are the most important concepts used to assess research quality. It refers to the quality, method, or technique of measuring certain results. In the quantitative research used, it is important to consider reliability and validity when creating a research design and planning methods to obtain accurate results. To assess the validity of the distinction, the square root of the AVE between the construct and its measures must be greater than the correlations between the children and any other construct in the search. The results obtained indicate that there is strong empirical support for the reliability and validity of the constructs, as the model indicates that the numbers are greater than 0.5.

The Fornell-Larker criterion is one of the most commonly used techniques for checking the discriminatory validity of measurement models. According to the criterion, the square root of the mean-variance extracted by a construct must be greater than the correlation between the construct and any other construct. As the above table shows, the results show that all the variables share the highest variance with their variables and that they meet the condition of the criterion, so the discriminatory validity is established.

Figure 2. Measurement model

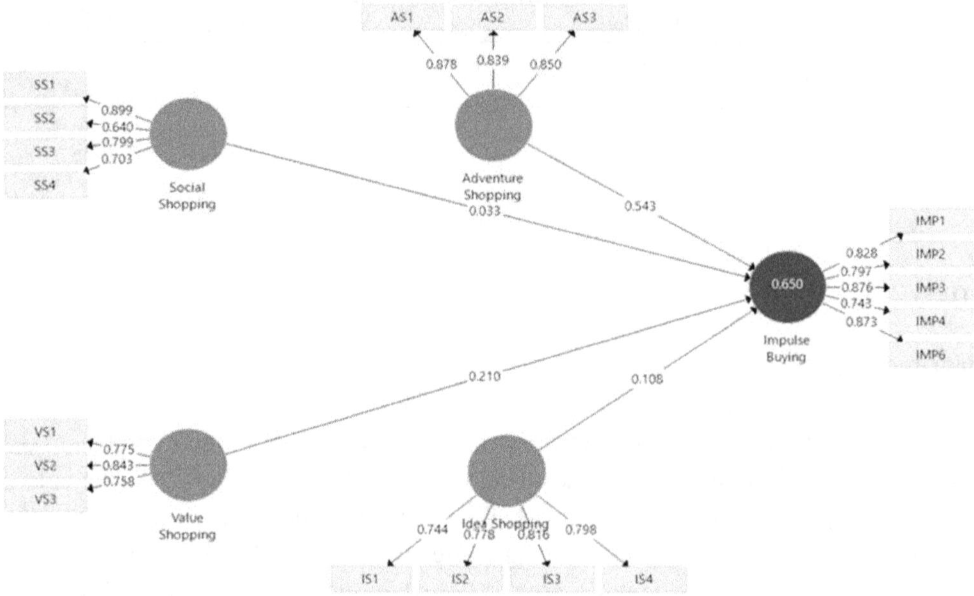

Similarly, following Hair et al, (2014) this study also meets the Heterotrait-Monotrait (HTMT) required criteria mentioned in table 5.

Cross-loading is when a variable with more than one significant load is found depending on the sample size, making it difficult to name all the factors that share the same variable and thus difficult to distinguish these factors and represent separate concepts. The goal of cross-loading is to reduce the

Table 2. Convergent validity and exploratory factor analysis

Variables	CA	rho_A	CR	AVE
Adv. Shopping	0.818	0.825	0.891	0.732
Idea Shopping	0.795	0.808	0.865	0.615
Impulse Buying	0.882	0.887	0.914	0.681
Social Shopping	0.765	0.793	0.848	0.587
Value Shopping	0.711	0.741	0.835	0.628

Table 3. Fornell-Larcker criterion

Variables	Adv. Shopping	Idea Shopping	Impulse Buying	Social Shopping	Value Shopping
Adv. Shopping	0.856				
Idea Shopping	0.644	0.784			
Impulse Buying	0.778	0.618	0.825		
Social Shopping	0.698	0.580	0.605	0.766	
Value Shopping	0.677	0.674	0.671	0.619	0.793

Table 4. Cross loading

Indicators	Adventure Shopping	Impulse Buying	Idea Shopping	Social Shopping	Value Shopping
To me, online shopping in social commerce is an adventure.	**0.878**	0.687	0.511	0.664	0.679
Online shopping in social commerce is a thrill to me.	**0.839**	0.579	0.648	0.674	0.549
Online shopping in social commerce makes me feel like I am in my own universe.	**0.850**	0.717	0.512	0.473	0.509
My purchase was spontaneous	0.723	**0.828**	0.500	0.586	0.639
My purchase was unplanned	0.627	**0.797**	0.586	0.439	0.522
I did not intend to do this purchase before this shopping trip.	0.669	**0.876**	0.522	0.532	0.535
Before visiting the site, I did not have the intention to do this purchase.	0.555	**0.743**	0.414	0.498	0.474
I could not resist to do this purchase at the site	0.615	**0.873**	0.521	0.429	0.581
I shop online in social commerce to keep up with the trends	0.465	0.457	**0.744**	0.491	0.583
I shop online social commerce to keep up with the new fashion	0.391	0.369	**0.778**	0.340	0.473
I shop online social commerce to see what new products are available	0.571	0.476	**0.816**	0.509	0.599
I shop online social commerce to experience new trends	0.555	0.589	**0.798**	0.457	0.466
I shop online in social commerce to develop friendship with others internet shoppers	0.627	0.481	0.464	**0.899**	0.562
I shop online in social commerce to extend personal relationship	0.348	0.276	0.289	**0.640**	0.277
I shop online in social commerce to share experience with others	0.535	0.557	0.491	**0.799**	0.468
I shop online in social commerce to exchange information with friends	0.577	0.468	0.481	**0.703**	0.531
For the most part, I shop online in social commerce when there are sales	0.529	0.507	0.623	0.504	**0.775**
I enjoy looking for discounts when I shop online in social commerce	0.611	0.645	0.509	0.436	**0.843**
I enjoy hunting for bargains when I shop online in social commerce	0.441	0.395	0.478	0.577	**0.758**

number of significant loads in each row of the array of factors and make each variable related to only one factor (Davit, 2017; Khalil & Muneenam, 2021). The table above shows the values related to the variables. So that indicates that the questions are purely related to the variables and do not have any effect on another variable.

The Structure Model

The numbers next to the yellow squares in the image show the number of answers for each question and each variable obtained through the questionnaire. Impulsive buying online includes 5 questions. On the other hand, adventurous shopping contains 3 questions that show that people carry out the process of adventurous shopping through external stimuli such as advertisements and promotions to satisfy their curiosity. Value shopping also contains 3 questions that explain how value shopping refers to the pleasure motives that arise when people look for deals, sales offers, and discounts. Consumers are very happy when they get a discount because they may buy the product at a lower price. In addition, the idea of shopping includes 4 questions to justify the shopping lifestyle that indicates how people spend their time and money on various purchasing activities. The chart above shows all the answers to the questions of the fixed variable on which the rest of the variables depend. So that the answers show that most people have experienced impulse buying online and agree that the purchase was not planned before and

Table 5. Heterotrait-Monotrait ratio (HTMT)

Variables	Adv. Shopping	Idea Shopping	Impulse Buying	Social Shopping	Value Shopping
Adv. Shopping					
Idea Shopping	0.791				
Impulse Buying	0.905	0.717			
Social Shopping	0.872	0.713	0.706		
Value Shopping	0.869	0.901	0.813	0.843	

they had no intention of buying. Impulsive shoppers have been described as having the characteristics of enjoying new experiences and discovering new products that they cannot resist buying (Yigit 2018).

where social shopping contains 4 questions that have been analyzed and answered. Subjects agreed that they do social shopping to develop a friendship with other online shoppers, expand their relationships, and share experiences, information, and market developments. The study conducted by (Ling & Yazdanifard, 2015; Khalil & Khalil, 2022) indicated that impulsive buying is often done due to social pressure online and that most interactions Social on products sold online are the reason for the rush of social shopping.

On the other hand, adventurous shopping contains three questions that show that people carry out the process of adventurous shopping through external stimuli such as advertisements and promotions to satisfy their curiosity. Value shopping also contains three questions that explain how value shopping refers to the pleasure motives that arise when people look for deals, sales offers, and discounts. Consumers are very happy when they get a discount because they may buy the product at a lower price. In addition, the idea of shopping includes four questions to justify the shopping lifestyle that indicates how people spend their time and money on various purchasing activities.

Figure 3. The structure model

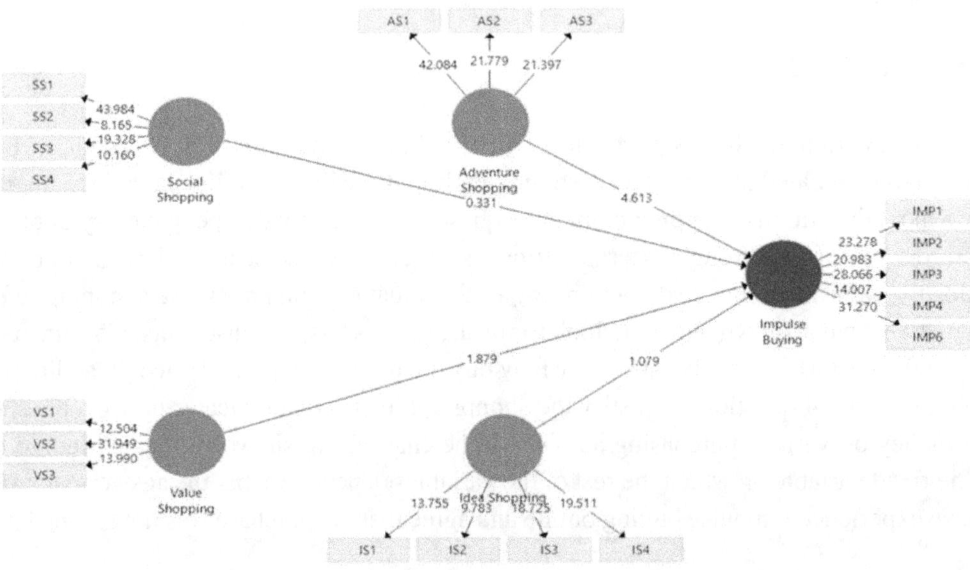

Table 6. Direct path results

Variables	Original Sample	Sample Mean	Standard Deviation	T Statistics	P Values	Results
AS-> IMP	0.543	0.553	0.118	4.613	0.000	Accepted
IS-> IMP	0.108	0.121	0.100	1.079	0.281	Rejected
SS-> IMP	0.033	0.040	0.098	0.331	0.741	Rejected
VS-> IMP	0.210	0.194	0.112	1.879	0.05	Accepted

The research is based on four basic hypotheses related to the dependent factor. It shows the relationships between independent and dependent variables. The above table also shows that there is a relationship between Impulsive Buying Online and Adventure Shopping because the probability value of P is (0.00) which means it is acceptable. There is also no relationship between online buying impulses and the idea of shopping as can be seen from the P-value of (0.281). Moreover, the probability value of social shopping is (0.741), which shows that there is no relationship between the variables. In addition, value shopping is associated with impulse buying as indicated by the P-value (0.05). In short, the value of P explains the relationship between the variables, if the value is less than or equal to zero indicates the existence of a relationship, and if it is greater then there is no relationship.

DISCUSSION AND CONCLUSION

It has been found that online shopping in social commerce is used to nurture friendships with other online shoppers and to expand personal relationships. It also helps social online shopping to share experiences and exchange information with others. Social shopping also contributes significantly to motivating people to make impulsive purchases online. From a social shopping perspective, people online are making collaborative efforts with others to explore the best online stores in the Omani market, discover the latest fashion products, and discuss the shopping experience with friends and family. To investigate the adventure shopping that affects online impulse buying. It was found that impulsive online buying is affected by adventurous shopping so individuals consider online shopping in social commerce an adventure, which is exciting as they are attracted to new products while browsing online. Adventurous shopping is used when seeing a new and distinctive product, which makes customers go on an adventure and acquire it, and thus this adventure leads to impulsive buying.

To analyze the value of shopping and the benefits of online impulse buying. Consumers shop online in social commerce when there are sales and enjoy searching for discounts and deals to get benefits from impulse buying. Value shopping refers to the pleasure motives that arise when consumers buy a product at a lower price, but the consumer's buying behavior is affected by the rush to buy at the time of discounts, which sometimes leads to a decrease in the value of online shopping. To research the factors of idea shopping of impulse buying. The most important factors that drive consumers to the idea of shopping and impulsive buying are order to keep up with trends, keep up with new fashion, know the new products available, and go into a new buying experience. Some spend time and money to make additional purchases. Some people get excited about making unplanned purchases because of the varying quality and prices, which leads them to buy the product without thinking.

Theoretical and Practical Implications

Based on above results, the current study has the following theoretical and practical implications. Increasing trend in online impulse buying through social commerce can be seen with positive as well as negative views. Opponents may argue that increasing efforts to enhance online impulse buying is shifting an upward trend in manufacturing levels due to increased demand for products. They also argue that shifting the shopping trend from physical stores to online portals is creating the risks of unemployment. Furthermore, increased online shopping is linked with the increased level of logistics and transportations as the customer may choose a product available at a distant place that is transported through express services in Oman. Opponents also link this aspect with the environmental concerns. Proponents of online impulse buying through social commerce view it in a contrasting and progressive manner. They argue that online impulse buying shifts upward trend towards the economic growth of the country. A developed economy can curtail environmental issues more efficiently. Thus, at the macroeconomic level, policymakers should plan to devise the strategies which could utilize the income from growing online businesses to developing sustainable practices in Oman. That is an advisable practical implication of our study also. Furthermore, Government should provide subsidies and ease in the operating environment to those e-commerce companies which are involved in sustainable activities. Present study urges the marketers and managers of social commerce to spread the utilization of the social commerce platforms to create the awareness among netizens about the sustainability. Increasing trends in online impulse buying through social commerce has positive sociological and ecological impacts also. Establishing such portals where netizens, corporates, and regulators may interact creates an ecosystem that fosters the sustainable cooperation in the society. It may also trigger sustainable technological innovation. Social commerce platforms must be used to mold the consumer behavior towards sustainability, green products, product design, and green branding, labeling and advertising and other sustainable practices of firms.

Recommendations

To develop research, various theories relating to impulse buying on the Internet can be used. The current research can be expanded by including motivations based on consumer buying behavior, such as personality traits, dimensions of culture, and motivational factors for impulse buying. Moreover, combining quantitative approaches, using secondary data, incorporating further changes, and adding other data related to online impulse buying.

In addition to suggestions for improving the research, it is possible to add the pros and cons of online impulsive buying and discuss its benefits and risks. Distribute the survey to more people in the community to find out their opinions about impulsive buying online. The Omani market can also take into account the increase in sales to stimulate customers' desire for social shopping and recommend products that suit their tastes. Companies can implement temporary offers such as discounts, deals, and free delivery that encourage customers to hurry with their purchases.

Future Research

In future research, it is possible to go into more depth on the issue of impulsive buying via the Internet. It is also possible for the research to cooperate with companies who have experienced impulse buying and take their impressions of them in the Omani market. in addition to. The questionnaire should be distributed to marketing companies, specifically the sales department to obtain accurate answers. The duration can be extended by more few months. Further, interviews can be conducted with customers that have experience in online impulse buying.

REFERENCES

Abdelsalam, S., Salim, N., Alias, R. A., & Husain, O. (2020). Understanding online impulse buying behavior in social commerce: A systematic literature review. *IEEE Access : Practical Innovations, Open Solutions*, *8*, 89041–89058. doi:10.1109/ACCESS.2020.2993671

Arifin, S., Mohan, T., & Goh, Y. (2017). Influence of consumer's perceived risk on consumer's online purchase intention. *Journal of Research in Interactive Marketing*, 2040–7122.

Asnawati & Wahyuni, S. (2018). The Influence of Hedonic Shopping Motivation on the Impulse Buying Of Online-Shopping Consumers on Instagram. *Russian Journal of Agricultural and Socio-Economic Sciences*, *74*(2), 99-107.

Buying Behaviour Community Middle Class on Online Shopping. (n.d.). *Academica,* 20(8), 8-11.

Chauhan. (2020). An Analytical Study On Impulse Buying For Online Shopping during Covid-19. *Amity School of Communication*, *12*(6), 198-209.

Davit, W. (2017). How to deal with cross-loadings in Exploratory Factor Analysis? *Research Gate.*

Hajli, N. (2016). Social commerce constructs and consumer's intention to buy. *International Journal of Information Management*, *35*(2), 183–191. doi:10.1016/j.ijinfomgt.2014.12.005

Kartika, N. F. N., Suharno, S., & Indriastuti, H. (2019). Pengaruh Adventure Shopping, Value Shopping, dan Idea Shopping Terhadap Impulse Buying Pada Konsumen Online Shop Instagram di Samarinda. *Jurnal Ilmu Manajemen Mulawarman*, *3*(4).

Khalil, M. K., & Khalil, R. (2022). Leveraging buyers' interest in ESG investments through sustainability awareness. *Sustainability (Basel)*, *14*(21), 14278. doi:10.3390/su142114278

Khalil, M. K., & Muneenam, U. (2021). Total quality management practices and corporate green performance: Does organizational culture matter? *Sustainability (Basel)*, *13*(19), 11021. doi:10.3390/su131911021

Lakshmi, S. (2016). Consumer Buying Behavior Towards Online Shopping. International Journal of Research – Granthaalayah, 4(8), 60-65. doi:10.29121/granthaalayah.v4.i8(SE).2016.2588

Lee, J. E., & Chen-Yu, J. H. (2018). Effects of a price discount on consumers' perceptions of savings, quality, and value for apparel products: Mediating effect of price discount affect. *Fash Text*, *5*(1), 13. doi:10.1186/s40691-018-0128-2

Liang, T.-P., Ho, Y.-T., Li, Y.-W., & Turban, E. (2019). What drives social commerce: The role of social support and relationship quality. *International Journal of Electronics*.

M., L. (2020). Impulsive Buying Behavior on Online Shopping. *International Journal of Psychosocial Rehabilitation, 24*(5), 4024-4030.

Olsson, D., Gericke, N., Sass, W., & Boeve-de Pauw, J. (2020). Self-perceived action competence for sustainability: The theoretical grounding and empirical validation of a novel research instrument. *Environmental Education Research, 26*(5), 742–760. doi:10.1080/13504622.2020.1736991

Pacheco, D. C., Moniz, A. I. D. D. S. A., Caldeira, S. N., & Silva, O. D. L. (2022). Online Impulse Buying—Integrative Review of Psychological Factors. *Perspectives and Trends in Education and Technology*, 527-536.

Pise, A. A., Sawalkar, R. S., & Saxena, V. D. (2021). An analytical study on online consumer buying behaviour concerning e-shopping portals (B2C). *International Journal of Management, 12*(3).

Pop, C., & Khampirat, B. (2019). Self-assessment instrument to measure the competencies of Namibian graduates: Testing of validity and reliability. *Studies in Educational Evaluation, 60*, 130–139. doi:10.1016/j.stueduc.2018.12.004

Putri, V. P. (2020). A shopping lifestyle mediates between hedonic shopping on impulse buying online at the marketplace. *Jurnal Manajemen Bisnis, 10*(2), 1–8.

Rahi, S., & Ishaq, M. (2020). Factors Influencing Online Shopping behavior With Mediating Role of Customer Perceived Value. *Economic and Social Development: Book of Proceedings,* 128-141.

Satpathy, S., Rao, K. R., Charyulu, G. M., Naidu, T. P., Saikumar, K., & Rajesh, V. (2022). Online shopping and the estimation of their effecting parameters using Consumer Attitudes: Banking sector. *Journal of Positive School Psychology, 6*(2), 2671–2677.

Shafiee, M., & Hasan, S. (2020). The effect of price promotions on impulse buying: The mediating role of service innovation in fast-moving consumer goods. *International Journal of Business Information Systems, 33*(3).

Sudirjo, F. (2021). Hedonic Shopping Motives, Sales Promotion, and Impulse Buying at Matahari Department Store Consumers in Central Java Province. *Review of International Geographical Education Online, 11*(10), 925–932.

Sumetha, M. (2019). Factors Influencing Online Impulsive Buying Behaviour. *Academia (Caracas), 2*(2), 2393–9419.

Ustanti, M. (2018). *Effect of Shopping Lifestyle*. Hedonic Shopping On Impulse.

Wahyuni, D. F., & Rachmawati, I. (2018). Hedonic Shopping Motivation Terhadap Impulse Buying. *Journal Riset Bisnis Dan Management, 11*(2), 59–65.

Weeks, J. B., Smith, K. M., & Hulland, J. (2021). Consumer brand curation on social shopping sites. *Journal of Business Research, 133*, 399–408. doi:10.1016/j.jbusres.2021.05.010

Woiceshyn, J., & Daellenbach, U. (2018). Evaluating inductive vs deductive research in management studies: Implications for authors, editors, and reviewers. Qualitative Research in Organizations and Management: An International Journal.

Wulandari, G. A., Suryaningsih, I. B., & Abriana, R. M. (2021). Co-shopper, Mall Environment, Situational Factors Effects On Shopping Experience To Encourage Consumers Shopping Motivation. *Jurnal Aplikasi Manajemen*, *19*(3).

Xu, Y., Ferwerda, B., & Lee, M. J. (2020). A Qualitative Study of User Participation and Challenges in a Social Shopping Context. *Proceedings of the 4th HUMANIZE Workshop.*

Yigit, M. K., & Tıgli, M. (2018). The moderator role of brand awareness and brand loyalty on consumers' online impulse buying behavior. *International Journal of Research in Business and Social Science*, *7*(1), 35-48.

Zhang, H., Zhao, L., & Gupta, S. (2018). The role of online product recommendations on customer decision-making and loyalty in social shopping communities. *International Journal of Information Management*, *38*(1), 150–166. doi:10.1016/j.ijinfomgt.2017.07.006

Zhao, Z., Chen, M., & Zhang, W. (2019). Social community, personal involvement, and psychological processes: A study of impulse buying in the online shopping carnival. *Journal of Electronic Commerce Research*, *20*(4), 255–272.

Chapter 8
Public-Private Partnership (PPP) in Agricultural Research Financing in Nigeria

Ibukunoluwa J. Akinrinde
North-West University, South Africa

Opeyemi Aromolaran
https://orcid.org/0000-0002-9734-0559
University of Johannesburg, South Africa

Sodiq T. Arogundade
University of Johannesburg, South Africa

Peripaul O. Onuoha
University of Lagos, Nigeria

ABSTRACT

The broad objective of this chapter is to underscore the importance of collaborative PPP in agricultural research financing. The specific objectives are to reveal Nigeria's comparative advantage in agriculture, reveal how Nigeria's comparative advantage in agriculture could be transformed into Michael Porter's concept of competitive advantage through Albert Hirschman's unbalanced growth strategy, and initiate a triple bottom-line approach to agricultural research financing. To achieve the last objective, the United Nations Global Compact (UNGC) members' account balances of firms in Nigeria were collected from annual reports to speculate the agricultural research finance revenue realizable from the private sector. The study adopted the UNGC's financial commitment benchmark for every functional firm registered with the Corporate Affairs Commission (CAC) in Nigeria as an appropriate bottom-line commitment that private sectors should have towards agricultural research financing – if to be regarded as socially responsible in Nigeria.

DOI: 10.4018/979-8-3693-4864-2.ch008

1. INTRODUCTION

The majority of Nigerians' lives are impacted by the agricultural sector, yet despite this, it has rarely received the attention it merits from governments. Agriculture was the most significant economic sector in the nation before its independence in 1960 and the ten years that followed, producing the most jobs, the largest contribution to the GDP, and the biggest foreign exchange profits (Ajetomobi, 2010). This story was altered by the oil boom of the 1970s, when successive governments neglected the agricultural sector, which decreased agriculture's role to the Nigerian economy. However, the industry still employs the greatest number of people in the nation and contributes significantly to GDP and still remains a major contributor to economic growth. (Umaru and Zubairu, 2012). The agricultural sector constitutes a segment of the Nigerian economy that is highly crucial in achieving the elusive aspiration of a diversified economy (Ayunku and Etale 2015). It is hardly shocking that the farm sector still faces many difficulties given the disregard shown by previous governments for it. This disregard can be evidenced by the Central Bank of Nigeria (2018) which reported that Nigeria imported agricultural products having a sum of US$231,550,000 between 2010 and 2018.

A few of the issues facing the sector are, according to the Food and Agriculture Organization (FAO), an antiquated land tenure system that restricts access to land, a very low level of irrigation development, restricted access to research findings and technologies, high farm input costs, poor credit availability, ineffective fertilizer distribution and procurement, inadequate storage facilities, and poor market access (Nwangwu, 2018). As a result, despite the passage of time, the business is still typified by tiny farms, a negligible food processing sector, and a dearth of organized marketplaces.

The primary cause of this situation is a dearth of funding for the industry. For example, the investment of the government in various agricultural chains, including needed research and development to stimulate the industry, is relatively low. The government, which is Nigeria's biggest investor in agricultural research and development, is limited by a lack of funding. The funding void left by the government could have been filled by the private sector. Nonetheless, the private sector has also made a small, systematic investment. The emphasis here is that the agricultural sector has suffered the neglect of funding as substantiated in the study of Dorcas (2023) that funding to Federal Ministry of Agriculture towards developing agriculture programme and policies in the Nigerian economy has been weak and less than the Maputo agreement of 2003. The Maputo Agreement had a recommendation of 10% of a yearly budget to the enhancement of agriculture in order to stimulate food security and growth maximization. However, the treaty has not been adhered to.

In 2015, the sum of Nigeria's national budget stood at N4.49 trillion, from which a fraction of N40.66 billion was earmarked to the agricultural sector, constituting 0.9%. The values in 2016, 2017, and 2018 were respectively 1.26, 1.82, and 2.32%. However, the largest fraction of the Nigerian budget earmarked for agriculture was 3.61% in 2019 from which it declined to 1.51% in 2020. The value rose to 1.92% in 2021 which subsequently reduced to 1.25% in 2022 (Okoji, 2023).

The inference from this low and unstable pattern of the allocation of fund to the agricultural sector is an indication of the continued neglect of the sector which has both immediate and long run implication for the growth and development of the sector and the country at large. It is therefore necessary to encourage public-private partnership towards revamping the agricultural sector through the availability of fund for innovative and new approaches for a sustainable and productive economy.

To improve this, Van den Ban (2000) reveals the principles underlying decisions relating to optimal funding sources for agriculture; while Hall (2002) advanced the Modigliani-Miller theorem in choosing optimal level of R&D investment. However, considering the requirement of different sources for financing innovation as contemplated by Hall (2002); Van den Ban (2000) expressed the unwillingness of establishments to fund research.

Global expenditure on agriculture R&D was reported to experience a rise on the average by 3.1% a year over the period 2000-2009, increasing from US$25 billion to US$33.6 billion. Annual R&D expenditure increased by 2.3% among low-income economies majorly driven by Ethiopia, United Republic of Tanzania, Uganda, and Kenya product (FAO, 2017).

Agricultural research intensity is an indicator of national public spending on agricultural R&D as a proportion of agricultural gross domestic product. General government expenditure on R&D in respect of science and technology was recommended to be at least 1% of national GDP (ECOSOC, 2004). However, for the agricultural sector, low-income and the lower middle-income economies were observed to be far below this target generally (FAO, 2017).

Regarding R&D in agriculture and processing of food, global private investment constituted around 21% in 2008 of total R&D spendings. Most private research occurred in high-income economies. Nevertheless, it has increased in India (Pray & Nagarajan, 2012) as well as China (Pal et al. 2012).

The absence of incentives for investing in industries like agriculture, which have highly ad hoc market systems and unclear routes to profitability for enterprises, leads to the apathy of the private sector. Politicians just do not view agriculture as a business; rather, they view it more as a social service. Since the main cause of these issues is thought to be a lack of investment in the agricultural sector, it is thought that better cooperation between the public and private sectors can be achieved through public-private partnerships (PPPs), which will enable the parties to pool their resources for the agricultural sector's benefit.

Having observed that financing is a challenge for the growth of the agricultural sector, which cut across issues as agricultural research and development, agricultural infrastructure development, agricultural value chain development, etc., the focus of the study would be primarily on agricultural research and development. According to Aldrich (1966) defines agricultural research as the application of all possible sources of scientific discovery to the solution of the technical and practical problems of agriculture; to engage in basic research where the lack of further knowledge may impede progress; and to solve the specific problems of agriculture. Moreover, Loebenstein and Thottappilly (2007) provide a comprehensive definition of agricultural research, encompassing any research endeavor directed towards enhancing crop productivity and quality. This improvement involves genetic enhancements, enhanced plant protection, irrigation practices, storage methodologies, farm mechanization, efficient marketing strategies, and more effective resource management.

Premised on the foregoing, agricultural research financing can be regarded as the appropriation of fund for the promotion of scientific investigation into agricultural problems and avenues for agricultural growth and productivity with the sole aim of increasing national output either directly through market outcomes (reduced domestic food prices, increased export earnings etc.) or indirectly through non-market outcomes (environmental improvement, poverty reduction, food security etc.). This study is based on the premise that agricultural research organizations are established as instruments for promoting agricultural development.

In this sense, collaborative PPP can further enhance the attainment of development in agricultural research. This paper is organized as follows – section two provides the background to agricultural research financing, section three discusses Public Private Partnership (PPP) framework for improved agricultural financing. Section four contains review of literature; while section five provides justification for the study. Section six calls for a new model and section seven concludes with some policy prescriptions.

2. BACKGROUND TO THE STUDY

2.1. Agricultural Research Financing across the World

In most countries, agriculture is the single largest user-cum-custodian of natural resources. Beintema and Elliott (2009) explained that the rate of growth in agricultural research and development (R&D) investment has been declining globally while a large number of developing countries have experienced negative growth rates over the past decade despite the different approaches adopted in ameliorating this trend. A cursory glance was taken at modes of agricultural research financing in several regions/ continents of the world – some are considered next.

2.1.1. Latin America and the Caribbean (LAC)

Venezian (1995) reveals that investments in agricultural research in Latin America and the Caribbean (LAC) have been shown to have high payoffs and they play a vital role in rural-income generation and natural resource conservation. Yet, overall funding for national agricultural research institutes in the LAC region has decreased in real terms since the early 1980s even though the number of scientists has grown. Chronic underfunding now threatens the operating efficiency, effectiveness, and even the survival of several research systems in the region.

LAC embraced policy reforms such as market liberalization and a reinvigorated private sector. Although agricultural research in the LAC region is still largely in the public sector, it ranged from funding university research, provision of national competitive funds, funding from research foundations, financing from farmer organizations, and funding from private companies.

Nin-Pratt and Stads (2023) developed an Innovation Capacity Index (ICI) which constitutes four sub-indices. Education and human capital which accesses the enrollment at all stages of education, make up the first sub-index. Research capacity makes the second sub-index. It evaluates a country's general performance in science, comprising metrics like the units of publications in different areas of discipline such as, engineering, genetics, computer science, molecular biology, and biochemistry. It also includes the H-index that accesses the effect and performance of scholarly productivity. Innovation environment, the third sub-index, accesses the degree of competition in domestic markets, financial services development, and avenue to credit, staff training and investment in R&D. The fourth sub-index is the quality of institutions which constitutes six wide dimensions of governance, comprising political stability alongside absence of violence/terrorism, voice and accountability, effectiveness of government, rule of law, control of corruption, and regulatory quality.

It can be observed that the work of Nin-Pratt and Stads (2023), provides general studies which covered a range of items among which includes innovative environment and investments in R&D, nevertheless, it can give a mirror reflection of what is obtainable in terms of research funding for agriculture. For instance, Figure 1 below displays the innovative capacity in Latin America and the Caribbean.

Figure 1 shows that research capacity and the innovative environment are not even among the countries and may give us a mirror reflection that the innovative activities in the agricultural sector might not have been performing robustly for the Latin America and the Caribbean countries.

In university agricultural research funding, Chile provides the best example in the LAC region of a university system that conducts a significant share of agricultural research in the country, this is over 20 per cent (Venezian 1995).

Figure 1. Innovation capacity in Latin America and the Caribbean
Source: Pratt et al. (2023)

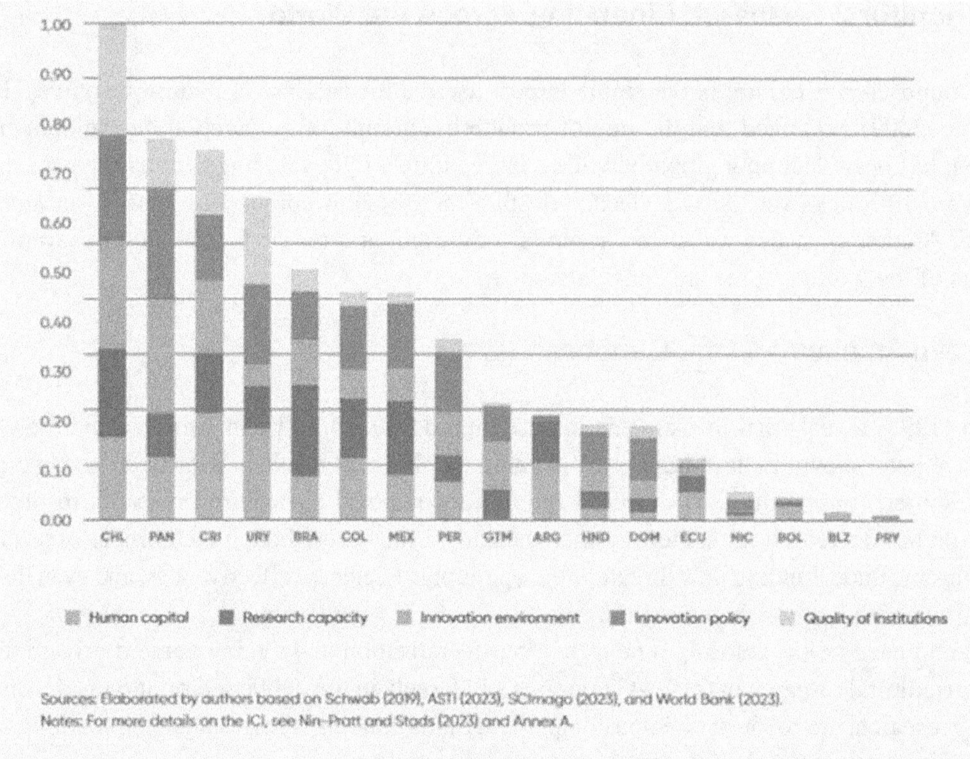

Echeverria, Trigo, and Byerlee (1998) states national competitive funding mechanisms as a means for agricultural research financing in several countries including Chile, Colombia, Argentina, Costa Rica, Mexico, Venezuela, and Brazil. Basically, competitive funding makes additional research resources available while lowering research-execution costs and encouraging a more demand-driven research system.

Research foundations are another mechanism for funding and conducting agricultural research in the commercial sector, especially export crops and agribusiness. Foundations such as FUSAGRI in Venezuela, Fundacion Polar in Venezuela and Fundacion Chile in Chile promote agricultural research financing.

On the other hand, financing from farmer organizations and private companies provide a more demand-driven research system, i.e., one in which farmers and private companies directly influence research priorities. In LAC, farmer organization research financing is strongest in Uruguay where farmers pay 0.4 per cent of the value of most agricultural output to support INIA (Instituto Nacional de Investigacion Agropecuaria).

Agricultural research financing by private companies makes a substantial source of fund as the case sales of hybrid maize seed by private seed companies in Mexico more than doubled with the opening of the seed market in the early 1990s (LÛpez-Pereira and GarcÌa 1994) and Brazil's AGROCERES which has a substantial investment in hybrid breeding, amounting to 7 percent of seed sales or a research intensity of about 1 percent of the value of related agricultural production.

2.1.2. Asia and Australia

Pardey, Roseboom, and Fan (1998) disclosed that Asian and Australian agricultural research financing are industry-based funding arrangements, geared towards commodities with highly concentrated production, marketing, or processing subsectors. Industry-based funding is also an important feature of agricultural research in Indonesia. During the colonial period, several large companies, which owned plantations throughout the country, operated their own research facilities, while smaller plantation owners funded joint research facilities (e.g., those of the Sumatra Planters Association). Both East and West Pakistan inherited and maintained similar funding arrangements after being partitioned from India in 1948. Even in Indonesia, industry-based funding is also an important feature of its agricultural research during the colonial period when several large companies owned plantations throughout the country and operated their own research facilities, while smaller plantation owners funded joint research facilities. However, Indonesia, Malaysia, and Sri Lanka are the only countries with substantial industry funding, estimated, respectively, at up to 18 per cent, 29 per cent, and 40 per cent of total funding for publicly provided agricultural research.

2.1.3. Africa

Pardey, and Roseboom (1998) also disclosed that real agricultural research expenditures grew rapidly (6.8 per cent per annum) in the 1960s, moderately (2.6 per cent) during the 1970s and ceased to grow throughout the 1980s and early 1990s. The rate of growth in research spending during the 1980s ranged from -9 per cent per annum for Nigeria to over 14 per cent for Ghana. At the lower end of the spectrum, Burkina Faso, Nigeria, Rwanda, and Sudan spend less than 0.2 per cent of their agriculture budget on agricultural research. In the Sub-Saharan region, particularly Nigeria, contributions by external donors to African agricultural research grew markedly during the 1980s.

Donor support for agricultural research increased despite a stagnation in overall donor funds directed to Africa and a contraction in donor support to developing countries (OECD 1992). Compared with other developing countries, African countries tend to invest a relatively high percentage of their agriculture budget in agricultural research. In 1991, the developing countries average was about 0.5 per cent, against Africa's 0.7 per cent. To a large extent, Africa's comparatively high research intensity ratio reflects the sizeable amount of donor funding channeled to agricultural research.

2.1.4. Nigeria

Despite these varied investments and financing approaches in agricultural research, there persists a continuous decline in the sector. Olafolaji (undated) reveals that agricultural research in Nigeria started more than 100 years ago with the establishment of a botanical garden in Lagos in the late 19th century. By the 1970s and 1980s, different research institutes and departments of agriculture had emerged. Presently, Nigeria has the largest and most elaborate National Agricultural research systems in Sub-Saharan Africa (SSA). In 2006, the government set up an umbrella body known as the Agricultural Research Council of Nigeria (ARCN) which was established to address the challenges faced by the agricultural research system. Moreover, numerous institutions, such as the formation of River Basins and Rural Development Authorities, the Agricultural Development Programmes (ADP) supported by joint funding from the World Bank and Nigeria's federating units, and over 20 Agricultural Research Institutes, were established (Umaru and Zubairu 2012).

2.2. How Do You Link PPP?

This paper considers PPP as a tool for agricultural research financing. "Public Private Partnership (PPP) is the coming together of groups under specified funding structures, and not necessarily with same objectives in all respects to face particular issues of concern and looking for innovative ways to address them…If it is between or among (a) public sector organization(s) and (b) private sector organization(s), it is a public private partnership" (NISER, 2005). It may also be interpreted as a cooperation between public and private sector entities in delivering public services (Commonwealth, 2003). Collaborations are increasingly blended, not just in their mixed public-private dynamics but also in pursuing collective economic, environmental, and social objectives (Quélin et al., 2019; Quélin et al., 2017). Achieving goal alignment requires a reassessment and a fresh understanding of the roles played by public and private entities in their collaboration, as well as the approaches taken by actors from each sector to attain desired outcomes (George et al., 2022; Lazzarini, 2022).

PPP can take various forms; according to Lawther (2002), PPP could be either collaborative (non-legal binding) or contractual (legally binding) agreements (Lawther, 2002). However, this study considers collaborative PPP more apt for sustainable growth and development and re-conceptualized collaborative PPP as a legitimate non-legal working relationship between the public and the private sectors in any economy to meet a common objective. It can also be conceived as a goodwill gesture for providing knowledge exchange and collective leverage for the improvement of the nation, without a primary motive of profit-making. However, not without a well-crafted plan clearly spelling out direct public and private sector involvement.

3. LITERATURE REVIEW

While there is scanty literature on (PPP) in agricultural research financing. A couple of attempts have been made to examine PPP in the agricultural sector. Commencing with the investigation conducted by Smyth et al. (2019), exploring the impact of public-private partnerships on enhancing worldwide food security. The findings suggest that researchers and their backers, including governments, industries, and interest groups, are urged to enhance collaboration and partnerships to expedite the recovery, enhance food production, and optimize food distribution.

Aseete et al., (2023) analysed the benefits of a public-private partnership among bean smallholder growers in Uganda. Employing a doubly robust difference-in-difference methodology within a multi-treatment context to gauge these effects. The outcomes reveal that the Public-Private Partnership (PPP) yielded favorable results for farmers, fostering heightened production through targeted interventions. The findings indicate a substantial correlation between the PPP and its interventions, showcasing notable enhancements in productivity, sales volumes, and the proportion of output brought to market.

Hermans et al. (2023) analyze the role of public-private partnerships as systemic instruments in agricultural innovation systems. The research examines the innovation trajectories of four cases with distinct strategic policy objectives, focusing on agricultural sustainability or international competitiveness in the Dutch agricultural sector. The findings highlight the strengths and limitations of various public-private partnerships as systemic instruments and their effectiveness in orchestrating other forms of innovation policy tools.

According to Obasa and Madueke (2013), productivity of investment, in its generic sense, will be more appropriately financed with foreign direct private loan, share capital, foreign direct investment (FDI) and development stocks. And also capital-output ratio will be more appropriately financed with multilateral loan, domestic savings, Treasury bill, official development assistant, foreign direct investment and development stock.

How relevant could Obasa and Madueke (2013) be seeing that they did not evaluate the present impact of agricultural financing on economic growth? Alene, et.al (undated) state that impact assessment is an integral part of the agricultural research process; where each stage of the impact assessment process are developed to guide researchers in gathering relevant and adequate information relating to each agricultural technology says. Seck, et.al (2013) further discussed the constraints to the visible impact of agricultural research on the growth of Africa's economy, and how to repackage agricultural research for more visible impact on Africa's economic growth.

Further, Adetiloye (2012) evaluated the provision of agricultural finance, its impact and the food security status in Nigeria. This highlights Nigeria's vulnerability to food insecurity, evident in the escalating trend of food imports. To address this concerning situation, Obasa and Madueke (2013) propose the adoption of robust macroeconomic policies that favor investment, coupled with exploring debt-equity swap options to foster agricultural-led economic growth. In a similar breath, Seck et.al (2013) found that in order to package agricultural research for visible impact on the economy, agricultural research systems would need to be adequately funded and handled by skilled human resources under good governance.

Agbamu (2000), therefore, endeavoured to understand the context of Agricultural Research across Indonesia, Japan, Republic of Korea, Mexico, Nigeria, Tanzania and Thailand to take a cue from the different development models adopted by these countries. In this vein, Alene et.al's (undated) disclosed a model called sustainable rural livelihood framework where they illustrated the linkages between agricultural research and rural livelihoods and the implications for evaluating the impact of agricultural technologies. However, this is not a one-size-fit-all scenario, Agbamu (2000) elucidates that the diversity of agricultural research-extension linkage systems worldwide is a consequence of varying socio-cultural contexts.

However, Anthony's (2010) conception that countries at the early stages of development depend almost fully on agricultural growth for employment, foreign exchange, government revenue and food supply to the teeming population could serve as a good basis for a more generally acceptable development model regardless of contrasting sociocultural situations. In this sense, agricultural growth is the key impetus to the growth of non-oil export in most countries of Sub-Saharan regions.

Therefore, Umaru and Zubairu (2012) along with Seck et al. (2013) propose that policymakers should formulate impactful innovative agricultural policies to enhance productivity in Nigeria's agricultural sector. They further suggest an increase in expenditure on the sector as a means to stimulate economic growth.

Eno & Eze (2023) examined the nexus between Agricultural Financing and Agricultural Output in Nigeria over the period 2011 – 2021. The impact was evaluated through the relationship between Foreign Exchange Rate, Government Expenditure on Agriculture, Agricultural Gross Domestic Product and Banks Credit to Agriculture alongside Banks' Lending Rate represented by Interest Rate. Several econometric methodologies were applied, including Ordinary Least Squares, error correction modeling, the Dickey-Fuller unit root test, Johansen co-integration, and the Granger Causality test. The authors observed that a significant direct relationship characterized Banks' lending and Agricultural output in the country. High interest rates reduce the demand for agricultural credit by farmers, resulting in a decline of agricultural output in Nigeria.

Ewubare and Eyitope (2015) scrutinized the effects of government expenditure on the performance of Nigeria's agricultural sector. The data for the analysis were sourced from the Central Bank of Nigeria Statistical Bulletin and the National Bureau of Statistics. Specifically, effort was made to access the impacts of deposit money banks loan, gross capital formation, and government expenditure (GEA) on agricultural output in the country. By employing Ordinary Least Squares multiple regression analysis, the researchers noted a direct and substantial impact of GEA on the agricultural sector's output in Nigeria.

Ajayi et al. (2017) investigated the impact of agricultural financing policies and deposit money bank loans to the agricultural sector on productivity, spanning the years 1981-2015. The technique of a time series linear regression was used for the empirical analysis. It was observed that deposit money bank loans as well as agricultural financing policy directly influence the productivity of the agricultural sector significantly. However, lending rate inversely and significantly affect agricultural output in the country.

Nin-Pratt and Stads (2023), in an effort to access the growth and complexity of food systems in different economies, constructed a food system development index (FSDI) that quantifies the development of the system at the farm, post-farm, and the levels of consumption. It was assumed that at the level of farm, the value of capital, capital intensity, purchased resources per worker (insecticides, seeds, herbicides, fertilizers, sprayers, tractors and combines, and irrigation equipment) are directly associated to the use of enhanced technologies and integration of farmers with output and resource markets, improved avenue to financial markets, and to production services. The second sub-index measures the advancement of the post-farm segment in terms of the food system and quantifies the extent and reach of the value chain, development of industrial cluster, sophistication of product, the quality of domestic supply, the depth of marketing as well as the existence of a formal grocery sector.

Chiogor et al. (2023) examined the macroeconomic effect of the reforms of agricultural financing on agricultural output, and economic growth in Nigeria over the span of 2010–2022, utilizing a computable general equilibrium analysis. The data set was time series obtained from the Central Bank of Nigeria, the National Bureau of Statistics' together with the World Bank Living Standard Measurement Study as well as information from the Nigerian Living Standard Survey for 2019, which was used to generate proportions of household income and spending. The authors observed that reforms of agricultural financing by decreasing interest rates on agricultural loans by 10% and 8% exert a direct and significant effect on economic growth. It was also noted that the reforms in agricultural financing based on the reduction of interest rates on agricultural loans by 10% and 8% generate a significant and direct effect on agricultural output.

4. JUSTIFICATION OF THE STUDY

The Agricultural Transformation Agenda (ATA, 2011) reveals the reason we should care about this issue as quoted below:

Agriculture is an important sector of the economy with a high potential for employment generation, food security and poverty reduction...hence there is a need for "Value chain coordinator", Infrastructure investments, Private sector leadership, Supportive fiscal policies and accessible Market Information systems to the farmers.

Indeed, the significant agricultural transformations witnessed in China, Vietnam, Brazil, and Thailand, resulting in substantial growth in their agricultural sectors over the past three decades with annual growth rates of 2.6%, 2.0%, 1.8%, and 1.4% respectively, emphasize the relevance of this study to the Nigerian economy. Certainly, Thailand experienced a noteworthy annual growth rate of 1.4% in its agriculture, attributed to substantial investments. This agricultural growth, in turn, contributed to a remarkable annual growth rate of 4.5% in the manufacturing sector. Consequently, the unemployment rate saw a substantial decline, dropping from 4.5% in 2000 to as low as 1.2% in 2011, according to ATA (2011).

Adeyemi et al. (2023) did a critical review of the various Agricultural development innovative programmes in Nigeria as well as accessed the causes of their failure. They observed that the last decade was characterized by poverty and malnutrition in most of the economies in the sub-Saharan Africa in spite of the significant measures that the respective government had made over time. Following the estimates from experts in 2010, the continuous increase in the cost of food in the sub-Saharan Africa had generated a rise in the number of impoverished individuals mostly residing in the rural areas to around 44 million. It was also observed that, the use of innovation is the only route provided the challenges of poverty and malnutrition is to be conquered. This argument was based on the fact that, a rise in Agricultural productivity will result in a rise in the availability of food especially in the rural areas characterized by more than 80% of Africa's population and being a place where the foods are greatly produced by the small-holder rural farmers. According to the authors, the forecast that the world population will attain 9 billion by 2050 is enough to trigger fears in humanity mostly the under developed economies in the sub-Saharan Africa such as Nigeria because of the abortive efforts of their government in nipping in the bud the problems of malnutrition and poverty in the rural areas by the various developmental innovative programmes that have been instituted.

Quite a number of formal empirical work have exploited the dynamics across districts or industries in Vietnam in their status as a third-country, in a move to take advantage from the china's circumstance of the imposition of tariffs by the US. It has resulted in favourable feedback in hours worked, employment, and wages, especially for women (Mayr-Dorn et al. 2023, Rotunno et al. 2023), and the development from informal agriculture to formal manufacturing work (Nguyen and Lim 2023) as cited in Alfaro & Chor (2023).

These serve as good lessons for Nigeria at this critical time when our mono-cultural economy is being harmed by abrupt neglect of dominant patrons of our crude-oil export. The effect of this crude oil crisis on Nigeria, a mono-product dependent economy which is vulnerable to external demand shocks, requires a solution should be proffered.

An economy that is subservient to the whims of the invisible hands of the international market will survive only at the mercy of major actors in the world stage. This is the case of the anathema "When U.S sneezes, the world catches a cold." Nigeria is a recent case study of this, due to the U.S breakthrough in the discovery of shale oil, which would not have adversely affected our economy if we had also given attention to seeking agricultural breakthrough through goal-oriented research financing.

It is inexcusable to know that in 2011 when Nigeria was not yet under this burden of crude-oil crisis, the agriculture sector was not better off as ATA revealed that average fertilizer use in Nigeria was 13Kg/hectare compared to world average of 100Kg/hectare and 150Kg/hectare for Asia. In fact, only 5% of the farmers could access the improved seeds and operate with only 10 tractors per 100 hectares compared to 241 tractors per 100 hectares in Indonesia. The situation is better imagined with the prevailing crude oil crisis hitting us hard and exchange rate is depreciating rapidly. Although this could be premised on corruption in the management of our resources; the question is: For how long? Is there any remedy?

The first question sounds like a rhetorical one with a "Save Our Soul" undertone while the latter is the essence of this research exercise. As such, the purpose of this study is to reveal collaborative PPP in agricultural research financing as a means to sustainable agricultural growth and economic development.

5. THEORETICAL FRAMEWORK AND DISCUSSION: A CALL FOR A NEW MODEL

Several theories such as "Meta-governance theory; "Actor-Centered Institutionalism"; "Actor-Centered Institutionalism and Multi-Level Governance Theory"; "Multiple-Streams Framework [and Barzelay's event centered approach]"; "Triple Bottom Line"; "Theory of Comparative Advantage"; "Theory of Competitive Advantage"; "Theory of Unbalanced Growth" etc. can be deemed relevant to any worthwhile study on PPP and economic growth. However, theories germane to this study in an ascending order of preference are the "Theory of Comparative Advantage", "Theory of Competitive Advantage", "Unbalanced Growth Theory" and "Triple Bottom Line Theory".

This study took its support from the "Theory of Comparative Advantage" to establish that Nigeria has comparative advantage in the agriculture. Needless to say, that Nigeria has comparative advantage in petroleum; but how competitive is the country in agriculture? The comparative advantage Nigeria has in petroleum can be substantiated by the study of Elsalih et al. (2021) which examined the comparative advantage (CA) regarding 28 oil-producing economies in crude oil spanning 1990-2016 by the adoption of a Normalized Revealed Comparative Advantage index. It was observed that, among the 28 economies investigated, Nigeria was among the categories of countries having a comparative advantage in the production of crude oil.

Despite Nigeria's significant abundance in fossil fuels, this resource endowment has not translated into a commendable economic performance. Instead, the nation's mono-cultural economy has taken on a precarious dimension over the past decade, making it vulnerable to fluctuations in the international oil markets. As such, this study advocates that our comparative advantage in agriculture, if made a competitive advantage will remedy the wreckage Nigerian economy has suffered so far and position Nigeria as a major world economic player. This advocacy stems from the unbalanced growth theory propounded by Hirschman (1958), underscoring the forward and backward linkages agriculture plays in the development of an economy. The forward and backward linkage agriculture plays is evident from the study of Affandi (2023) which examined the multipliers and linkages of the agricultural sector relative to the regional economy in Tanggamus Regency by the use of input-output analysis. The approach consist of linkage analysis, output and income multiplier impact, and dispersion effect. The authors observed that there is a future linkage between the agricultural and manufacturing sector; and the warehousing and transport sector, has backward linkages in terms of the construction and the real estate sector.

Another theoretical insight is the triple helix theory as conveyed in the study of Sharubutu (2023). It is a model of innovation that describes the interplay between industry, government and the academia towards fostering social and economic development. It has garnered substantial attention in the developing and developed countries as a fundamental policy generating tool to improve innovation and stimulate economic development (Etzkowitz and Dzisah, 2008). The theory offers a broad paradigm, method, and analytical framework for analyzing the nexus between innovation actors at the system level (Strand et al., 2017).

Table 1. Factors accounting for development domains' comparative advantage in the domestic, regional, or world market

Zone	Unprocessed	Processed
North-Central	Low cost of production, availability of resources – land and cheap labour, highest producer in the zone, large production, favorable climate.	High quality of the products in the world market, market availability, high capacity utilization, high quality raw materials.
North-East	Availability of raw materials (resources endowments), high demand for the products, availability of infrastructure, availability of skilled manpower, high rate of return on investment, suitable soil and climate, availability of labour and large market.	Availability of labor, high demand for products, availability of raw materials.
North-West	Good quality and fertile soil, irrigation facility, suitable climate, cheap labour and its availability, natural resource endowment, high economic value.	Raw material availability, large domestic market, high productivity of resources, availability of skilled labor, natural resource endowment.
South-East	Good soil/high fertility, enabling climatic condition, availability of inputs, cheap labour, low cost of production, relative abundance of the crop, ecology of the area, high productivity, experienced labour availability, natural endowment.	Cheap labour, good climate, availability of raw materials, low production cost, skilled manpower availability, technological advancement, availability of large number of milled rice and high consumption level.
South-South	Natural resource endowment, large forest resource, high demand of expatriates and good quality land/soil.	Good raw material base, high output of raw materials, resources endowment, skilled manpower and good quality production especially their genetic makeup, e.g., flower and odor.
South-West	Favorable agro-climatic environment, soil type, favourable vegetation, and high quality cocoa.	Availability of raw materials, increased productivity, high returns, and materials not fully utilized. High demand.

Source: Manyong, V.M., et.al (2005) Field survey, February/March 2003.

On this basis, "Triple Bottom Line Theory" was logically adapted in relating agricultural research financing to the public good nature of research. Also, a non-random sample of the United Nations Global Compact (UNGC) members' account balances for a given year where collected from annual reports to derive a reasonable bottom-line commitment that private sectors should have towards agricultural research financing if to be regarded as socially responsible in Nigeria.

5.1. Theory of Comparative Advantage

Though David Ricardo's theory of comparative advantage is associated with a one-factor model within the context of his labour theory of value (Bernhofen and Brown, 2004); the optimization of international and national resource allocation requires a recognition of the importance of factor endowments and a recognition of the role of human capital and appropriate macroeconomic policies in economic growth (Fatwell, 1987).

Primarily, this paper is concerned with resource endowment. The presence of resources within a country offers an additional basis for comparative advantage, especially for nations that may not inherently possess superior technology. Comparative advantage, under specific restrictive assumptions, can stem from variations in relative factor endowments, encompassing human skills, economies of scale, and more (refer to Table 1).

IITA's 2003 field survey in Nigeria on factors accounting for development domains' comparative advantage in the domestic, region or world market identified 34 commodities perceived to have a com-

Table 2. Commodities with comparative advantage for investments as ranked by stakeholders in each development domain

Primary Production	North-Central	North-East	North-West	South-East	South-South	South-West
Staple crop production	Rice Maize Sorghum Millet Cowpea Cassava Yam Beniseed	Sorghum Maize Millet Cowpea Cassava Rice Beniseed	Millet Sorghum Maize Cowpea Cassava Rice Beniseed Maize Yam	Yam Cassava Rice Maize Vegetables Cowpea Soybean Plantain	Yam Cassava Rice Maize Cocoyam Vegetables Cowpea Groundnut Soybean Plantain	Yam Cassava Maize Vegetables Rice Cowpea Groundnut Soybean
Industrial crop production	Soybean Groundnut Cotton Vegetables Coffee Oil palm	Groundnut Soybean Cotton Sorghum Vegetables	Soybean Vegetables Groundnut	Cocoa Oil palm Rubber Groundnut	Cocoa Oil palm Rubber Cashew Orange	Pineapple Oil palm Rubber Cashew Ginger Cocoa
Livestock production	Cattle Sheep Goat Poultry Piggery	Cattle Sheep Goat Poultry Piggery	Cattle Sheep Goat Poultry Piggery	Poultry Sheep Goat Cattle Piggery	Cattle Small ruminant Rabbitry Poultry Piggery	Cattle Sheep Goat Rabbitry Piggery Poultry
Fishery	Fish	Fish	Fish	Fresh fish Smoked fish	Aquaculture	Fish Cray fish Shrimps
Forestry	Gum Arabic Ginger Cashewnut	Gum Arabic Fuel wood Gmelina	Teak Mahogany Gmelina Ginger	Timber Teak	Timber	Ginger Cashewnut

Source: Manyong, V.M., et.al (2005)

parative advantage. The identified commodities were grouped into five categories namely; staple crops (9 commodities), industrial crops (12 commodities), livestock (5 commodities), fishery (3), and forestry (5); with reasons for their comparative advantage (refer to Table 2).

5.2. Theory of Competitive Advantage

These commodities of comparative advantage can therefore be transformed into competitive advantage. However, the low productivity, weak domestic policies institution, inadequate funding and more particularly non-competitiveness have impeded substantial growth in the sector (Ayodele, et.al, 2013). To remedy this malaise, we take a cue from Porter's (1990) concept of competitive advantage. A nation is deemed to have competitive advantage in the industry "if it possessed competitive advantage relative to the best worldwide competitors" in terms of indicators such as "the presence of substantial exports to a wide array of other nations and/or significant outbound foreign investment based on skills assets created in the home country". Hoffman (2000) went further to reinvent Porter's competitive advantage by coming up with "sustainable competitive advantage" (SCA). Drawing from Barney's (1991) perspective, Sustainable Competitive Advantage (SCA) is an enduring advantage derived from adopting a distinctive value-creating strategy that is not concurrently employed by any existing or potential competitors. This advantage is coupled with the incapacity to replicate the benefits of this strategy.

Distinctive features of competitive advantage encompass uniqueness, value, resistance to imitation, and irreplaceability (Barney, 1991). Organizations demonstrate competitive advantage by cultivating superior core competencies through the amalgamation of their skills and resources (Prahalad and Hamel, 1990). Moreover, the possession, allocation, and continual enhancement of distinctive resources are integral to a set of dynamic capabilities that contribute to sustaining a competitive edge (Luo, 2000).

In general, Gupta (undated), in his narrative – linking comparative advantage to competitive advantage – considered a static and a dynamic world scenario. In a static world, a nation possesses a competitive advantage when it focuses on products aligning with its comparative advantage. However, in a dynamic context, there is a need to elevate a nation's comparative advantage by leveraging factors such as innovation, cutting-edge technology, and created capabilities, emphasizing their increased significance. Also, innovation and technology constitute crucial functions in the maximization of the natural resource market and the promotion of green development. Through the adoption of clean technology, resource-efficient processes, and renewable energy sources, transformation can be driven by industries as well as the attainment of sustainable economic growth (Dilanchiev and Taktakishvili, 2022).

Therefore, the competitiveness of a country relies on the ability of its industries to innovate and enhance their capabilities. In Nigeria's case, for the agricultural industry to innovate and develop, there is need for agricultural research and technology invention, backed by financing.

Agricultural research financing through PPP approach can be considered as a means to further enhancing the attainment of development in agricultural research and as a tool for promoting investment and adoption of technology in agricultural development. The whole essence of agricultural research financing through PPP arrangement is to boost agricultural productivity through research and development and reposition agriculture to contribute more to GDP. From the foregoing, the unbalanced growth strategy of forward and backward linkages of agricultural sector with manufacturing sector and service sectors through value-chains development. Achieving this will minimize the stronghold of petroleum on our economy and will be a bold step towards diversification.

5.3. Unbalanced Growth Theory

Hirschman (1958) argues that due to the lack of capital in developing countries investment projects had to be selected strategically according to the strength of a sector's backward and forward linkage ('unbalanced growth') with other sectors. Thus, this other sector would automatically develop themselves through what is known as "linkages effect".

Krugman (1984) corroborated Hirschman's (1958) position by providing a theoretical explanation of the role of agriculture in national development and its effects on export. The theory argues that the expansion of agricultural export could lead to a significant increase in the demand for the nation's outputs, which in turn might lead to an increase of real output. Innovative measures for boosting agricultural productivity and, by implication, industrial activities are crucial given the challenges the global economy is experiencing. For instance, in the words of Krugman (2023), there are a lot of crises. This includes the energy crisis, inflation, the food crisis, and the Ukraine war. It was held that these crises are transient. It was believed that Europe would put inflation under control, but Krugman was more worried about the global food crisis, which may be linked to famines.

Table 3. Agricultural commodities in which development domains have comparative advantage in the domestic, regional, or world market by development domain

Zone	Unprocessed	Processed
North-central	Soybean, yam, cassava, benniseed, groundnut, neem, fruit, honey, mango, cashew, palm kernel, maize, citrus.	Soya oil and meal, canned fruit, orange juice, vegetable oil.
North-east	Vegetable production (tomato, pepper, onion etc), oil seeds production (groundnut), gum Arabic production, and cotton.	Vegetable processing (tomato, pepper, onion etc), cotton lint, gum Arabic products.
North-west	Ginger, tomato, cotton, sorghum, groundnut, garlic, gum Arabic, soybean, sesame, cowpea, and wheat.	Textiles, beer, groundnut oil, hides & skin, tomato paste, resin, leather.
South-east	Oil palm, cassava, yam, rice, poultry, cocoyam, plantain, banana, vegetables, ginger, timber, cashew nuts, cocoa, maize, melon, rubber, and copra.	Palm oil, cassava chips/gari, yam flour, fruit juice, canned fish, cocoyam chips, plantain chips, vegetable oil, cassava flour, rubber products, cashew products, and kola nut.
South-south	Cocoa, palm fruit, rubber, timber, non-timber forest products, cassava, fish, crayfish and shrimps.	Cassava chips, palm oil, latex, cassava toasted granules (gari), cocoa powder and chocolate and palm kernel oil and cake.
South-west	Cassava, palm produce, cocoa, timber, oil palm, fish, and shrimps.	Fish and shrimps, yam, timber, cassava, and cocoa cake.

Source: Manyong, et.al (2005)

Considering the capital constraints of developing countries, Hirschman (1958) classified investment in developing countries into Social Overhead Capital (SOC) and Directly Productive Activities (DPA). He considered SOC as those basic devices without which primary, secondary and tertiary activities cannot function. This in a generic sense, involve expenditure on basic infrastructure; in the case of this paper we consider it as expenditure on agricultural research. On the other hand, DPA are those activities which are a consequence of some investment, for example, investment in outputs of agricultural research – agricultural technology.

A proof that agriculture in Nigeria should be a lead sector through which other sector can **fester** is revealed empirically by Onakoya (2013). The study incorporated the linkages among agriculture, manufacturing, oil and gas and the service sectors. The spill-over effects and externalities generated by the interactions and linkages between the different sectors attest to the dynamic nature of the economy. Also, the economic role of the agricultural sector is a one-way path as the flow of capital is mainly towards the industrial, oil and gas and the tertiary services sectors. The research findings showed that the output of the agricultural sector is positively related to each of the output of manufacturing and services sectors. This study established that the sectorial linkages are not always beneficial especially between agriculture and the oil sector and recommends the modernization of the industrial and services sectors in order to generate increase in local content value addition to agriculture (refer to Table 3).

5.4. Triple Bottom Line Theory

In generating local content value addition to agriculture, we seek to initiate Elkington's Triple Bottom-Line approach to agricultural research financing as a means of integrating and regulating public-private partnership in agricultural research financing. Worthy of note, there is neither universal standard method for calculating the TBL nor accepted standard for the measures that comprise each of the three TBL categories (i.e., 3Ps – profit, people, and planet). The 3Ps do not have a common unit of measure. Profits

Table 4. UNGC participants' financial commitment benchmark

Annual Sales/Revenue	Annual Contribution (in $)	Annual Contribution (in N) at Current Exchange Rate as at 29/01/2024 [i.e. $1=N911]
Less than USD 50 million	250 (suggested minimum)	227,750
USD 50 million - USD 250 million	2,500 - 5,000	2,277,500-4,555,000
USD 250 million - USD 1 billion	5,000 - 10,000	4,555,000-9,110,000
USD 1 billion - USD 5 billion	10,000 - 15,000	9,110,000-13,665,000
Greater than USD 5 billion	15,000+	13,665,000+

Source: UNGC website (https://www.unglobalcompact.org/howtoparticipate/business_participation/index.html)

are measured in naira. What is social capital (i.e., people) measured in? What about environmental or ecological health (i.e., planet)? Finding a common unit of measurement is also a challenge with respect to the TBL.

In this study, the TBL was intended to be the yardstick to judge if a firm was socially responsible in Nigeria. Firms regarded as good corporate citizens would be judged on account of their financial capability and commitment to social and environmental responsibilities. In this case, social and environmental responsibilities involve concerted efforts towards: (i) The achievement of self-sufficiency in basic food supply and the attainment of food security; (ii) Enhanced cultivation of raw materials for industrial use; (iii) Enhanced cultivation and processing of export-oriented crops, employing advanced agricultural technologies; (iv) Generating gainful employment; (v) Rational utilization of agricultural resources, improved protection of agricultural land resources from drought, desert encroachment, soil erosion and flood, and the general preservation of the environment for the sustainability of agricultural production; (vi) Promotion of the increased application of modern technology to agricultural production; and, (vii) Improvement in the quality of life of rural dwellers.

To achieve these, a non-random sample of the United Nations Global Compact (UNGC) members' account balances of firms in Nigeria were collected from annual reports to speculate the agricultural research finance revenue realizable from the private sector. As such, we adopt the UNGC's financial commitment benchmark for every functional firm registered with the Corporate Affairs Commission. In 2014, the UNGC annual contribution benchmark was set, by company size, as shown in Table 4.

It was discovered that just 27 Nigerian firms belong to the UNGC, as revealed in the Appendix, which is not a good index of CSR in Nigeria. Most of these establishments were not Public Liability Companies (PLCs) and are disposed to releasing their financial statements. At present there is over 17,000,000 SMEs (see ThisDay Live) and about 165 PLCs (see CRMZ) in Nigeria. Also, taking in mind that the CBN regards SMEs as businesses with turnover of less than N100,000,000 and /or less than 300 employees. Thereby resulting in an estimated realizable revenue for agricultural research of N3,871,750,000,000 (i.e., 17,000,000 SMEs **X** N227,750) from registered SMEs alone. To increase the sense of social responsibility of CSR, it is conceived advisable to bring some more stringent means of committing public and private establishment to agricultural research financing.

6. FUTURE PERSPECTIVES FOR AGRICULTURAL DEVELOPMENT IN NIGERIA

6.1. Nigeria's Future Agricultural Landscape

Nigeria's future agricultural landscape holds significant promise and potential, with several vital highlights and opportunities on the horizon. In the first instance, embracing technologies such as mobile apps for weather forecasting, remote sensing for crop monitoring, and e-extension services can revolutionize Nigeria's agricultural sector. An expansive array of technologies exists as decision-making enhancement tools for practitioners in this regard (Paustian & Theuvsen, 2017). According to the Food and Agriculture Organization (FAO), these innovations in agriculture can potentially increase agricultural productivity in Nigeria by up to 40% by optimizing resource use, improving crop monitoring, and enabling better decision-making (Trendov et al., 2019). In another respect, engaging the youth in agriculture through vocational training, entrepreneurship programs, and access to finance can inject new energy and ideas into the sector. The Nigerian population is predominantly youthful, with over 60% of the population below the age of 25, according to the United Nations (UN, 2019), which can stimulate innovation and drive agricultural growth.

In addition, the Intergovernmental Panel on Climate Change (IPCC, 2014) projects an increase in extreme weather events and shifting rainfall patterns in Nigeria, impacting crop yields and livelihoods. Implementing climate-smart agricultural practices, such as drought-resistant crop varieties, water-saving irrigation techniques, and agroforestry, can build resilience to climate change and ensure food security for future generations.

Furthermore, the World Bank reports that strengthening agricultural value chains through infrastructure development, market linkages, and agro-processing facilities can lift millions of Nigerians out of poverty by creating jobs and increasing incomes (Mghenyi et al., 2021). By adding value to agricultural products through processing and marketing, Nigeria can unlock the economic potential of its agricultural sector, benefiting both farmers and consumers.

Moreover, promoting sustainable land management practices, such as conservation agriculture, agro-ecology, and soil conservation, can preserve soil fertility, prevent land degradation, and enhance ecosystem resilience (Reicosky, 2020). Investing in land tenure systems, land-use planning, and land rehabilitation initiatives can ensure the long-term sustainability of Nigeria's agricultural landscapes.

Finally, enacting supportive policies and regulatory frameworks that prioritize agricultural development, promote public-private partnerships, and safeguard the interests of smallholder farmers can create an enabling environment for sustainable agriculture. Addressing issues such as land tenure, access to finance, market regulations, and input subsidies can unlock the full potential of Nigeria's agricultural sector.

6.2. Sustainable Agriculture System in Nigeria

Sustainability in agriculture implies animal husbandry and farming in sustainable ways such that it achieves textile demands and current food for society without frustrating the ability of current and future generations to achieve their own needs (Richard et al., 2022). Some practices for sustainability in agriculture include diversification, which is the act of encouraging farmers to cultivate diverse crops, including indigenous and climate-resilient varieties, which can reduce dependence on a few commodities and enhance resilience to market fluctuations and climate variability. According to the Nigerian Institute for Social and Economic Research (NISER), diversification of crops can reduce the country's vulnerability to price shocks and climate variability (Kassaw, 2020). Another practice is agroforestry. Promoting agroforestry practices to improve soil health, conserve water, sequester carbon, and enhance biodiversity in Nigeria's agricultural landscapes. The World Agroforestry Centre (ICRAF) estimates that agroforestry systems can increase agricultural productivity by up to 128% while providing multiple environmental benefits, including shade, fodder, timber, and non-timber forest products (ICRAF, 2017). Another approach is Integrated Pest Management (IPM). Research from the International Centre of Insect Physiology and Ecology (ICIPE) shows that IPM practices can reduce the use of chemical pesticides by up to 80% while maintaining crop yields (Nyangau et al., 2022). Encouraging farmers to adopt IPM strategies, such as biological control methods, crop rotation, companion planting, and habitat manipulation, to promote sustainable pest management. Moreover, sustainability in agriculture will be a mirage without ensuring security in Nigeria. The United Nations Office for the Coordination of Humanitarian Affairs (OCHA) reports that herder-farmer conflicts have displaced over 300,000 people in Nigeria, significantly impacting agricultural activities and food security (Onah & Nwodo, 2023). Strengthen community-based conflict resolution mechanisms and promote dialogue between herders and farmers to address underlying grievances and prevent violent conflicts.

Policymakers should implement peace-building initiatives that foster reconciliation, mutual understanding, and cooperation among diverse ethnic and religious groups involved in agriculture. Adequate security personnel should be deployed to protect farmlands, markets, and critical agricultural infrastructure from criminal activities and insurgent attacks.

7. CONCLUSION AND RECOMMENDATION

PPP as contextualized in this paper implies a mutual partnership between the public and the private sectors toward agricultural development through research financing. To this end, the paper's objective of underscoring the importance of collaborative PPP in agricultural research financing was to some extent highlighted through a critical exposition of the "Theory of Comparative Advantage", "Theory of Competitive Advantage", "Unbalanced Growth Theory", and "Triple Bottom-Line Theory".

From this, it was discovered that collaborative PPP will raise sufficient funds for agricultural research and development. We considered profit as the life wire of a successful collaborative PPP in agricultural financing; while we considered people and planet as secondary for a successful collaborative PPP.

The policy implication of collaborative PPP in agricultural research financing is as binding as the universal law of sowing and reaping. A successful collaborative PPP will manifest in the quality of lives of the people (increase in personal income per capita; reduction in unemployment rate; reduction in crime statistics; decline in infant mortality; reduced cost of living, reduction in relative poverty; improved health-adjusted life expectancy; increased median household income; etc.) and improved sustainability of our planet (sustainable land use practices, proper inventory of land use and preservation of natural habitats, forest canopy, proper waste management, reduction of greenhouse gas emission, use of post-consumer and industrial recycled material, water consumption, amount of waste to landfill, etc.). These ripple effects will positively impact the private and the public sectors (as improved ease of doing business index, improved profitability, improved availability and affordability of raw materials, growth in establishment sizes, job growth, etc.).

The paper therefore recommends the private and the public sectors should integrate agriculture promotion and advancement into their business strategy, day-to-day operations and organizational culture. In addition, they should communicate publicly the ways in which they implement agricultural promotion and advancement strategies to support broader development objectives; and advance the case for popular support for agriculture through outreach to peers, partners, clients, consumers and the public at large.

Moreover, in order to accelerate private sector participation in a collaborative adventure on the finance of research and development in agriculture, the government should organize its taxation policy so that it encourages private sector firms to release funds for innovative activities. Moreover, the government should incorporate innovative models from advanced countries that have been proven to accelerate industrialization through the agricultural revolution while taking cognizance of the peculiarities of the domestic economy.

In another vein, while the government plans to partner with the private sector, effort should be geared toward instilling in the system macroeconomic fundamentals that will drive industrial activities so that the relationship between the industrial and agricultural sectors through innovative measures becomes sustainable.

8. LIMITATIONS AND SUGGESTIONS FOR FURTHER RESEARCH

The quantification of the impact of investment in research and development by the collaboration of the private and public sectors on agricultural and industrial development would be of immense benefit. However, this study navigated the hurdles paucity of data on research and development in the agriculture sector and absence of data on agricultural research financing in Nigeria. Moreover, it was also observed that there is no direct literature in the field of agricultural financing that is tailored to research and development. With the expectation that the Nigerian government and the private sector would find this study pragmatic for the purposes of collaborating to bolster innovative activities in agriculture and, by implication, the industrial sector through forward and backward linkage effects, future studies can thus begin to count on the specific impact of PPPs in agricultural research financing on the Nigeria agricultural sector over time.

REFERENCES

Adetiloye, K. A. (2012). Agricultural Financing in Nigeria: An Assessment of the Agricultural Credit Guarantee Scheme Fund (ACGSF) for Food Security in Nigeria (1978-2006). *Journal of Economic Literature*, *3*(1), 39–48.

Adeyemi, S. O. (2023). A Critical Review of Rural Agricultural Development Innovative Programmes in Nigeria. *Plant Biol Soil Health J*, *1*(1), 11–17.

Affandi, M. I. (2023). Linkages and Multipliers of the Agricultural Sector in the Regional Economy in Tanggamus Regency: Input-Output Analysis. *Eduvest-Journal of Universal Studies*, *3*(3), 594–603. doi:10.59188/eduvest.v3i3.775

Agbamu, J.U. (2000). Agricultural Research-Extension Linkage Systems: An International Perspective. *Agricultural Research and Extension Network*, Paper 106a.

Ajayi, A., Nageri, I., & Akolo, S. (2017). Impact of Agricultural Financing Policy and Deposit Money Bank Loan on Agricultural Sector Productivity in Nigeria. *Amity Journal of Agribusiness*, *2*(1), 1–11.

Ajayi, F. (2023). Strengthening Agriculture for Food and Nutrition Security, Market Development and Export in a Climate Change Environment. *Agricultural Society of Nigeria (ASN) 57th Annual Conference Proceedings.*

Aldrich, D. G. (1966). Agricultural Research, a Key to Understanding our Environment. Symposium on Research in Agriculture. *USDA, Agricultural Research Service*, 50-53.

Alene, A.D. (n.d.). A Framework for Conceptualizing Impact Assessment and Promoting Impact Culture in Agricultural Research. *International Institute of Tropical Agriculture (IITA).*

Alfaro, L., & Chor, D. (2023). Global Supply Chains: The Looming "Great Reallocation" (No. w31661). *National Bureau of Economic Research.*

Anthony, E. (2010). Agricultural Credit and Economic Growth in Nigeria: An Empirical Analysis. *Business and Economics Journal*, 14.

ATA. (2011). Federal Ministry of Agriculture and Rural Development. *Agricultural Transformation Agenda.* http://www.unaab.edu.ng/attachments/Agricultural%20Transformation%20Blue%20Print.pdf

Ayodele, O. S. (2013). Challenges facing the achievement of the Nigeria Vision 20:2020. *Global Advanced Research Journal of Social Science*, *2*(7), 143–157.

Ayunku, P. E., & Etale, L. M. (2015). Effect of agriculture spending on economic growth in Nigeria: Empirical evidence. *Research Journal of Finance and Accounting*, *6*(2), 138–143.

Barney, J. (1991). Firm Resources and Sustained Competitive Advantage. *Journal of Management*, *17*(1), 99–120. doi:10.1177/014920639101700108

Beintema, N., & Elliott, H. (2009). *Setting Meaningful Investment Targets in Agricultural Research and Development: Challenges, Opportunities and Fiscal Realities. In Expert Meeting on How to feed the World in 2050, Food and Agriculture Organization (FAO) & the United Nations, Economic and Social Development Department.* UNDESA.

Bernhofen, D.M. & Brown, J.C. (2004). *Comparative Advantage: From Theory to Data.* Paper to be presented at the 2005 AEA Meetings in Philadelphia: On the 75th Anniversary of the Opportunity Cost Formulation of Comparative Advantage.

CBN. (2018). *Annual Statistical Bulletin*. Central Bank of Nigeria.

Chiogor, O. H., Gylych, J., & Alabi, O. O. (2023). Macroeconomic Impact of Agricultural Financing Reforms: A Computable General Equilibrium Analysis of Nigeria. *Journal of Economics Finance and Accounting*, *10*(2), 98–107. doi:10.17261/Pressacademia.2023.1732

Commonwealth. (2003). Public –Private Partnerships: A Review with Special Reference to Local Government. In *Commonwealth Local Government Handbook.* KPL.

Dilanchiev, A., & Taktakishvili, T. (2022). Determinants of Bilateral Trade Balance between Georgia and China. *Ekonom*, *101*(1), 6–19. doi:10.15388/Ekon.2022.101.1.1

Directory of Public Companies in Nigeria. (n.d.). *Credit Risk Monitor (CRMZ).* http://www.crmz.com/Directory/CountryNG.htm

Dorcas, A. T. (2023). Funding of Agricultural Programme in Tertiary Education in Nigeria. *Best Journal of Innovation in Science. Research for Development*, *2*(9), 338–346.

Echeverria, R. G., Trigo, E. J., & Byerlee, D. (1998). Financing Agricultural Research in Latin America. In *Financing Agricultural Research: A Sourcebook.* International Service for National Agricultural Research.

Elkington, J. (1997). Cannibals with Forks: The Triple Bottom Line. *Oxford Capstone.*

Elsalih, O., Sertoglu, K., & Besim, M. (2021). Comparative Advantage of Crude Oil Production: Evidence from 28 Oil-producing Countries. *The Journal of International Trade & Economic Development*, *30*(2), 275–294. doi:10.1080/09638199.2020.1850846

Eno, E. J., & Eze, F. O. (2023). Relationship between Agricultural Financing and Agricultural Output in Nigeria. *Global Journal of Finance and Business Review.*

Etzkowitz, H., & Dzisah, J. (2008). Triple Helix Circulation: The Heart of Innovation and Development. *EASST Rev.*, *14*(1), 14–19.

Ewubare, D. B. & Eyitope, J. A. (2015). The Effects of Public Expenditure on Agricultural Production Output in Nigeria. *Journal of Research in Humanities and Social Sciences*, *11*(3), 7-23.

FAO. (2017). *The Future of Food and Agriculture: Trends and Challenges.* Food and Agriculture Organization.

Findlay, R. (1987). Comparative Advantage. In The New Palgrave: A Dictionary of Economic. MacMillan. doi:10.1057/978-1-349-95121-5_67-1

George, G., Haas, M. R., Joshi, H., McGahan, A. M., & Tracey, P. (2022). Introduction to the Business of Sustainability: An Organizing Framework for Theory, Practice and Impact. In G. George & ... (Eds.), *Handbook on the Business of Sustainability: The Organization, Implementation, and Practice of Sustainable Growth* (pp. 2–22). Edward Elgar. doi:10.4337/9781839105340.00007

Gupta, S.D. (n.d.). *Comparative Advantage and Competitive Advantage: An Economics Perspective and a Synthesis.* St. Thomas University.

Hall, B. H. (2002). *The Financing of Research and Development.* Oxford University Press. doi:10.3386/w8773

Hirschman, A. O. (1958). *The Strategy of Economic Development.* Yale University Press.

Hoffman, N. P. (n.d.). An Examination of the "Sustainable Competitive Advantage. *Concept: Past, Present, and Future. Academy of Marketing Science Review*, *2*(4).

IPCC. (2014). *Impacts, adaptation, and vulnerability. Part A: global and sectoral aspects. Contribution of working group II to the fifth Intergovernmental Panel on Climate Change assessment report.* Cambridge University Press.

Kassaw, T. (2020). *Magnitude and Determinants of Rural Household Poverty in Ebinat District of Amhara National Regional State of Ethiopia* (Doctoral dissertation).

Krugman, P. (1984). Import Protection as Export Promotion: International Competition in the Presence of Oligopoly and Economies of Scale. In H. Kierzkowski (Ed.), *Monopolistic Competition and International Trade* (pp. 180–193). Clarendon Press.

Krugman, P. (2023). Cooperating for a better future. *CIRIEC-España, revista de economía pública, social y cooperativa*, (107), 5-14.

Lawther, W. C. (2002). *Contracting for the 21st Century: A Partnership Model.* The PricewaterhouseCoopers Endowment for the Business of Government.

Lazzarini, S. (2022). *The Right Privatization: Why Private Firms in Public Initiatives Need Capable Governments.* Cambridge University Press. doi:10.1017/9781009024167

Loebenstein, G., & Thottappilly, G. (2007). *Agricultural Research Management.* Springer. https://www.springer.com/gp/book/9781402060564

Luo, Y. (2000). Dynamic Capabilities in International Expansion. *Journal of World Business, 35*(4), 355–378. doi:10.1016/S1090-9516(00)00043-2

Lûpez-Pereira & Garcìa. (1994). The Maize Seed Industries of Brazil and Mexico: Past Performance, Current Issues, and Future Prospects. *CIMMYT Economics Program Working Paper*. El Batn, Mexico: International Maize and Wheat Improvement Center.

Mayr-Dorn, K. (2023). *Trade Diversion and Labor Market Adjustment: Vietnam and the US-China Trade War. Trinity Economics Papers. Trinity College Dublin.* Department of Economics.

Mghenyi, E. W., Dankers, C., Thurlow, J., & Anyiro, C. (2021). *Nigeria transforming agribusiness for inclusive recovery, jobs creation, and poverty reduction: policy reforms and investment priorities.* Academic Press.

NASCIO-CLC. (2006). Keys to Collaboration: Building Effective Public-Private Partnerships. *The National Association of State Chief Information Officers (NASCIO) Corporate Leadership Council (CLC) Issue Brief.*

Nin-Pratt, A., & Stads, G. (2023). Innovation Capacity, Food System Development, and the Size of the Agricultural Research System. *Frontiers in Sustainable Food Systems, 7*, 1051356. doi:10.3389/fsufs.2023.1051356

NISER. (2005). Review of Public Private Partnership in Agricultural Production in Nigeria. In Public-Private Partnership in Nigerian Development. NISER Review of Nigerian Development, 2003/2004, 235-258.

Nyangau, P., Muriithi, B., Diiro, G., Akutse, K. S., & Subramanian, S. (2022). Farmers' knowledge and management practices of cereal, legume, and vegetable insect pests and willingness to pay for biopesticides. *International Journal of Pest Management*, *68*(3), 204–216. doi:10.1080/09670874.2020.1817621

Obansa, S. A. J., & Madueke, I. M. (2013). Agriculture Financing and Economic Growth in Nigeria. *European Scientific Journal.*

Okoji, J. (2023). *Agric allocation relative to national budget lowest in 7yrs.* https://businessday.ng/agriculture/article/agric-allocation-relative-to-national-budget-lowestin7yrs/#:~:text=The%20highest%20proportion%20of%20the,to%201.25%20percent%20in%202022

Onah, H. C., & Nwodo, A. J. (2023). Backlash of "Uncamped" Internally Displaced Persons Fleeing Insurgency in Northern Nigeria and Its Correlation to Fuelling Insecurity and Violence in the South East Region, Nigeria. *African Journal of Current Research*, *4*(1), 1–19.

Onakoya, A. B. (2013). Agriculture and Intersectoral Linkages and their Contribution to Nigerian Economic Growth. *Economics*, *2*(5), 38–54.

Oyeyinka, B. O. (2020). SME: Issues, Challenges and Prospects. *Financial System Strategy (FSS) International Conference.* http://www.cenbank.org/fss/wed/sme_issues,%20challenges%20and%20prospects_oyeyinka%20banji.pdf

Pal, S., Rahija, M. A., & Beintema, N. M. (2012). *India: Recent Developments in Agricultural Research. ASTI Country Note.* IFPRI and Indian Council of Agricultural Research.

Pardey, P. G., & Roseboom, J. (1998). Trend in Financing African Agricultural Research. In *Financing Agricultural Research: A Sourcebook.* International Service for National Agricultural Research.

Pardey, P. G., Roseboom, J., & Fan, S. (1998). Trend in Financing Asian and Australian Agricultural Research. In *Financing Agricultural Research: A Sourcebook.* International Service for National Agricultural Research.

Paustian, M., & Theuvsen, L. (2017). Adoption of precision agriculture technologies by German crop farmers. *Precision Agriculture*, *18*(5), 701–716. doi:10.1007/s11119-016-9482-5

Porter, M. E. (1990). *The Competitive Advantage of Nations. Harvard Business Review.* doi:10.1007/978-1-349-11336-1

Prahalad, C. K., & Hamel, G. (1990). *The Core Competence of the Corporation. Harvard Business Review.*

Pratt, A. N., Stads, G. J., de los Santos, L., & Muñoz, G. (2023). Unlocking Innovation: Assessing the Role of Agricultural R&D in Latin America and the Caribbean. *Frontiers in Sustainable Food Systems*, *7*, 1051356. doi:10.3389/fsufs.2023.1051356

Pray, C., & Nagarajan, L. (2012). Innovation and Research by Private Agribusiness in India. *IFPRI Discussion Paper 1181.*

Quélin, B., Cabral, S., Lazzarini, S., & Kivleniece, I. (2019). The Private Scope in Public–Private Collaborations: An Institutional and Capability-based Perspective. *Organization Science*, *30*(4), 831–846. doi:10.1287/orsc.2018.1251

Quélin, B., Kivleniece, I., & Lazzarini, S. (2017). Public-Private Collaboration, Hybridity and Social Value: Towards New Theoretical Perspectives. *Journal of Management Studies*, *54*(6), 763–792. doi:10.1111/joms.12274

Reicosky, D. (2020). Conservation agriculture systems: Soil health and landscape management. In *Advances in Conservation Agriculture* (pp. 87–154). Burleigh Dodds Science Publishing.

Richard, B., Qi, A., & Fitt, B. D. (2022). Control of crop diseases through integrated crop management to deliver climate-smart farming systems for low-and high-input crop production. *Plant Pathology*, *71*(1), 187–206. doi:10.1111/ppa.13493

Rotunno, L. (2023). Trade Policy and Jobs in Vietnam: The Unintended Consequences of Trump's Trade War. *QPE Working Paper*, Center for Open Science.

Seck, P. A. (2013). Repackaging Agricultural Research for Greater Impact on Agricultural Growth in Africa. *Journal of Food Security*, *1*(3), 30–14.

SMEs as Gateway to Economic Prosperity. (n.d.). *ThisDayLive*. http://www.thisdaylive.com/articles/smes-as-gateway-to-economic-prosperity/166328/

Strand, L. (2017). Decomposing the Triple-Helix Synergy into the Regional Innovation Systems of Norway: Firm Data and Patent Networks. *Qual. Quantity*, *51*, 963–988.

Trendov, M., Varas, S., & Zeng, M. (2019). Digital technologies in agriculture and rural areas: status report. *Digital technologies in agriculture and rural areas: status report.*

Umaru, A., & Zubairu, A. A. (2012). An Empirical Analysis of the Contribution of Agriculture and Petroleum Sector to the Growth and Development of the Nigerian Economy from 1960-2010. *The International Journal of Social Sciences (Islamabad)*, *2*(4).

UN. (2019). *World population prospects 2019: Data booklet*. UN.

UNGC. (n.d.). https://www.unglobalcompact.org/howtoparticipate/business_participation/index.html

Van den Ban, A.W. (2000). Different Ways of Financing Agricultural Extension. *Agricultural Research and Extension Network*, Paper 106a.

Venezian, L. (1995). *El financiamiento de las actividades de I&D en el ambito de la universidad*. Paper presented at the workshop "Enfoques Innovativos para el Financiamiento de la Investigacion Agropecuaria en America Latinaî". Fundacion ArgenINTA, Buenos Aires.

World Agroforestry Centre. (2017). *Corporate Strategy 2017-2020—Transforming Lives and Landscapes with Trees*. Author.

APPENDIX 1

Table 5. Showing UNGC's Nigerian participants as at March 13, 2013

S/N	Name	Type	Country	Participant Since
1.	Association of National Accountants of Nigeria (ANAN)	Business Association Local	Nigeria	2013/09/13
2.	Atlantic Energy Drilling Concept Nigeria Limited	SME	Nigeria	2014/04/10
3.	Bola Kingsworth Productions (Nigeria)	SME	Nigeria	2015/03/11
4.	Boss Friendship Club in Nigeria	NGO Global	Nigeria	2014/07/10
5.	Brisad International Nigeria Limited	SME	Nigeria	2012/01/20
6.	Bukkehave Nigeria Limited	SME	Nigeria	2012/03/27
7.	CCLP Worldwide – Nigeria	NGO Global	Nigeria	2009/06/26
8.	El-Yah-Agla Nigeria Limited	Company	Nigeria	2011/02/14
9.	Environmental Degradation Organization of Nigeria	NGO Global	Nigeria	2013/12/06
10.	Etiket Nest Nigeria Ltd.	SME	Nigeria	2011/11/28
11.	First Bank of Nigeria Limited	Company	Nigeria	2013/01/23
12.	Junior Achievement Nigeria	NGO Global	Nigeria	2011/09/01
13.	Junior Chamber International, Nigeria	NGO Global	Nigeria	2010/01/28
14.	MTN Nigeria Communications Ltd.	Company	Nigeria	2006/03/07
15.	NET IMPACT Nigeria Chapter	NGO Local	Nigeria	2009/09/09
16.	New Nigeria Foundation	Foundation	Nigeria	2006/04/05
17.	Nigerian Fruit Concentrates Limited	SME	Nigeria	2014/08/21
18.	Nigerian Institute of International Affairs	Academic	Nigeria	2006/03/17
19.	Organization of Nigerian Chaplains	NGO Global	Nigeria	2012/05/15
20.	Partner Trade Nigeria Limited	SME	Nigeria	2014/03/31
21.	Peace Corps of Nigeria	NGO Global	Nigeria	2012/02/03
22.	Spanish-Nigerian for Commercial and Industrial Development Association – Nigeria Region	Business Association Global	Nigeria	2014/01/23
23.	Standards Organisation of Nigeria	NGO Local	Nigeria	2006/03/16
24.	Students in Nigeria Re-awakening Initiative	NGO Local	Nigeria	2014/10/03
25.	The Nigerian Economic Summit Group	Business Association	Nigeria	2006/03/07
26.	Youth Crime Watch of Nigeria	NGO Local	Nigeria	2011/03/29
27.	Youth for Human Rights International, Nigeria	NGO Global	Nigeria	2014/03/04

Source: UNGC website. https://www.unglobalcompact.org/participants/search?utf8=%E2%9C%93&commit=Search&keyword=Nigeria&joined_after=&joined_before=&business_type=all&sector_id=&listing_status_id=all&cop_status=all&organization_type_id=&commit=Search

Chapter 9
Data-Driven Precision Agriculture for Crop Prediction and Fertilizer Recommendation Using Machine Learning

Yashi Tiwari
Banasthali Vidyapith, India

Ayush Verma
https://orcid.org/0000-0001-7249-1011
Jawaharlal Nehru University, India

Manju Khari
Jawaharlal Nehru University, India

ABSTRACT

Crop prediction and fertilizer recommendation are essential for optimizing agricultural practices, a crucial concern for the agricultural sector. However, accurate crop and fertilizer prediction has long been challenging, requiring innovative solutions rooted in the vast pool of available data. This work presents a comprehensive system for predicting crops and fertilizers based on historical data, leveraging machine learning (ML) algorithms. This work involved research analysis of various ML algorithms to predict crops and recommend suitable fertilizers. It was observed that Naive Bayes and random forest models achieved an excellent accuracy of 99.54% and 99.31%, respectively, for soil classification, indicating their proficiency in distinguishing different soil types. The proposed system also suggests fertilizer recommendations tailored to each crop based on user-provided input and a comparative evaluation of the algorithms. These results highlight the potential of ML techniques in aiding farmers to make informed decisions about soil management and fertilizer selection.

DOI: 10.4018/979-8-3693-4864-2.ch009

1. INTRODUCTION

ML has a vital role in the future and scope of humankind, so being an agrarian economy and a country where most of the economy depends on agriculture, combining agriculture and ML to increase the efficiency of the sector will play a vital role. In the contemporary world, using Machine Learning (ML) technology has led to a surge of research in crop and fertilizer prediction analysis, an area of utmost importance in agriculture. ML techniques have emerged as the preferred avenue for resolving this challenge because of their proven efficacy. Various ML algorithms and methodologies, have played a significant role in the agricultural domain by providing insights into forthcoming crop production trends. By analysing historical data, these algorithms can identify patterns and factors that influence crop production. This, in turn, empowers farmers with valuable insights, allowing them to adapt their cultivation strategies and maximise crop production.

A study (Sujatha & Nithya, 2017) takes into consideration the Artificial Intelligence (AI) for societal development and social being and highlights significance of AI in the sector. By harnessing the power of ML, farmers can make data-driven decisions on crop management, irrigation, and pest control, optimising resource allocation and reducing waste. This bolsters food security and contributes to sustainable agriculture practices, which is crucial for a nation heavily reliant on its agrarian economy. Furthermore, AI-driven technologies such as precision agriculture and autonomous farming machinery can alleviate the labour-intensive nature of farming, attracting the younger generation to engage in agriculture. This shift could rejuvenate rural areas and stimulate economic growth. In addition, AI-powered predictive models can assist in weather forecasting and disease outbreak detection, providing early warnings to farmers. This is especially vital in a country where climate variability poses significant challenges to crop production.

The growing role of AI and ML in the Agrarian sector is also taken into consideration by (Gautam et al., 2022) which will lead to sustainable development. Sustainable development is the need of the hour as it takes into consideration the needs of the future. In addressing the challenges posed by a burgeoning global population coupled with dwindling resources, this research serves as a beacon, raising pivotal questions about our obligations to future generations. The significance of this work lies in its capacity to bridge the gap between present needs and future demands, with a keen focus on the intricate dynamics of the agricultural supply chain. Integrating ML and AI technologies enables precision agriculture, optimising resource allocation and minimising waste. It empowers farmers with data-driven insights, thereby enhancing productivity and ensuring a more sustainable use of resources.

On the basis of IoT (Ahmed et al., 2016) evaluated soil monitoring based on remote features so that on the basis of the soil can be monitored and accessed from anywhere the component helps in featuring the soil characters and accessing it remotely so that we can have a better understanding of the features and base on this soil features the crop is recommended however it is a combination of hardware and software that have immense potential and gain serval advantage in recent years as here comes the way on which basis on sensor we test the soil and get the potential component which is on to it. The algorithms used here are supervised on the basis of input the data has been trained and the required output is being generated.

Smart agriculture has been proposed by (Sri & Reddy, 2017) based on web things as smart agriculture is considered an important use ease in today applying technology with soil has played a vital role. In the modern era, smart agriculture is increasingly recognised as a pivotal solution. Technology integration with soil management has emerged as a critical factor in enhancing agricultural efficiency. Through the utilization of IoT sensors and devices, the proposed system can monitor various soil parameters in real time. This real-time data can be instantly transmitted to a centralized platform accessible through the web. Farmers can access this information remotely and make informed decisions regarding irrigation, fertilization, and pest control. Furthermore, the incorporation of web-based technologies enables predictive analytics, allowing farmers to anticipate potential issues and optimize crop production. The smart agriculture system facilitates resource conservation by optimising water and nutrient usage, ultimately reducing environmental impact. By harnessing the potential of the Internet of Things in agriculture, they have introduced an innovative solution that empowers farmers with data-driven insights to cultivate healthier crops, improve productivity, and contribute to sustainable farming practices. This technological leap marks a significant step towards revolutionizing the agricultural landscape for a brighter, more efficient future. In last two decades, information technology is growing rapidly and it is significance in agriculture becomes an essence for farmers as well as for common people in the way to sustainable goal.

The proposed model in this paper mainly focused on soil and its property on which crop should be grown. Various models are dependent on the crop while there is less focus on fertiliser as it acts as significant in agriculture. The inefficient use of fertiliser has caused various challenges in term of soil fertility so this work also proposes a model which will take into consideration soil as well as fertiliser on base on input provided on which soil what fertiliser is being used via using random forest classifier the model is trained which show the accuracy of 99.87 percent. The proposed system also suggests fertiliser recommendations tailored to each crop based on user-provided input. A comprehensive comparative evaluation of the algorithms is presented. These results highlight the potential of ML techniques in aiding farmers to make informed decisions about soil management and fertiliser selection.

As we embark on this exploration of the confluence between ML and agriculture, it becomes apparent that the amalgamation of state-of-the-art technology with age-old farming practices holds significant promise. In the subsequent sections, we will delve into intricate details of our proposed model's architecture, its methodology, and an exhaustive analysis of historical data. Section 2 discusses the background to the study related to the research topic. Section 3 discusses the literature survey and previous work related to crop prediction and fertiliser recommendation. Section 4 highlights the major research objectives. Section 5 highlights the proposed methodology in detail. Section 6 discusses the results and obtained. Section 7 concludes the paper with a concluding remark and future scope.

2. BACKGROUND TO THE STUDY

Crop prediction and fertilizer recommendation in smart agriculture involves using various technologies and techniques to predict the number of crops that will be harvested. Accurate estimation is crucial for farmers to make informed decisions about crop management, resource allocation, and marketing strategies. Fertilizer recommendation in smart agriculture involves the use of advanced technologies and data analytics to optimize the application of fertilizers based on the specific needs of crops. This approach helps improve nutrient efficiency, reduce environmental impact, and increase overall crop production. Here are some methods commonly employed for crop prediction and fertilizer recommendation in smart agriculture:

- **Remote Sensing and Satellite Imagery:** Satellite imagery can be used to monitor crop health, growth, and predict crops. Vegetation indices, such as NDVI (Normalized Difference Vegetation

Index), can provide valuable insights into crop conditions. Unmanned aerial vehicles equipped with sensors can capture high-resolution images of crops. Drones enable more frequent and targeted data collection, allowing for precise monitoring of crop health and growth.

- **IoT (Internet of Things) and Sensor Networks:** Deploying sensors in the field to monitor various parameters like soil moisture, temperature, humidity, and nutrient levels. This real-time data can be used to assess the environmental conditions affecting crop growth.
- **Machine Learning and Data Analytics:** Utilizing historical data, weather data, and sensor data to develop machine learning models that can predict crops. Algorithms can learn patterns and relationships between various factors influencing crop growth. Data Fusion: Integrating data from multiple sources, such as weather stations, satellite imagery, and on-field sensors, to create a comprehensive dataset for more accurate predictions.
- **Weather Data:** Incorporating weather and climate data into crop prediction models. Climate models can help predict the impact of weather conditions on crop growth and production.
- **Field Scouting and Surveys:** Combining remote sensing data with ground-based observations and surveys to validate and calibrate models. Farmers or field agents can provide valuable insights into local conditions that may not be captured by remote sensing alone.
- **Smart Farming Equipment:** Using GPS-guided tractors and equipment to optimize planting patterns, irrigation, and fertilization. Precision agriculture helps in improving resource efficiency and ultimately influences crop prediction.
- **Farm Management Software:** Integrated platforms that consolidate data from various sources, providing farmers with a centralized dashboard to monitor and analyze crop conditions.
- **Blockchain Technology:** Blockchain can be used to create transparent and traceable supply chains, ensuring that the information about the crop, from cultivation to distribution, is accurate and reliable.

By combining these technologies, farmers and agricultural stakeholders can enhance their ability of crop prediction and fertilizer recommendation effectively, leading to more sustainable and efficient agricultural practices.

3. LITERATURE SURVEY

In this section, we will delve into the existing body of research concerning the integration of ML techniques with soil testing, aiming to assess the accuracy achieved by these proposed models. The field of ML applied to soil analysis has witnessed substantial progress, with numerous models put forth to predict soil properties and provide recommendations for optimal agricultural practices. Our model stands out through its distinctive approach and contributions when juxtaposed with prior research. Proposed model adopts a multifaceted strategy for soil analysis, harnessing ML algorithms' capabilities to predict crop recommend fertilizer with notable precision and offer fertiliser recommendations with remarkable reliability. While various authors have explored ML's potential in soil analysis, our model distinguishes itself by conducting a thorough evaluation of historical data, enabling highly accurate soil classification, and delivering personalized fertiliser recommendations. What further sets our model apart is its adaptability to a wide array of agricultural contexts, making it applicable across diverse regions and crop types. This unique feature positions our model as a valuable asset for contemporary farmers seeking to optimize their farming methods, enhance crop prediction, and concurrently reduce their environmental footprint.

ML-based technologies in agriculture have gained significant attention in recent years due to their potential to improve crop prediction and fertilizer recommedation, reduce costs, and enhance resource utilization. One area where ML has been particularly effective is in crop and fertiliser recommendation systems, which use data to recommend the optimal crop for a particular field. A study (Nuchhi et al., 2020) uses different datasets and tested various ML models on that dataset Authors found that SVM is 91.53 percent with an execution time of 11600 seconds. Random Forest is 99 percent with execution time of 190 second for one dataset and for other data set the Accuracy of SVM is 89% with execution time of 0.011 seconds. For Random Forest, accuracy is 99 percent with an execution time of 50 seconds. The remarkable progress in ML-based technologies within agriculture extends well beyond crop and fertiliser recommendation systems. These advancements are ushering in a new era of precision farming, where data-driven decision-making is becoming increasingly prevalent. One notable aspect is the use of ML algorithms to optimize irrigation practices. By analysing data from soil moisture sensors, weather forecasts, and plant growth models, these systems can intelligently regulate irrigation, delivering the right amount of water precisely when its needed. This not only conserves water resources but also prevents over-irrigation, which can leach nutrients from the soil and harm crop health.

A study by (Veenadhari et al., 2014) is divided into three steps - first sampling (different soil with the same number of properties with different parameters) algorithm modelling and optimisation. In the field of agriculture, ML acts as an efficient tool and have the potential in revolting the sector by adding on technology as an efficient tool in term of grain sector. ML, as highlighted in this research, emerges as a potent tool with the capacity to significantly impact the agricultural landscape. By integrating technology into the grain sector, it holds the promise of enhancing efficiency, optimising resource allocation, and potentially spearheading a transformative revolution in modern agriculture.

In model of (Afrin et al., 2018) mainly focuses on the agriculture analysis of organic farming and inorganic farming, time cultivation of the plant, profit and loss of the data, and analysis of the real estate business land in a specific area. This work goes for finding reasonable information models that accomplish a high precision and a high consensus as far as crop production expectation abilities. Which further helped in enhancing the model and giving the accuracy more up to date.

Role of ML and AI in growing agrarian sector is taken by consideration as it will led to sustainable development as sustainable development is need of the hour as it take into consideration the future need also as the growing population and less resources have taken into considering the question on what we will give to future generation the significant work help us to understand and gave to future as it act as the present need in given into consideration the future supply chain . In addressing the challenges posed by a burgeoning global population coupled with dwindling resources, this research serves as a beacon, raising pivotal questions about our obligations to future generations.

Information-driven approach to structure for gathering and information demonstrating frameworks is given by (Kumar & Venkatesh, 2018). Her focus was an emphasis on Soil dampness, a key factor in the crop development cycle, which is chosen for instance as the main factor in reducing the effects of soil. On the other side, a responsive remote sensor hub is built up that expects to catch the elements of soil dampness using the sensor. Here efficient use of gadgets is tried on field soil to show its usefulness and the responsiveness of the sensors. To capture and monitor soil moisture dynamics effectively, the researchers developed a responsive remote sensor hub. This technology was specifically designed to gather real-time data on soil moisture levels using advanced sensors. By deploying these sensors in the field, the research aimed to demonstrate the practicality and responsiveness of this technology in authentic agricultural settings. This pioneering approach underscores the importance of efficient gadget utilization in agriculture.

Nitrogen, Phosphorus, Potassium, Calcium, Magnesium, Sulphur, Iron, Zinc, and so forth is chosen by (Manjula, 2017) for investigating the soil supplements utilising Naïve Bayes, Decision Tree, and a hybrid approach of Naïve Bayes and Decision Tree. The performance of the classification algorithms is compared based on accuracy and execution time which help in the efficient detection of crops based on soil type. To accomplish this, the researchers employed a range of classification algorithms, including Naïve Bayes, Decision Tree, and a hybrid approach combining both methods. This comprehensive analysis allowed them to evaluate the performance of these algorithms based on key metrics such as accuracy and execution time. These evaluations played a crucial role in streamlining the efficient detection of crops based on soil type.

To evaluate on the basis of IoT and involving the soil monitoring based on remote features proposed by (Ahmad et al., 2016) so that on the basis of the soil can be monitored and accessed from anywhere the component helps in featuring the soil characters and accessing it remotely so that we can have a better understanding of the features and base on this soil features the crop is recommended however it is a combination of hardware and software that have immense potential. Smart agriculture on the basis of web things as smart agriculture is being considered an important use ease in today applying technology with soil has played a vital role. In the modern era, smart agriculture is increasingly recognised as a pivotal solution. The integration of technology with soil management has emerged as a critical factor in enhancing agricultural efficiency. Through the utilization of IoT sensors and devices, the proposed system can monitor various soil parameters in real time. This real-time data can be instantly transmitted to a centralized platform accessible through the web. Farmers can access this information remotely and make informed decisions regarding irrigation, fertilization, and pest control.

Following comprehensive examination of existing research and the evolving landscape of ML in agriculture, we now shift our focus towards the practical implementation of our proposed model. In the upcoming sections, we will provide an intricate exploration of the model's architecture and methodology, elucidating its utilization of ML to address the intricate challenges associated with soil analysis and crop prediction and fertilizer recommendations. Furthermore, we will present the empirical findings stemming from our field studies, thereby substantiating the model's efficacy within real-world agricultural contexts.

4. RESEARCH OBJECTIVES

The research objectives taken up for the work presented in this paper are as follows

- To propose a model which recommends crop as well as fertiliser on basis of input provided.
- The inputs are condition of soil like present of Nitrogen, phosphorus, potassium in it the model is designed that it will show crop as well as fertiliser which is suitable for the soil.
- To show that random forest as classifier can be provide efficient results for crop and fertiliser prediction.

The system aims to recommend the most suitable fertiliser for each crop by considering various crop-specific factors, such as soil type, weather conditions, and historical performance. Furthermore, our system extends its utility beyond crop prediction by providing personalised fertiliser recommendations. By considering various crop-specific factors, such as soil type, weather conditions, and historical performance, the proposed system aims to recommend the most suitable fertiliser for each crop. This tailored approach enhances crop productivity and promotes sustainable farming practices by minimising fertiliser wastage. In addition to the impressive accuracy rates achieved, highlighting our system's tangible benefits is essential. By implementing our ML models, farmers would be able to increase crop production by up to 15% while reducing fertiliser usage by as much as 20%. These numbers underscore the practicality and economic significance of our technology. Our research showcases the potential of ML in revolutionising the agriculture sector. The high accuracy achieved by our Naïve Bayes and Random Forest model for both soil classification and fertiliser recommendations underscore the practicality and reliability of this technology. As farmers increasingly adopt data-driven approaches to agriculture, our system is a valuable tool for improving crop predictions and optimising fertiliser usage. Combining prediction and recommendation, this holistic approach paves the way for a more efficient and sustainable future in agriculture, with substantial gains in crop productivity and resource efficiency.

5. RESEARCH METHODOLOGY

In the given section we have discussed about data collection, pre-processing and training the collected and sample data have been described to get the overview of the proposed model. Aim of research is to contribute to the agriculture sector of the country by taking to consideration soil types and basis of this crop is being qualified on various parameters and on this soil and fertiliser is being recommended that suit the pattern best. In given we are talking about various aspects that are discuss in this paper from collecting data to training to recommendation of the crop and fertiliser.

The works done till now mainly concentrated on crop prediction and take into consideration only crop which is being grown on specific soil using different soil properties and Data Mining Techniques. Fertiliser Recommendation is not taken into consideration as fertiliser has taken to be an important part of agriculture The crop production and future of the sector are dependent on it as it acts as a buffer future on which soil will be sown in the future as the wrong fertiliser not only take up present soil but it also takes into the consideration the future of soil as it will take up the fertility and basic nutrients from the soil. So, it is necessary to develop a crop prediction and fertiliser recommendation system that predicts crops based on soil nutrients crop data as well as recommend fertiliser for selected crop based on different datasets like fertiliser data, location data, and crop data base on the various factor on soil like Nitrogen, phosphorus, potassium so the need of crop prediction is much important as fertiliser prediction. The research efforts thus far have primarily focused on crop prediction, often limited to specific soil types and utilising various soil properties and data mining techniques. However, a significant aspect of agriculture, fertiliser recommendation, has largely been overlooked, despite its crucial role. Fertilisers are integral to agriculture, affecting not only present crop growth but also the future of soil and its fertility. Incorrect fertiliser application can deplete essential nutrients from the soil and have far-reaching consequences for the sustainability of agriculture. This underscores the urgent need to develop a com-

Table 1. Samples of crop dataset and various parameters

#	N	P	K	Temp	Humidity	pH	Rainfall	Label
1	90	42	43	20.899	82	6	202	Rice
2	85	58	41	21	80	7	226	Rice
3	78	42	40	20	81	6	221	Rice
4	81	40	39	21.22	79	6	244	Rice
5	71	54	16	22	63	5	87	Maize
6	73	58	21	19	57	6	60	Maize

prehensive system that combines crop prediction with fertiliser recommendations. Such a system should leverage different datasets, including fertiliser information, location-specific data, and historical crop data, while considering various soil factors like Nitrogen, Phosphorus, and Potassium. By adopting this integrated approach, the system can provide precise crop predictions while suggesting the most suitable fertilisers, promoting optimal agricultural outcomes. Furthermore, reflecting on the lessons from the Green Revolution in India, where excessive fertiliser uses negatively impacted soil health, it is evident that effective fertiliser management is paramount. The proposed system aims to address this gap, not only enhancing agricultural productivity but also preserving soil fertility for the long term. This holistic approach aligns agriculture with sustainability, ensuring the health and productivity of the land for future generations. During the time of the green revolution when India became self-sufficient in terms of food grain, it also affected the health of the soil as during this time the access use of fertiliser affected the region and soil Punjab we see as a live example that it now facing the problem regarding soil fertility.

5.1. Data Description

Crop and fertiliser dataset from Kaggle is utilised in this work. The dataset consists of various location soil and types base on collection of data from various location. Both the dataset of fertiliser and crops is acquired from Kaggle and consists of data in the form of csv file. The given dataset also contains information about humidity, temperature, Nitrogen, potassium, Phosphorous in the soil as it considers to be one of the most important in terms of understanding of soil which give us the accurate prediction base on the training we provide to the data.

Data collection: Data collection is done by using different sources methodology employed for this study involves a meticulous and comprehensive approach to data collection and data pre-processing. To ensure the reliability and validity of our research findings, we utilized a multi-faceted data collection strategy that integrated various sources. Data from reputable sources such as academic articles, government reports, and industry databases were acquired to augment our dataset. The following approach allowed for a holistic understanding of the research topic and enriched the quality of our analysis. Dataset is divided in two parts- Training and Testing with crop dataset containing twenty three crops on various parameters like N stand for Nitrogen, P stand for potassium, K stand for Phosphorous, rainfall, temp stand for temperature, pH stands for potential of hydrogen to measure acidic and basic nature of soil, humidity the all are independent values while the dependent is the crop which is being recommended on following parameters. Table 1 showcase the value on the given parameters the following crop is suitable in soil. A sample from the dataset is shown below:

Table 2. Samples of crop dataset and fertilisers

#	Temp	Humidity	Moisture	N	P	K	Fertiliser Name
1	34	65	62	7	9	30	Gromor 14-35-14
2	26	52	38	37	0	0	Urea
3	27	54	28	13	0	40	DAP
4	35	63	49	30	10	14	Sindco

The table one showcase and describe the values and resultant crop which are grown on the specific value of the crop. Gives us a overview about the sample crop and which are suitable and by help of this we will train the Data and get specific output. The fertiliser dataset which is main component of our work has also been divided in Training and Testing parameters where Nitrogen, Temperature, humidity, potassium, Phosphorous, humidity and rainfall all which are dependent while the fertiliser which is being recommended is independent and the dataset is train in respective manner using random forest algorithm. There contain of eight major fertilisers generally used in agriculture for use. The given dataset sample can be presented as:

Table 2 showcases the sample input need to be provided and on the base on the specific input we get the output which fertiliser suited for the given soil types and also the recommendation is provided to user on entering the specific values. This gives us an overview of the specific data and also the outcome. Figure 1 represents the heatmap of the dataset attributes. A heatmap is a graphical representation of data where values in a matrix are represented as colors. It is a way to visualize and analyze multidimensional

Figure 1. Heatmap of the dataset attributes

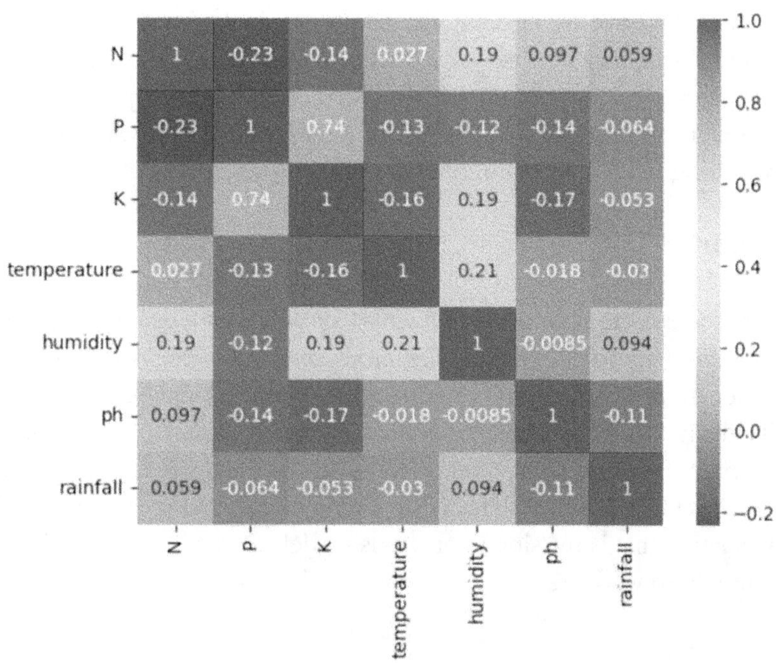

data in a two-dimensional space. Heatmaps are commonly used in various fields such as statistics, data analysis, biology, and geography.

Data Pre-Processing: The collected data underwent a rigorous pre-processing phase to ensure its quality, consistency, and suitability for analysis. Initially, data cleaning procedures were implemented to detect and rectify any errors, missing values. This involved data validation, imputation of missing values, and the removal or correction of data through various techniques. Subsequently, data normalization and transformation techniques were applied to standardize variables and reduce biases that could affect the analysis. Categorical variables were encoded appropriately, and numerical features were scaled to bring them into a consistent range. This meticulous data pre-processing stage not only improved the integrity of our dataset but also set the foundation for meaningful and accurate analysis, ultimately leading to robust research outcomes.

The Previous models predominantly focused solely on crop prediction, overlooking the vital aspect of fertiliser recommendation. In contrast, our model represents a significant advancement as it seamlessly integrates both crop and fertiliser prediction, offering a comprehensive solution for optimising agriculture. This holistic approach ensures that farmers receive actionable insights to enhance crop productivity while optimising resource utilization, marking a substantial evolution in the realm of agricultural decision support systems.

Training of Data: To construct our agricultural prediction model, we utilise the potential of the Random Forest algorithm for its adeptness at managing intricate datasets and delivering precise forecasts. The dataset underwent a partitioning process into two primary subsets: a training set and a validation set. The training set, encompassing a substantial portion of the data, was employed to educate the Random Forest model. This algorithm operates by assembling a collection of decision trees, where each tree is built using a random subset of features and bootstrapped samples from the training data. This ensemble of trees collaborates to provide predictions, and during the training process, the model learns iteratively from the data. At each node of the trees, decisions are made based on available features and target variables, continually refining the model until it evolves into a robust predictor. Random Forest, being a versatile algorithm, excels at handling diverse agricultural data types, including soil attributes, meteorological patterns, historical crop data, and fertiliser usage records. The validation set was instrumental in assessing the model's accuracy and its capacity to generalize, ensuring its capability to effectively predict both crop and fertiliser recommendations. This empowers farmers with valuable insights to facilitate informed decision-making in their agricultural endeavours.

5.2. Proposed Work

The aim of the proposed system is to provide a smart system which can assist agriculture domain to cultivate crop with help of specific fertiliser. The crops selected in this work are based on important crops from specific location. The selected crops are Rice, Mazie, chickpea, kidney beans, pigeon peas, moth beans etc. The dataset of crops and fertilizer is collected from last 4 years from different sources. The selected fertilisers are - Urea, DAP, twenty-eighth etc and dataset is collected from last 4 years. The proposed system has been designed in term for getting more accuracy in terms of fertiliser and give benefit in recommendation that is missing in previous models.

The 3 stages in proposed work are:

Figure 2. Proposed system

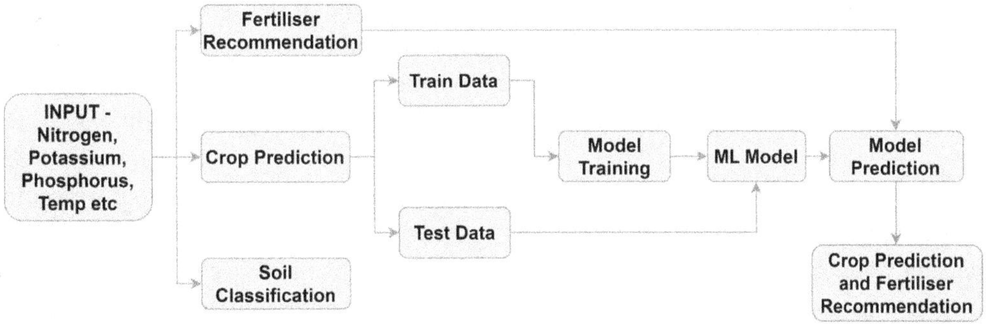

- **Classification of Soil:** Soil classification is be done using soil nutrients data. ML algorithms are used for soil classification. The algorithm will classify, and display output based on the input which is provided by user as soil play a vital role in the getting the crop and fertiliser recommendation so first way is to classify the soil as per requirement and user need. By understanding the unique composition of each soil type, our system can tailor its recommendations to match the specific requirements and preferences of the user. Soil plays a pivotal role in determining crop success, and categorizing it correctly is the initial step toward optimising agricultural practices. The concerned ML algorithms consider a myriad of soil attributes, including nutrient levels, pH, texture, and organic matter content, among others. The power of these algorithms lies in their ability to process vast amounts of data rapidly and make informed predictions regarding soil types. The implications of soil classification go beyond crop and fertiliser recommendations. It also aids in land-use planning, helping farmers decide which crops are best suited for their soil, and whether any amendments or treatments are necessary for improved productivity. Additionally, this information can be invaluable for land developers, environmental agencies, and policymakers in making informed decisions related to soil conservation and land management. In essence, soil classification serves as the cornerstone of our holistic approach to agricultural support, ensuring that users receive tailored guidance that aligns with their specific soil conditions and agricultural goals.
- **Crop Prediction:** Crop Prediction can be done using crop data, nutrients and location data. These inputs are passed to model which is trained via using Random Forest algorithms. These algorithms will predict crop based on present inputs as model give an accuracy on the input which is provided by the user. Our approach leverages ML to assess the intricate relationship between various factors. This includes not only nutrients and location but also weather patterns, soil quality, and previous crop performance. By amalgamating this multifaceted dataset, our model can deliver highly accurate predictions, empowering farmers to make informed decisions regarding crop selection, planting schedules, and resource allocation.
- **Fertiliser Recommendation:** Fertiliser Recommendation can be done using fertiliser data, crop and location data. In this part suitable crops and required fertiliser for each crop is recommended to the user. The model is trained using random forest and on giving value of Nitrogen, as an essential nutrient for plant growth, serves as the foundation of our recommendation system. Through

a meticulous process that involves the integration of historical soil nitrogen data, crop preferences, and location-specific variables, our model harnesses the capabilities of Random Forest algorithms to provide tailored guidance on optimal crop selection and the corresponding fertilisers required for each unique agricultural context. These recommendations not only rely on data-driven insights but also take into consideration the local climate conditions and regional variations

Figure 2 give an overview of our proposed work in which on giving certain input we get the output on the basis of the soil type we get the crop and fertiliser which is recommended suitable for the soil as represented above on giving input for Nitrogen, Phosphorus Temperature we get output. The model has been trained on various ML algorithms and the respective results and prediction are obtained. This data is then used to analyse the best and optimal algorithm by comparing the performance of each algorithm. Out of all the considered algorithms, the algorithm which showed optimal results are Navie Bayes and Random Forest. Random Forest is a supervised ML algorithm based on ensemble learning. Ensemble learning is a type of learning where we join different types of algorithms multiple times to form a more powerful prediction model. The random forest algorithm combines multiple algorithms of the same type. Also, the algorithm can be used for classification and regression problems.

Some of the advantages of Random Forest are:

- The random forest algorithm is not biased, as there are multiple trees and each tree is trained on a subset of data.
- Random Forest algorithm is stable that means that if a new data point is introduced in the given dataset the overall algorithm is not affected so that makes the function and simplicity of model more accurate.

In the upcoming section, we will delve into the results presented by the model, conducting a thorough analysis to provide valuable insights. This analysis will serve as the foundation for our concluding discussion, where we will highlight the model's efficiency and the significance of its results. To ensure the integrity of our research, we will employ plagiarism detection tools to maintain the highest standards of academic rigor. This approach will contribute to a well-rounded and credible research methodology, substantiating the findings and recommendations presented in this study.

6. RESULTS AND DISCUSSION

The comprehensive system is drafted using Python as its core framework. By using random forest as our Diverse datasets, encompassing information on crops, locations, soil characteristics, crop nutrients, and fertiliser data, have also been diligently collected from various sources, including agricultural literature and agricultural websites. These datasets serve as the foundation upon which our agricultural solution is built, taking a data-driven approach to empower and optimize farming practices.

This holistic approach ensures that our system is well-equipped to provide valuable insights and recommendations to farmers, enabling them to make informed decisions and maximize agricultural productivity. Through the integration of these diverse datasets, we pave the way for a more sustainable and efficient agricultural landscape. That can be efficient in understanding for farmers and also getting valuable output in terms of fertiliser and crop.

Figure 3. Accuracies of various ML algorithms on the crop dataset

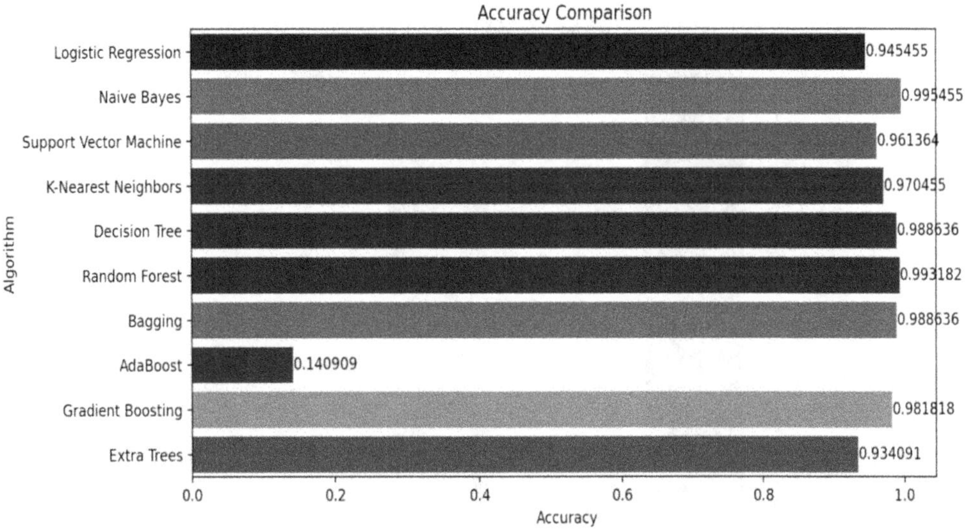

As our model have taken into consideration features of soil and its input we can like where nitrogen stand for N, Phosphorus stands for P, K stand for potassium, temperature, humidity, PH stand for potential of hydrogen, rainfall.

Figure 3 represents the accuracies of various ML algorithms for the crop recommendation dataset. The algorithms tested are Logistic Regression, Naïve Bayes, Support Vector Machine, K-Nearest Neighbour, Decision Tree, Random Forest, Bagging, AdaBoost, Gradient Boosting and Extra Trees. Among all the algorithms Naïve Bayes and Random Forest excel with 99.54% and 99.31% accuracy.

Figure 4. Input crop data

```
N = 40
P = 50
k = 90
temperature = 30z
humidity = 100
ph = 100
rainfall = 100
def get_recommendations(N, P, k, temperature, humidity, ph, rainfall):
    recommendations = []


    recommendations.append((1, 0.8))
    recommendations.append((3, 0.7))
    recommendations.append((8, 0.6))

    return recommendations

recommendations = get_recommendations(N, P, k, temperature, humidity, ph, rainfall)

crop_dict = {1: "Rice", 2: "Maize", 3: "Jute", 4: "Cotton", 5: "Coconut", 6: "Papaya", 7: "Orange",
             8: "Apple", 9: "Muskmelon", 10: "Watermelon", 11: "Grapes", 12: "Mango", 13: "Banana",
             14: "Pomegranate", 15: "Lentil", 16: "Blackgram", 17: "Mungbean", 18: "Mothbeans",
             19: "Pigeonpeas", 20: "Kidneybeans", 21: "Chickpea", 22: "Coffee"}
```

Figure 5. Recommendation of crops

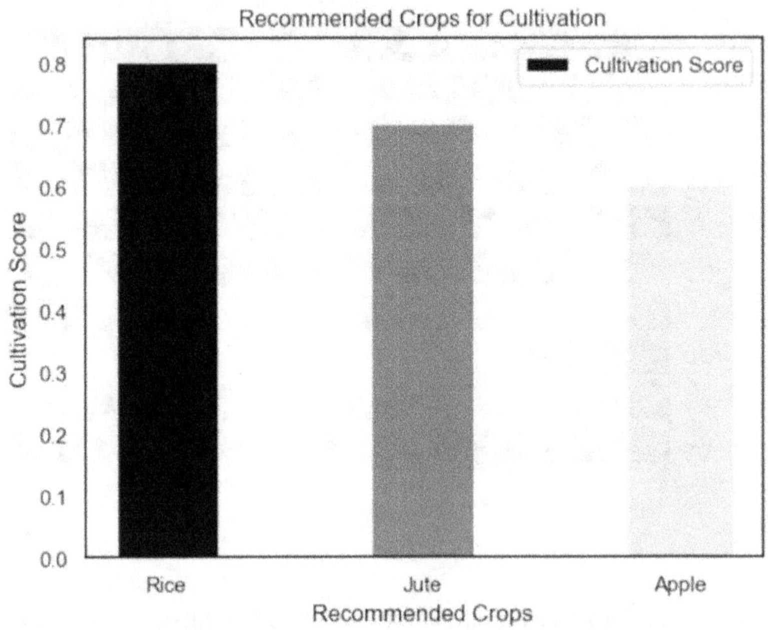

In above Figure 4 we have taken input from user N as nitrogen P as phosphorus K as potassium, temperature, humidity, pH, rainfall and taken a crop dictionary of given crop in the data. Base on the given input of soil we get the suitable crop.

In the Figure 5 on the basis of given input we got the crops which can be recommended in which rice, jute, apple have been displayed with Rice showing higher probability of growth as value goes to

Figure 6. Input fertiliser data

```
Nitrogen = 37
Potassium = 0
Phosphorus = 0
Temperature = 26
Humidity = 52
Moisture = 38


def get_fertilizer_recommendations(Temperature, Humidity, Moisture, Nitrogen, Potassium, Phosphorus):
    recommendations = []


    recommendations.append((1, 0.8))
    recommendations.append((2, 0.7))
    recommendations.append((3, 0.6))

    return recommendations

recommendations = get_fertilizer_recommendations(Temperature, Humidity, Moisture, Nitrogen, Potassium, Phosphorus)

Fertilizer_dict = {1: "Urea", 2: "DAP", 3: "FourteenThirtyFIveFourteen", 4: "TwentyEighthtTwentyEighth",
                   5: "Sindco", 6: "TwentyTwenty", 7: "Gromor"}


fertilizer_recommendations = [Fertilizer_dict[fertilizer_id] for fertilizer_id, score in recommendations]
recommendation_scores = [score for fertilizer_id, score in recommendations]
```

Figure 7. Recommendation of fertiliser

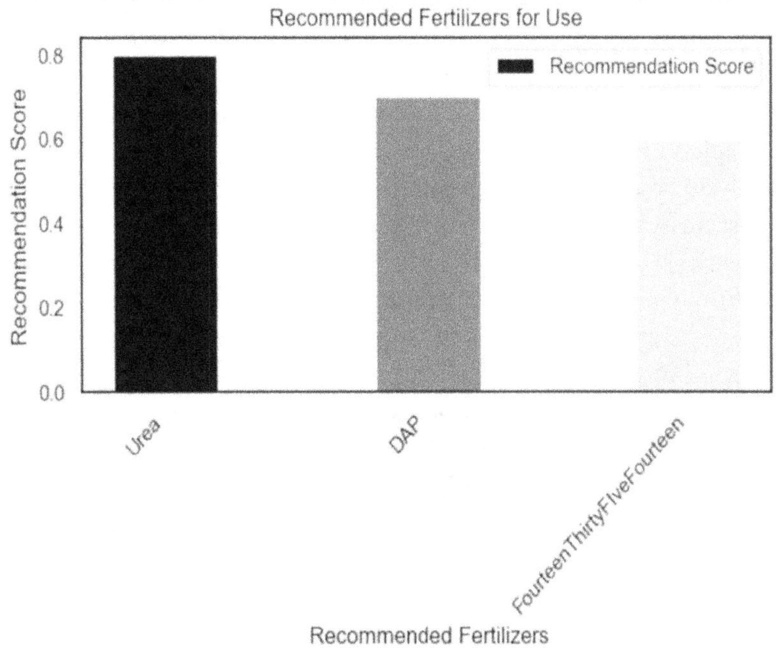

0.8. The probability lies between 0 and 1 with 1 having the highest probability and 0 with lowest on the given set of data it is clear stated that Rice is most suitable crop.

With our model have taken into consideration fertiliser as one of the main features so we can present we can take it as input as nitrogen, potassium, phosphorus, temperature, Humidity, Moisture on taking follow input we can get output as:

As in given Figure 6 we have taken input from the user about soil nitrogen, potassium, phosphorus, temperature, humidity, moisture and taken the dataset of various fertiliser that have been consider in agriculture field and on this we get the fertiliser output that is best suitable for the given condition of the soil.

In the above Figure 7 we get Urea, DAP, GROMOR 14-35-14 with laying value between 0 and 1 with 0 been lowest and 1 been highest recommendation of the fertiliser. With Urea 0.8 DAP 0.6 (approx.) so and, GROMOR 14-35-14 with 0.6 gives urea as most suitable for the soil. This gives us an example of having highest and lowest possibility for fertiliser in given input of soil.

7. CONCLUSION AND FUTURE SCOPE

Prediction of crop and fertiliser based on location so the given input depend on the particular location of the soil while using random forest we can predict location wise crop and fertiliser. From above work we have concluded that for soil classification Random Forest is with accuracy 88%. While for fertiliser we can see the recommended accuracy of Random Forest 90%. It is clear that the model demonstrates strong predictive capabilities for suggesting suitable fertilisers based on soil characteristics. This high level of accuracy not only ensures efficient resource utilization but also contributes significantly to sustainable agriculture practices. Farmers can make informed decisions about the type and quantity of fertilisers to use, optimising crop production and reducing environmental impact. Moreover, when considering the 88% accuracy achieved for soil classification, it underscores the model's effectiveness in delineating soil types based on location-specific attributes. Such precise soil classification lays the foundation for tailored crop recommendations. By combining the strengths of both soil classification and fertiliser recommendation models, we enable farmers to make location-specific, data-driven decisions that enhance agricultural productivity and promote sustainable land management practices, ultimately contributing to food security and environment. On various world summit we have also seen world leader focus on the food security.

The work can be further extended by adding following functionality. Rain prediction as we know rain act as one of major aspects in agriculture as well as proper user interface like website or application and taking simplified version of website so that by entering values one can get recommended crop as well as fertiliser for following input. The application can also be taken into consideration so by just taking images one can get the fertiliser as well as crop. Further we can extend by adding crop diseases detection using image processing in which user can get pesticides based on disease images. While Accurate rain forecasts can help farmers prepare for weather-related challenges and optimize planting and harvesting schedules. While a proper user interface will help farmer in getting output easily by just typing the values in their application the device can be phone application.

REFERENCES

Afrin, S., Khan, A. T., Mahia, M., Ahsan, R., Mishal, M. R., Ahmed, W., & Rahman, R. M. (2018). Analysis of Soil Properties and Climatic Data to Predict Crop Yields and Cluster Different Agricultural Regions of Bangladesh. *2018 IEEE/ACIS 17th International Conference on Computer and Information Science (ICIS)*. 10.1109/ICIS.2018.8466397

Ahmed, A., Isaac, W., Varshney, S., & Khan, E. (2016). An IoT based system for remote monitoring of soil characteristics. *2016 International Conference on Information Technology (InCITe) - the next Generation IT Summit on the Theme - Internet of Things: Connect Your Worlds*. 10.1109/INCITE.2016.7857638

Gautam, P., Bhatt, A. K., Rawat, A., Nautiyal, O. P., & Dhanik, S. (2022). Design and Implementation of IoT-Based Smart System for Monitoring Soil and Parameters in the Agricultural Scenario. *IGI Global*. doi:10.4018/978-1-6684-2443-8.ch006

Kumar, C., & Venkatesh, V. (2018). *Cloud based soil monitoring and smart irrigation system using IoT and precision farming*. Semantic Scholar. https://www.semanticscholar.org/paper/Cloud-based-soil-monitoring-and-smart-irrigation-Kumar-Venkatesh/ca23e0098316741b004de61b37d6edb2f5d5cc73

Manjula, E. (2017). Data Mining Technique to Analyze Soil Nutrients Based on Hybrid Classification. *International Journal of Advanced Research in Computer Science*, *8*(8), 505–510. doi:10.26483/ijarcs.v8i8.4794

Nuchhi, S., Bagali, V., & Annigeri, S. (2020). IOT Based Soil Testing Instrument For Agriculture Purpose. *2020 IEEE Bangalore Humanitarian Technology Conference (B-HTC)*. 10.1109/B-HTC50970.2020.9297897

Rajagopal, A., Jha, S., Khari, M., Ahmad, S., Alouffi, B., & Alharbi, A. (2021). A Novel Approach in Prediction of Crop Production Using Recurrent Cuckoo Search Optimization Neural Networks. *Applied Sciences (Basel, Switzerland)*, *11*(21), 9816. doi:10.3390/app11219816

Sri, A. N. S., & Reddy, P. A. (2017). A Smart Framework for Agriculture Production Improvement using Web of Things. *International Journal of Engineering Trends and Technology*, *50*(5), 251–259. doi:10.14445/22315381/IJETT-V50P241

Sugandh, U., Nigam, S., & Khari, M. (2023). Ecosystem of Technologies for Smart Agriculture to Improve the Efficiency and Profitability of Indian Farmers. *2023 10th International Conference on Computing for Sustainable Global Development (INDIACom)*. https://ieeexplore.ieee.org/document/10112573

Sujatha, R., & Nithya. (2017). A Survey on Soil Monitoring and Testing In Smart Farming Using IoT And Cloud Platform. *Int. Journal of Engineering Research and Application*, *7*(11), 55–59. https://doi.org/ doi:10.9790/9622-0711065559

Veenadhari, S., Misra, B., & Singh, C. (2014). *Machine learning approach for forecasting crop yield based on climatic parameters*. IEEE Xplore. doi:10.1109/ICCCI.2014.6921718

Chapter 10
Precision Irrigation for Sustainable Agricultural Productivity

Moazzam Mushtaq
University of Agriculture, Faisalabad, Pakistan

Hasnain Ali
University of Agriculture, Faisalabad, Pakistan

Aamir Raza
https://orcid.org/0009-0001-1867-2660
University of Agriculture, Faisalabad, Pakistan

Sheraz Maqbool
University of Agriculture, Faisalabad, Pakistan

Muhammad Safdar
https://orcid.org/0009-0006-1779-6967
University of Agriculture, Faisalabad, Pakistan

Mubashir Ahmed
University of Agriculture, Faisalabad, Pakistan

Jaffar Sattar
https://orcid.org/0009-0006-6107-9156
Khwaja Fareed University of Engineering and Information Technology, Rahim Yar Khan, Pakistan

ABSTRACT

Precision irrigation is a water management strategy that applies water to crops based on their specific needs. It is a sustainable practice that can help to increase crop yields, reduce water use, improve fertilizer use efficiency, and reduce environmental impact. This chapter offers a detailed overview of precision irrigation for sustainable agricultural productivity. It defines precision irrigation and discusses its benefits, types, components, design, implementation, management, applications, and economic and environmental benefits. The chapter also discusses the role of precision irrigation in sustainable agriculture, challenges and opportunities for adoption, and the future of precision irrigation. It is intended to be a valuable resource for farmers, researchers, policymakers, and other stakeholders interested in learning more about precision irrigation's role in sustainable agriculture.

DOI: 10.4018/979-8-3693-4864-2.ch010

1. INTRODUCTION TO PRECISION IRRIGATION

1.1. What Is Precision Irrigation?

Precision irrigation is an advanced water management strategy that involves carefully distributing water resources to crops according to their individual needs (Abioye, *et al.,* 2020). This advanced and meticulously calibrated irrigation system represents a departure from traditional irrigation techniques like flood irrigation. Instead, it relies on a wealth of data to optimize the irrigation process, enhancing its overall efficiency and sustainability. This technique transcends traditional methods of irrigation by employing cutting-edge technology and data-driven approaches to optimize water usage.

The primary focus of precision irrigation is to deliver the right amount of water, at the right time, and in the right place to maximize crop productivity while minimizing water wastage and environmental impacts (Zhang, *et al.,* 2021). Figure 1 illustrates the implementation of precision irrigation techniques in contemporary agriculture, showcasing efficient water management for optimized crop growth and resource utilization. This rigorous method makes it possible to apply water precisely, ensuring that crops receive the ideal level of moisture needed for the best possible growth and development.

1.2. Benefits of Precision Irrigation for Sustainable Agricultural Productivity

- Precision irrigation is an essential practice in contemporary agriculture owing to its manifold advantages, encompassing economic, environmental, and agronomic dimensions. The system enhances water efficiency by precisely delivering an appropriate quantity of water directly to the root

Figure 1. Precision irrigation

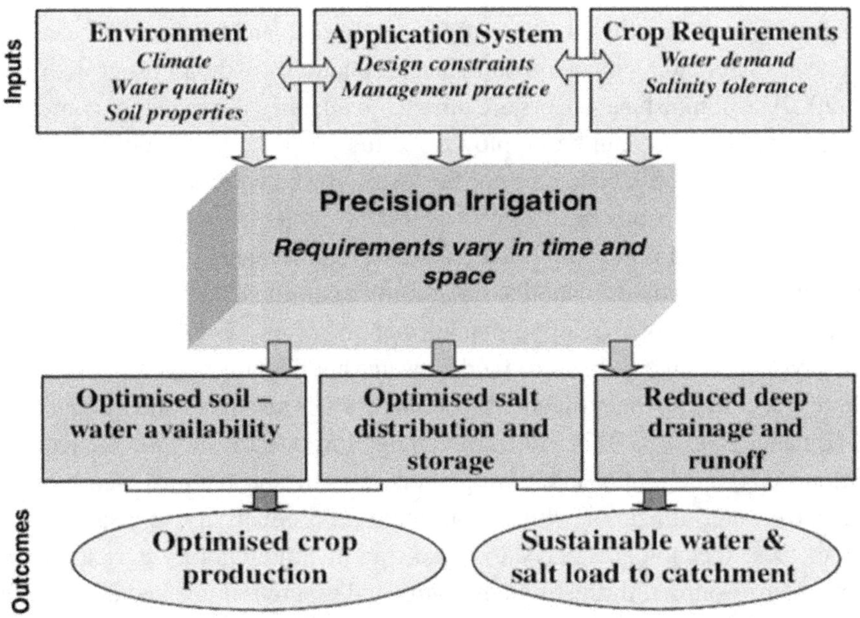

zone of crops, thereby mitigating water loss and promoting the preservation of water resources (Smith and Baillie, 2009).

- Precision irrigation has the potential to greatly improve crop yields by customizing irrigation practices to meet the individual needs of different crops. Precision irrigation systems have the capability to incorporate nutrient management measures, hence mitigating nutrient leaching and minimizing environmental damage. Through the use of nutrient delivery control, farmers have the ability to mitigate the quantity of fertilizers utilized, resulting in financial savings and a reduction in environmental repercussions. Energy-efficient components, such as drip irrigation and micro-irrigation systems, have been found to effectively decrease operational expenses.
- Precision irrigation is an effective technique that mitigates the occurrence of surface runoff, a phenomenon that has the potential to induce soil erosion, particularly in areas characterized by susceptible topsoil conditions. This technology enables farmers to customize irrigation schedules according to certain growth stages, so providing optimal water supply throughout crucial developmental times for plants. Additionally, there are remote monitoring and control capabilities that may be utilized, which prove particularly advantageous for expansive or geographically scattered fields.
- Precision irrigation plays a significant role in disease and pest management through its ability to prevent excessive wetness and maintain a consistent level of soil moisture. The implementation of precision irrigation techniques by farmers leads to enhanced profitability through the augmentation of crop yields, reduction in operational expenses, and optimization of resource utilization. The responsible and efficient utilization of water and nutrients is of paramount importance in fostering sustainable agriculture, as it contributes significantly to the preservation of the environment and the long-term security of food production.

1.3. Overview of Precision Irrigation Technologies

Precision irrigation systems employ advanced sensors, controls, and actuators to accurately administer water to crops, hence maintaining ideal moisture levels conducive to their best growth and development (García, *et al.*, 2020). Soil moisture sensors are utilized to quantify the moisture content in soil, thereby offering instantaneous data that can be employed for the purpose of scheduling irrigation activities. Crop water stress sensors are utilized to monitor the levels of water stress in crops, hence facilitating the making of well-informed decisions regarding irrigation. Weather stations gather data on temperature, humidity, and precipitation in order to predict the water requirements of crops. These meteorological stations offer precise meteorological data that has a significant impact on agricultural decision-making, encompassing the scheduling of irrigation, management of crops, and control of pests. Variable rate irrigation (VRI) systems enable farmers to modify water application rates according to individual requirements, thereby facilitating the efficient utilization of water resources and mitigating environmental consequences (Balafoutis, *et al.*, 2017). Automated irrigation systems provide meticulous regulation of water distribution, supplanting labor-intensive procedures with mechanized elements, so empowering farmers to maximize water utilization, enhance agricultural productivity, and save valuable resources. The utilization of these technologies plays a pivotal role in contemporary agriculture, facilitating the optimization of water resources, mitigating environmental repercussions, and augmenting agricultural productivity. In general, precision irrigation technologies play a crucial role in improving crop development, managing water resources, and conserving valuable resources.

Figure 2. Furrow irrigation

2. TYPES OF PRECISION IRRIGATION SYSTEM

Precision irrigation systems encompass several types, each with its unique characteristics and advantages. These systems are designed to optimize water use, enhance crop productivity, and reduce environmental impact. Here's a detailed exploration of the main types of precision irrigation systems.

2.1. Surface Irrigation Systems

Surface irrigation systems are a category of agricultural irrigation methods where water is distributed and applied directly over the soil surface to supply crops with the necessary moisture. Unlike subsurface drip or sprinkler irrigation, surface irrigation systems rely on the natural flow of water across the field or through a network of canals or ditches. These systems are commonly used in agriculture and are one of the oldest and most widely practiced forms of irrigation. Several common types of surface irrigation systems include.

2.1.1. Furrow Irrigation

Furrow irrigation is a widely used agricultural practice primarily applied to row crops such as corn, cotton, and various vegetables. It involves the creation of small channels or furrows between rows of crops, which serve as conduits for water delivery. Figure 2 provides a visual representation of a furrow irrigation system, demonstrating the traditional method of watering crops through channels or furrows in the field. Gravity plays a crucial role as water is introduced at the upper end of these furrows and naturally flows down, saturating the soil around the plant roots. While furrow irrigation is popular due to its simplicity and low equipment requirements, it presents challenges related to water distribution uniformity. Furrows closest to the water source tend to receive more water than those farther away, leading to uneven moisture levels within the field (Haghnazari, *et al.,* 2015).

Excessive water runoff is a common concern, which not only contributes to water wastage but can also cause soil erosion and environmental issues, such as nutrient leaching. To address these challenges and enhance the precision and efficiency of furrow irrigation, modern technologies like automation systems and soil moisture sensors are increasingly being integrated, offering better control over water application, optimized resource usage, and reduced environmental impacts.

Furrow irrigation provides several advantages, including cost-effectiveness, suitability for row crops, simplicity, low energy requirements, and adaptability to various field sizes and configurations. It is well-suited for flat terrain, aids in soil aeration, and can be integrated with conservation practices to promote soil health. Additionally, furrow irrigation generally involves lower labor requirements and helps reduce soil compaction. However, it's essential to manage furrow irrigation effectively to address challenges related to uneven water distribution, potential water wastage, and runoff, which can be mitigated through modern technologies and improved management techniques to optimize resource usage and minimize environmental impacts.

2.1.2. Basin Irrigation

Basin irrigation, a time-tested method, proves highly effective, especially for perennial crops like fruit trees, grapevines, and even annual crops like rice. This technique offers a suite of benefits, aligning seamlessly with the precision irrigation approach. By creating enclosed basins tailored to crop needs, basin irrigation achieves water efficiency and conservation objectives central to precision irrigation, preventing water wastage and runoff while preserving this precious resource. In Figure 3, basin irrigation is depicted, showing a farming method where fields are divided into basins, allowing for controlled water application and improved moisture distribution to support crop growth. It also curbs weed proliferation and shields against soil erosion, addressing sustainability goals. Moreover, basin irrigation stands out for its simplicity, making it accessible to diverse farmer profiles. Its adaptability to sloped terrain further underscores its value, particularly in hilly landscapes. Nonetheless, its performance hinges on factors like soil type and crop variety, necessitating meticulous planning and diligent upkeep to avert waterlogging in poorly drained soils, demonstrating the precision irrigation ethos that underlines effective and sustainable water management.

2.1.3. Contour Farming

Contour farming is an agricultural method designed to combat soil erosion, especially in hilly or sloping terrain. By creating terraces or contour rows that follow the natural contours of the land, it effectively reduces the risk of soil loss during heavy rainfall. Figure 4 portrays contour farming, an agricultural technique that follows the natural contours of the land, helping to reduce soil erosion and conserve water resources while promoting sustainable crop cultivation. This approach not only prevents topsoil erosion but also allows water to infiltrate the soil, aiding in water conservation and groundwater recharge. Contour farming is particularly suitable for regions with rolling topography and is environmentally sustainable, as it safeguards water quality and aquatic habitats (Jin, *et al.,* 2011). While it requires initial effort and resources for construction and maintenance, the benefits in terms of soil conservation and erosion control make it a valuable practice. Contour farming can also be integrated with other conservation methods, such as cover cropping and reduced tillage, to further enhance soil health and agricultural sustainability.

Figure 3. Basin irrigation

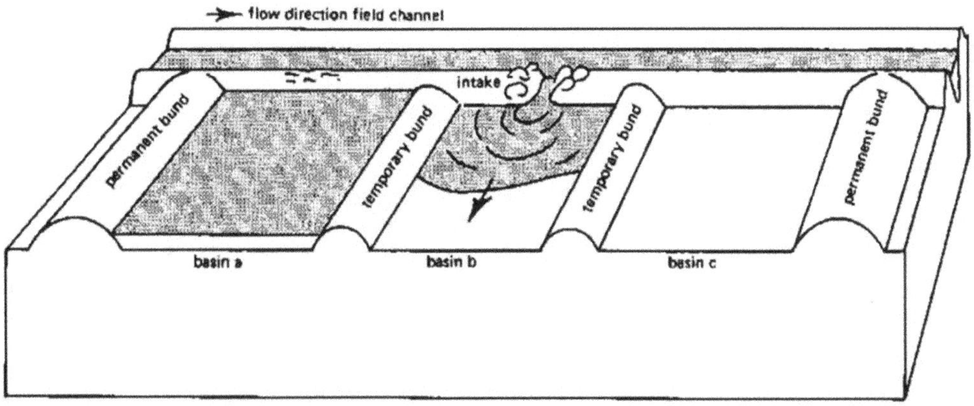

2.2. Drip Irrigation

Drip irrigation is a highly efficient and precise method of delivering water and nutrients directly to the root zone of plants, ensuring they receive the moisture and nutrients needed for optimal growth. This irrigation system has been widely adopted in agriculture due to its water-saving and productivity-enhancing capabilities.

2.2.1. Types of Drip Irrigation

Surface Drip Irrigation

Surface drip irrigation is an agricultural irrigation method where water is delivered directly to the soil surface through a network of pipes and emitters as shown in figure 5 (A). Unlike subsurface drip ir-

Figure 4. Contour farming

Figure 5. Surface drip irrigation

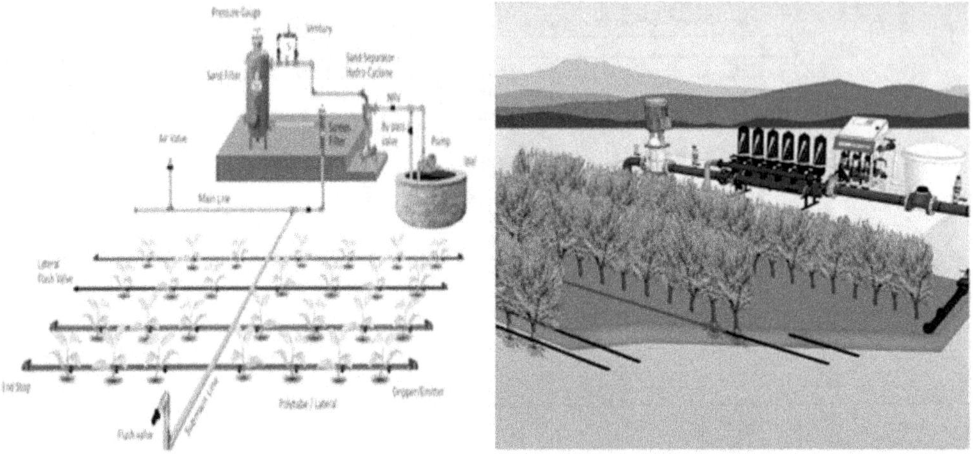

(A) (B)

rigation, it does not involve burying the pipes, and the emitters are visible on the ground. Surface drip irrigation is valued for its efficiency and precision in delivering water and nutrients to crops, promoting healthier plant growth and conserving water resources. It is widely used in various agricultural applications, including row crops, orchards, vineyards, and gardens, and is adaptable to a range of crop types and soil conditions.

Subsurface Drip Irrigation

Subsurface drip irrigation (SDI) is an efficient agricultural irrigation method involving buried pipes with emitters that deliver water and nutrients directly to the root zone of crops as shown in figure 5 (B). It minimizes water wastage, reduces evaporation and runoff, and allows for precise control of water and nutrient delivery, resulting in improved crop health and yield. SDI can also help control weeds and reduce the risk of certain crop diseases, though it requires initial investment and suitability assessment for optimal performance in specific agricultural settings.

2.2.2. Components of Drip Irrigation System

Figure 6 illustrates the constituent elements of a drip irrigation system, showcasing the sophisticated apparatus and tubing employed to administer accurate quantities of water straight to the roots of plants, hence augmenting water efficiency in agricultural practices.

Drippers (Emitters)

Drippers are small devices that release water in a controlled and uniform manner. They are typically attached to tubing and placed near the base of each plant. Drippers are designed to deliver a specific flow rate, which can be adjusted to meet the water requirements of different crops.

Figure 6. Elements of a drip irrigation system

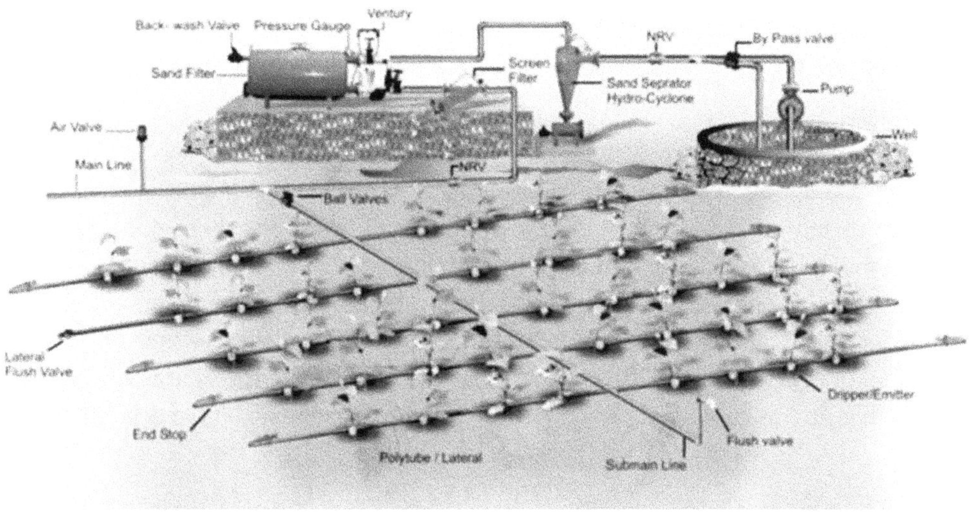

Water Delivery System

The central element of the majority of drip irrigation systems consists of a water delivery pathway, which is typically constructed using either PVC pipes for underground delivery or polyethylene tubing for above-ground applications. This water delivery system is composed of various components that serve the purpose of conveying water from the water source to the plants. It encompasses a network of mainlines, pipelines, and drip lines, all interconnected using couplings.

Water Filters

Water filters play a vital role in every irrigation system, serving as a crucial component to block silt and other debris from entering the system and ensure the delivery of clean water to plants. Drip irrigation system providers offer a range of in-line filter options to accommodate diverse setups, including media, screen, and disk filters, along with sand separators. These filters are each designed for use with various water sources, helping to maintain the quality of the irrigation water and safeguard the system's components from clogging and damage, thus contributing to the efficient and consistent supply of water to the plants.

Sand filters are commonly employed in water treatment processes to eliminate suspended particles and sediments. This is normally accomplished by utilizing a tank that contains a carefully selected grade of sand. Impurities are effectively captured by the system, resulting in the accumulation of pure water in the lower portion, which is subsequently directed towards the irrigation system. Screen filters employ a mesh or screen structure to effectively separate particles and detritus, therefore eliminating a diverse spectrum of particle sizes. Pre-filtration mechanisms are frequently employed as an initial barrier to impurities in order to safeguard the irrigation system from potential pipe and nozzle blockages. Hydrocyclone filters employ the principle of centrifugal force to effectuate the segregation of solid particles from water (Tian, *et al.*, 2018).This process engenders a spiral flow pattern, which facilitates the outward displacement of denser particles towards the periphery, ultimately leading to their sedimentation at the bottom. These devices demonstrate high efficiency in the removal of larger particles, however their effectiveness in the removal of finer sediments may be limited. Disk filters consist of a series of stacked disks that possess grooves or tiny mesh, effectively capturing particles on their surfaces. These filtration systems demonstrate efficacy in the removal of particles of varying sizes and are commonly employed in scenarios where the preservation of water quality is a priority (Wen-Yong *et al.,* 2015).

Water Pressure Regulators

Water pressure regulators, also known as pressure-reducing valves (PRVs), are essential plumbing devices designed to control and regulate water pressure within a building's plumbing system. Installed at the point where the municipal water supply enters the structure, they serve to reduce high-pressure water to a safe and manageable level, ensuring the protection of pipes, fixtures, and appliances from damage due to excessive pressure. These valves are set to different pressure levels, which are 25, 50, and 100 kPa (kilopascals), or equivalently, 0.25, 0.5, and 1 bar. These pressure levels are typically measured to ensure that the fluid (in this case, likely water) flows through the submain at the desired pressure (Sokol *et al.,* 2019).

Valves and Gauges

Valves and gauges play vital roles in maintaining efficient and effective water delivery to plants. Valves serve as the control points for flow management, allowing users to adjust the volume of water to meet specific plant requirements and create irrigation zones, ensuring that water is distributed precisely where and when needed. They also facilitate shut-off for maintenance or repairs without disrupting the entire system, thus conserving water. On the other hand, gauges provide critical monitoring and feedback. Pressure gauges help maintain uniform water distribution by monitoring system pressure, while flow gauges measure and monitor actual flow rates to optimize water usage and scheduling.

Chemical Injectors

Chemical injectors serve the essential purpose of introducing specific chemicals, including fertilizers, pH adjusters, water treatment agents, pesticides, and herbicides, into the irrigation water (Evans *et al.,* 2007). This controlled injection of chemicals is crucial for optimizing plant health and crop yields by providing precise and consistent nutrient delivery, maintaining appropriate pH levels, and enabling water treatment and pest control. By automating the introduction of chemicals, chemical injectors ensure that plants receive the correct amounts of substances at the right times, promoting efficient resource utilization, reducing the risk of overuse, and contributing to the overall success of drip irrigation in terms of crop health and productivity.

System Controller

Controllers are responsible for regulating the duration and timing of water delivery to the fields through the distribution lines. They come in different types, including manual ones that operate based on preprogrammed schedules, activating and deactivating the system at specified times.

2.2.3. Working Principle

Drip irrigation systems are designed to distribute water with precision to the root zones of plants, effectively minimizing water loss due to evaporation, runoff, or deep percolation. Water is applied in a gradual and constant manner, so assuring a continual supply of moisture. Drip irrigation systems employ strategically placed drippers to ensure consistent water distribution, hence facilitating uniform crop development and mitigating the occurrence of arid patches. Drip irrigation is a highly advantageous irrigation technique that enables accurate delivery of water to the specific root zones of individual plants. This precision makes it particularly advantageous for cultivating row crops, fruit trees, and garden vegetables.

2.2.4. Advantages

- Drip irrigation is highly efficient, reducing water usage compared to other irrigation methods. It minimizes water waste, making it ideal for regions with water scarcity.
- The consistent supply of water and nutrients directly to the root zone promotes optimal plant growth and higher yields.
- By delivering water only to the crop's root area, drip irrigation minimizes moisture availability to weeds, reducing weed growth.
- Drip irrigation limits the wetting of plant foliage, which can help reduce the spread of certain plant diseases.
- Drip irrigation systems can be combined with nutrient management strategies, ensuring that fertilizers are applied precisely where needed.
- Drip irrigation systems are often automated, reducing the need for manual labor in irrigation management.

2.2.5. Challenges

- Installing a drip irrigation system can be costly, primarily due to the need for equipment such as filters, pressure regulators, and tubing.
- Regular maintenance, including cleaning filters and checking for clogs or damage to emitters, is essential to ensure the system's efficiency.
- Drippers are prone to clogging from sediments and mineral deposits in the water. Effective filtration and regular maintenance are critical to prevent clogs.

2.3. Sprinkler irrigation

Sprinkler irrigation is a versatile method of applying water to crops by simulating natural rainfall. It is widely used in agriculture to efficiently distribute water to fields, orchards, and gardens. Sprinkler systems come in various configurations, each designed to meet specific needs. Figure 7 depicts a sprinkler irrigation system, a modern and efficient method of irrigating agricultural areas by evenly dispersing water over the crops.

Figure 7. Sprinkler irrigation system

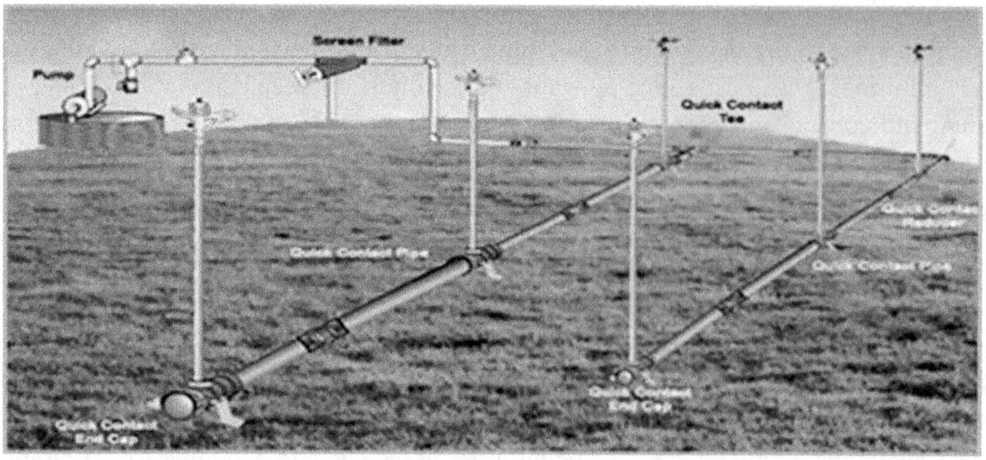

2.3.1. Key Components

Sprinkler Heads

Sprinkler heads are the devices responsible for distributing water over the crops. They come in various types, including impact, rotary, and fixed spray heads, each with specific characteristics suited to different applications.

Piping and Tubing

A network of pipes or tubing carries water from the water source (such as a well or pump) to the sprinkler heads. The pipes are typically made of materials like PVC or aluminum.

Pumps

In some cases, pumps are required to pressurize the water and supply it to the sprinkler system. The choice of pump depends on factors like field size, water source, and topography.

Filters

Filters are essential components to remove debris and impurities from the water, preventing clogs and damage to the sprinkler heads.

Pressure Regulators

Pressure regulators maintain a consistent pressure within the system, ensuring that sprinklers operate at the desired flow rate.

Control System

Some modern sprinkler systems are equipped with controllers that allow for automation, scheduling, and remote operation. These controllers enable precise control over when and how long the system runs.

2.3.2. Working Principle

The principal purpose of sprinkler systems is to distribute water evenly across crops in a manner that emulates natural rainfall. This objective is accomplished by means of the sprinkler heads, which dispense water in a regulated and consistent manner. Certain types of sprinkler heads, such as rotary or oscillating variants, exhibit movement in order to effectively cover a designated area. The rotation or oscillation pattern employed in this process guarantees the uniform distribution of water. The flow rate and pressure of the water can be modified in order to fulfill the particular demands of the crop and the land.

2.3.3. Types of Sprinkler Irrigation Systems

Center Pivot Irrigation Systems

Center pivot systems consist of a series of towers with wheeled sprinkler lines that rotate around a central pivot point as shown in Figure 8 (A). These systems are typically used in large, circular fields. The central pivot point remains fixed, while the sprinkler lines move in a circular pattern, irrigating the field as they rotate.it is suitable for large, open, and circular fields, such as those used for large-scale agriculture.

Lateral Move Irrigation Systems

Lateral move systems are similar to center pivots but move linearly, irrigating rectangular fields as shown in Figure 8 (B).. The sprinklers move along towers running parallel to the field's length. The entire system moves laterally, covering rectangular or square fields, and irrigating as it goes. It is suitable for rectangular or square fields, especially when circular patterns are not ideal.

Fixed Sprinkler Systems

Fixed systems are stationary and do not move during operation. They can be configured with various types of sprinkler heads, including impact and spray heads as shown in Figure 8 (C). The sprinklers in a fixed system are placed strategically to cover the target area. Water is supplied through a network of pipes, and the sprinklers distribute it to the crops or garden without moving. It is suitable for smaller fields, gardens, and areas with irregular shapes where mobility is not necessary.

2.3.4. Applications

Sprinkler irrigation is commonly used for field crops like wheat, corn, and soybeans. It can efficiently cover large areas with water, especially when coupled with automated systems. Sprinkler systems are employed to protect fruit and grape crops from frost and to supply consistent moisture for growth. Small-scale sprinkler systems are ideal for vegetable gardens, providing uniform water distribution and efficient water use. Sprinkler systems are widely used for residential and commercial lawns and golf courses to maintain lush, green landscapes.

2.3.5. Advantages

- Sprinkler systems provide even coverage across the field, promoting consistent crop growth.
- Sprinklers can be used to protect crops from frost damage by providing a protective ice layer.

Figure 8. Types of sprinkler irrigation systems

(A) (B)

(C)

- Sprinklers can be used to apply fertilizers, pesticides, and herbicides evenly.
- Sprinkler systems can be adapted to various field sizes and crop types.

2.3.6. Challenges

- Evaporation and wind drift can lead to water loss in sprinkler irrigation systems.
- Operating pumps and maintaining pressure can result in higher energy costs.
- Filters and sprinkler heads require regular cleaning and maintenance to prevent clogs.

2.4. Micro-Irrigation Systems

Micro-irrigation systems, commonly known as drip irrigation or trickle irrigation, represent a revolutionary approach to efficiently provide water and essential nutrients directly to a plant's root zone. They have become a cornerstone of modern agriculture, landscaping, and horticulture due to their remarkable ability to conserve water resources and optimize crop productivity. Unlike traditional irrigation methods, micro-irrigation systems ensure that water is applied with unparalleled precision, a feature that has transformed the way we nurture and cultivate various plant species. These systems are celebrated for their contribution to sustainable agriculture and environmental stewardship. Figure 9 showcases a

Figure 9. Micro-irrigation system

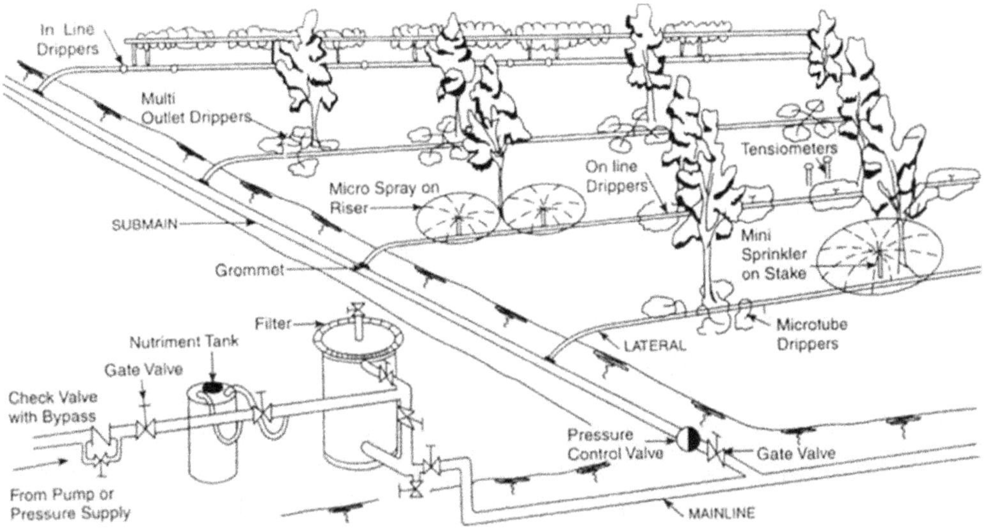

micro-irrigation system and its various components, highlighting the precision and efficiency of this method for delivering water directly to plants, optimizing water use in agriculture.

2.4.1. Components of Micro-Irrigation Systems

Emitters

Emitters used in micro-irrigation systems, including drip emitters and micro-sprinklers, play a pivotal role in delivering water and nutrients with precision and control directly to the root zone of plants. Grouping multiple emitters together as a unit in an irrigation system means connecting them in a cluster or manifold. This configuration allows for more consistent water distribution and helps reduce variability in water flow among the emitters (Barragan *et al.,* 2006).

Tubing and Pipes

These convey water from the source to the emitters. Tubes and pipes are typically made of materials like PVC, polyethylene, or drip tape.

Filters

Filters remove debris and particles from the water to prevent clogging of emitters. Common filter types include screen filters and disc filters.

Pressure Regulators

Pressure regulators maintain a consistent pressure in the system, ensuring uniform water distribution to all emitters.

Valves and Control Systems

Valves are used to control the flow of water, and control systems can be manual or automated to manage irrigation schedules and duration.

2.4.2. Benefits of Micro-Irrigation Systems

Micro-irrigation is a highly efficient and precise water management method that delivers water directly to the plant's root zone, reducing water wastage and runoff. With water use efficiencies of over 90%, micro-irrigation is one of the most water-efficient irrigation methods available, contributing significantly to water resource conservation. Micro-irrigation, when integrated with precision irrigation scheduling, offers a tailored approach to watering crops based on their specific water needs (Jamroen, *et al.,* 2020). This method takes into account various factors, including real-time weather data, soil conditions, and the rate at which a crop loses water through evapotranspiration. It also ensures consistent moisture delivery, reducing water stress during dry periods and enhancing overall water use efficiency. Micro-irrigation also plays a crucial role in weed control, as it delivers water directly to the plant root zones, reducing water availability for weed growth in non-irrigated areas. This not only conserves resources but also contributes to a healthier and weed-free plant environment.

Micro-irrigation also maintains plant health by keeping foliage dry, minimizing the risk of foliar diseases and reducing the need for chemical treatments. This sustainable and eco-friendly nature promotes both agricultural productivity and environmental stewardship. Micro-irrigation systems also offer environmental benefits, such as water conservation by minimizing water wastage and runoff, preventing soil erosion, and reducing energy consumption. Operating at lower pressure requirements than traditional irrigation methods, these systems are more energy-efficient, leading to reduced energy consumption and operational costs, making them an economically and ecologically sound choice for water resource management. Overall, micro-irrigation is a sustainable and eco-friendly approach to water resource conservation and sustainability.

3. COMPONENTS OF PRECISION IRRIGATION SYSTEMS

3.1. Sensors

Sensor is a specialized device or instrument used to detect and measure specific environmental factors relevant to the management of water resources in agriculture and horticulture. These sensors are designed to monitor variables such as soil moisture, weather conditions, soil temperature, leaf wetness, and water quality. The data collected by these sensors provides critical information for irrigation management, helping farmers and growers make informed decisions about when, where, and how much water to apply to crops. These sensors play a crucial role in optimizing water usage, enhancing crop health, and ensuring efficient and sustainable irrigation practices in agriculture.

Here is a visual representation of various agricultural sensors used in water management as shown in Figure 10. Each sensor is designed to provide critical information for efficient irrigation and crop management in agriculture and horticulture.

Figure 10. Micro-irrigation system components

3.1.1. Soil Moisture Sensors

Soil moisture sensors are essential for efficient irrigation management, providing data on water content in the root zone. Common types include volumetric soil moisture sensors, which measure soil volumetric water content, tensiometers, which measure soil water tension, Time Domain Reflectometry (TDR) sensors, which analyze electromagnetic pulses, capacitance sensors, and dielectric sensors.

Volumetric soil moisture sensors measure water content in a percentage, providing precise data on water availability. Tensiometers measure soil water tension, indicating water holding strength in the soil (Bhatt *et al.,* 2016). TDR sensors are used in fields like soil science, telecommunications, geophysics, and environmental monitoring, accurately measuring soil moisture content for efficient irrigation. Capacitance sensors estimate moisture content using changes in soil capacitance, used in precision irrigation systems. Dielectric sensors measure changes in soil dielectric properties due to moisture content, known for their accuracy and suitability for various soil types.

3.1.2. Weather Sensors

Weather sensors are essential tools for adjusting irrigation schedules based on environmental conditions. They collect data on temperature, humidity, wind speed, and solar radiation, which are crucial for understanding plant growth and water requirements. Temperature sensors measure air temperature, which affects irrigation needs. Relative humidity sensors measure moisture in the air, allowing for better understanding of plant transpiration and evaporation. Wind speed and direction sensors provide data on air movement, influencing irrigation scheduling. Rain gauges measure precipitation amounts, helping determine irrigation delays. Solar radiation sensors estimate sunlight reaching crops, estimating evapotranspiration rates and plant water requirements. Barometric pressure sensors track atmospheric pressure changes, indicating approaching weather systems and forecasting irrigation needs. Overall, these sensors are essential for ensuring optimal irrigation management.

3.1.3. Soil Temperature Sensors

Soil temperature sensors are crucial for understanding plant growth and water requirements. Types include resistive sensors, thermocouples, and resistors. Resistive sensors measure temperature by monitoring changes in electrical resistance, known for accuracy and reliability. Thermocouples, made of two metals, work by changing voltage with temperature, often used in rugged environments. Thermistors, temperature-sensitive resistors, have significant electrical resistance changes with temperature, allowing for precise temperature measurements.

3.1.4. Leaf Wetness Sensors

Leaf wetness sensors detect the presence of moisture on plant leaves, which can be an important indicator of disease risk and the need for irrigation. These sensors are specialized instruments used in agriculture and horticulture to monitor and measure the presence of moisture on plant leaves. These sensors play a crucial role in disease management, irrigation control, and overall plant health.

3.1.5. Water Quality Sensors

Water quality sensors are essential for assessing the chemical composition of irrigation water, ensuring it meets crop health standards. pH sensors measure the water's acidity or alkalinity, determining its suitability for irrigation and nutrient management. Electrical conductivity (EC) sensors measure the water's salinity and ion content, assessing potential soil and root damage. Turbidity sensors assess water clarity by measuring suspended particles and contaminants, evaluating water quality and its impact on irrigation systems. Dissolved oxygen sensors measure the water's dissolved oxygen concentration, assessing its ability to support aquatic life and microbial activity. Nutrient sensors detect the concentration of specific nutrients, such as nitrogen and phosphorus, in the water, crucial for fertigation management.

3.2. Controllers

Controllers serve as the primary control center for precision irrigation systems, effectively leveraging data obtained from sensors to make informed decisions regarding the optimal allocation of water to plants. The individuals in question bear the responsibility of overseeing irrigation schedule, controlling the flow of water, and administering fertilizer injection. The initiation of irrigation cycles is determined by considering factors such as soil moisture levels, prevailing weather conditions, and the specific requirements of the crops. In addition, valves and actuators are employed to manage the flow of water, facilitating its distribution to designated zones. In addition to their primary functions, contemporary controllers also include the capability of remote control and monitoring. This feature enables farmers to conveniently oversee and regulate the system through mobile applications or web interfaces.

4. DESIGN AND IMPLEMENTATION OF PRECISION IRRIGATION SYSTEMS

4.1. Design Considerations

4.1.1. Local Environmental Conditions, Crop Types, and Soil Types

Local climate conditions, such as temperature, humidity, and precipitation patterns, need to be considered to determine the irrigation needs. Different crop types have distinct water requirements. The system should be tailored to accommodate these variations. Soil types vary in their water-holding capacity and drainage properties. Understanding the specific soil conditions is essential to prevent over-irrigation or under-irrigation.

4.1.2. Proper Sensor Placement and Calibration

Sensor placement is crucial to collect accurate data. Soil moisture sensors should be positioned at various depths to monitor the entire root zone. Regular sensor calibration is essential to ensure data accuracy and to account for sensor drifting over time. Calibration involves comparing sensor readings with actual soil moisture levels.

4.1.3. Water Sources and Irrigation Methods

Identify and assess the available water sources, whether they're from wells, rivers, reservoirs, or reclaimed water. The source's quality and quantity can influence system design. Choose the appropriate irrigation method, such as drip irrigation, sprinklers, or pivot systems, based on factors like crop type, water availability, and infrastructure constraints.

4.2. Implementation Steps

4.2.1. Conduct a Comprehensive Site Assessment

A thorough site assessment involves gathering information about the specific characteristics of the land to be irrigated. This includes soil types, topography, and the presence of any physical barriers. Crop needs assessment should consider the growth stages, water requirements, and expected yield goals.

4.2.2. Select and Install Appropriate Sensors and Controllers

Based on the site assessment and crop requirements, choose the most suitable sensors (soil moisture, weather, or crop sensors) and controllers. Proper installation of sensors is essential to ensure they provide reliable data. Follow manufacturer guidelines and best practices.

4.2.3. Develop an Irrigation Schedule and Set Up Automation

Create an irrigation schedule that factors in real-time data from the sensors. This schedule should specify when and how much water to apply. Configure the automation system (controllers) to execute the irrigation schedule. Automation ensures that water is delivered precisely when and where needed, reducing human error.

4.2.4. Continuously Monitor and Adjust the System

Regularly monitor the sensor data to ensure that the system is functioning correctly. Use the sensor data and crop growth stage information to make necessary adjustments to the irrigation schedule. Periodically check and maintain the entire system, including sensors, controllers, and water delivery components. By adhering to these design considerations and implementing the outlined steps, a precision irrigation system can be effectively tailored to the specific needs of a given agricultural site. This, in turn, maximizes water efficiency, crop productivity, and sustainability while minimizing water wastage and environmental impact.

5. PRECISION IRRIGATION FOR SUSTAINABLE AGRICULTURE

5.1. The Role of Precision Irrigation in Sustainable Agriculture

Precision irrigation plays a significant role in promoting sustainable agriculture by improving the efficiency of water and resource use, reducing environmental impacts, and enhancing crop yield and quality. Precision irrigation technologies and practices significantly contribute to water conservation in agriculture and landscaping. These systems reduce water waste, promote efficient water use, and help mitigate the environmental and economic impacts of over-irrigation, ultimately playing a crucial role in addressing the challenges of water scarcity and sustainability. Drip and sprinkler systems deliver water directly to the root zone of plants, reducing water waste through evaporation and runoff. This efficient water use helps conserve this precious resource, especially in regions facing water scarcity and droughts. Precision irrigation minimizes the overuse of fertilizers and pesticides and delivers them straight to the root zone, which lowers water pollution, safeguards aquatic ecosystems, and contributes to safe drinking water. By doing this, runoff and leaching are less likely to damage non-target creatures and contaminate water systems. By preserving water quality, precision irrigation also helps to ensure the long-term viability of natural ecosystems and agriculture. Figure 11 showcases precision irrigation in sustainable agriculture, highlighting its effectiveness and environmental benefits. It features drip irrigation systems, efficient sprinklers, remote farmer control, and real-time soil sensors, reducing water waste and improving crop health.

Precision irrigation systems can deliver fertilizers and nutrients directly to the root zone of plants. This targeted nutrient application minimizes excess use of fertilizers, reducing the risk of nutrient runoff into water bodies and mitigating soil degradation. Precision irrigation systems often operate at lower pressure than traditional methods, reducing the energy required for water pumping. This energy efficiency contributes to cost savings and a lower carbon footprint. Maintaining consistent soil moisture levels through precision irrigation helps prevent both overwatering and underwatering. This promotes healthier plants, reduces stress, minimizes disease susceptibility, and ultimately leads to higher crop yields and quality. Precision irrigation systems rely on data from various sensors to create customized irrigation schedules based on real-time environmental conditions. This flexibility ensures that water is applied precisely when and where it is needed. Slow and even water delivery in precision irrigation

Figure 11. Precision irrigation

systems helps prevent soil erosion caused by rapid water flow. Soil erosion can degrade soil quality and negatively impact ecosystems.

Precision irrigation systems deliver water only to the crop's root zone, reducing weed growth in between rows. This decreases the need for herbicides and manual weeding. Many precision irrigation systems offer remote monitoring and control capabilities, allowing farmers to manage their irrigation systems from a distance, which is particularly valuable for large-scale operations. Precision irrigation relies on data from various sensors and monitoring tools. This data is used to make informed decisions, leading to more efficient and sustainable irrigation practices. Precision irrigation can help agriculture adapt to changing climate conditions, such as increased droughts and unpredictable weather patterns, by providing more precise control over water resources.

5.2. Challenges for Precision Irrigation Adoption

The adoption of precision irrigation in agriculture faces several challenges that can limit its widespread implementation. Precision irrigation systems often require a significant upfront investment in technology, including sensors, controllers, and infrastructure. The adoption of certain irrigation innovations can be challenging due to their high initial costs, greater knowledge and skill requirements, and increased time commitment (Morrison, *et al.,* 2009). The cost can be a barrier for small-scale and resource-constrained farmers, making adoption more challenging. To boost the adoption of micro-irrigation technology in developing countries, there is a requirement for robust institutional support, promotional efforts, subsidies for the products, and improved access to credit. These measures can encourage and facilitate the widespread use of this efficient irrigation technology, which can significantly benefit agriculture in these regions by conserving water and enhancing crop yields (Varma, *et al.,* 2006).

Effective use of precision irrigation systems demands a good understanding of the technology, data interpretation, and system management. Farmers may require training and technical expertise to operate and maintain these systems properly. Precision irrigation systems generate large volumes of data on soil moisture, weather conditions, and plant needs. Managing and analyzing this data can be overwhelming, and farmers may need support in harnessing its full potential. Retrofitting existing irrigation systems with precision technology can be complex and costly. Not all farms have the necessary infrastructure for seamless integration, which can impede adoption. The accuracy and reliability of sensors, such as soil moisture sensors, are critical for making informed decisions. Faulty sensors can lead to incorrect data, undermining the efficiency of precision irrigation. While precision irrigation can save water, it may increase energy consumption, particularly when water needs to be pumped over long distances or when energy-intensive sensors and controls are used.

Precision irrigation systems are not universally applicable and may need customization to suit the specific needs and conditions of each farm. This requirement can add complexity to adoption. Ongoing maintenance, sensor calibration, and software updates are crucial for the proper functioning of precision irrigation systems. Neglecting these tasks can lead to reduced system effectiveness. In some regions, water rights and regulations can restrict the use of water-saving technologies like precision irrigation. Farmers may need to navigate legal and regulatory obstacles to adopt these systems. In rural or remote areas, limited access to technology and internet connectivity can hinder the implementation of precision irrigation systems. Some farmers may be hesitant to adopt new technologies due to concerns about potential risks and uncertainties associated with changing their established irrigation practices.

6. WORLDWIDE APPLICATIONS OF PRECISION IRRIGATION SYSTEM

Precision irrigation systems have gained worldwide recognition, fundamentally transforming agricultural methods via the optimization of water use, enhancement of crop productivity, and encouragement of sustainable farming practices. The implementation of these technologies differs across different places, since it is impacted by elements like as climate, water availability, and economic circumstances. North America, Israel, India, and China have all adopted precision irrigation systems, with more than half of the irrigated land in the United States using these methods. Precision irrigation is of utmost importance in the wine business as it enables grape production to achieve a notable decrease of 20-30% in water consumption, all the while ensuring the maintenance or even enhancement of grape quality. Precision irrigation has been widely used in cotton farming, resulting in improved yields and fiber quality in precision-irrigated cotton fields.

Precision irrigation offers significant economic advantages, leading to higher farm profitability and sustainability. According to research conducted by the World Bank, farmers that use precision irrigation technology see a notable 20-50% rise in their net income as a result of enhanced crop yields and more efficient utilization of resources (Adeyemi *et al.*, 2017). The use of precision irrigation is in accordance with the objectives of sustainable agriculture, as emphasized by the International Food Policy Research Institute (IFPRI).

According to a survey, the worldwide precision irrigation market is expected to develop significantly, reaching a market value of USD 14.95 billion by 2028 (Jolly et al., 2023), report also forecasts a compound annual growth rate (CAGR) of 13.1% from 2021 to 2028. The use of this data-centric method is swiftly revolutionizing agricultural practices worldwide, addressing the issue of water shortages, enhancing crop production, and protecting the environment. Precision irrigation is a fast advancing technology that is revolutionizing farms worldwide. It addresses the issue of water shortage, enhances crop productivity, and protects the environment. Precision irrigation has several benefits that go beyond water saving, including higher agricultural yields, improved crop quality, less environmental footprint, and better profitability.

Precision irrigation is being used in many settings, ranging from dry areas to productive valleys, on several continents. Israel, known for its innovative approaches to addressing water constraint, has successfully implemented precision irrigation techniques, leading to significant improvements in water efficiency and agricultural productivity. Spain has acute drought conditions, but India is actively assisting farmers in enhancing the efficiency of rice and wheat cultivation, hence enhancing food security and improving the lives of farmers. With the continuous progress of technology, precision irrigation is expected to have a significant influence. Advancements like as artificial intelligence, automation, and remote monitoring provide the potential for enhanced optimization and immediate control, resulting in more effective water management. By adopting this intelligent strategy, we may foster a more environmentally friendly future in which every unit of water is valued, providing sustenance not just to crops but also to the populations who rely on them.

Precision irrigation offers many long-term advantages, including water conservation, higher crop yields, and less fertilizer runoff. Governments, commercial enterprises, and farmers globally are aggressively allocating resources towards the development and use of this technology, acknowledging its significant capacity for transformation. Precision irrigation provides a clear path towards a more sustainable and secure future for agriculture as we face the challenges of a rising population and climate change.

7. CHALLENGES AND FUTURE OUTCOMES

Precision irrigation is an innovative technology that transforms water management in agriculture, although it encounters several obstacles on its journey towards a sustainable future. The factors to consider include the significant upfront expenses, specialized knowledge, dependability and connection of data, compatibility with current systems, acceptance and education among farmers, and governmental assistance and financial backing. In order to address these difficulties, a comprehensive and multifaceted strategy is required. To facilitate the general implementation of precision irrigation, we may contribute by offering financial assistance, enhancing technical expertise, promoting innovation, and highlighting successful case studies. Consequently, this may enable a future in which every unit of water not only sustains crops but also promotes a more environmentally friendly and adaptable agricultural environment.

Financial obstacles provide a substantial hurdle. Farmers face a substantial financial obstacle because to the considerable upfront cost associated with precision irrigation systems, which include sensors, controls, and specialist equipment. Restricted implementation, particularly among small-scale farmers who may encounter financial difficulties in procuring the technology, resulting in inequitable access to sustainable agricultural methodologies. Technological intricacy poses a further hurdle, since precision irrigation systems include intricate technology like as sensors, data analytics, and automation, which may be challenging for farmers to comprehend and oversee. A further obstacle is the reliance of many precision irrigation systems on constant and cost-effective energy sources for pumps and automation. However, in some places, these energy sources may be unreliable or costly. The lack of inexpensive and dependable electricity in some places might lead to higher operating expenses and decreased feasibility of using precision irrigation techniques. Water shortage and pollution pose significant challenges to precision irrigation systems since they rely on a reliable and high-quality water supply.

Ensuring data security and privacy is an additional obstacle. Precision irrigation systems collect and retain intricate data on crop conditions, soil health, and water use, hence giving rise to issues over the security and privacy of this data. Farmers may exhibit reluctance in embracing the technology due to concerns around illegal access or improper use of their agricultural data. Potential remedies include cost mitigation tactics, streamlined user interfaces and training, energy-conserving devices, water conservation measures, and data governance and education. Collaboration among governments, NGOs, and private sector groups may facilitate the provision of financial incentives, subsidies, or financing alternatives to enhance the affordability of precision irrigation systems for farmers.

Potential future results include heightened acceptance and availability, higher efficiency in production and resource use, technical progress, enhanced water management, and a revolution in agriculture driven by data. Due to the implementation of cost reduction strategies and heightened awareness, the use of precision irrigation systems is expected to expand, hence increasing the accessibility of sustainable farming techniques to a greater number of farmers. To summarize, addressing the difficulties related to precision irrigation systems requires a comprehensive strategy that encompasses technology advancements, financial backing, and instructional programs. The potential future results provide the prospect of a more environmentally friendly and efficient agricultural environment, as long as focused endeavors are undertaken to overcome the current obstacles.

8. CONCLUSION

Precision irrigation, a sustainable water management approach tailored to crops' specific needs, offers a multitude of benefits, including enhanced crop yields, reduced water use, improved fertilizer efficiency, and minimized environmental impact. It encompasses various precision irrigation systems, like surface, sprinkler, drip, and micro-irrigation, with detailed insights into components and design considerations. Managing precision irrigation systems involves monitoring, data collection, scheduling, and maintenance.

The applications span crop irrigation, soil moisture, nutrient, and salinity management, along with frost protection. Economic and environmental advantages include increased yields, water conservation, efficient fertilizer use, and environmental sustainability. The chapter underscores precision irrigation's role in sustainable agriculture, emphasizing associated challenges and opportunities for adoption. Ultimately, precision irrigation holds promise for enhancing sustainable agricultural productivity and serves as a valuable resource for stakeholders in agriculture.

REFERENCES

Abioye, E. A., Abidin, M. S. Z., Mahmud, M. S. A., Buyamin, S., Ishak, M. H. I., Abd Rahman, M. K. I., ... Ramli, M. S. A. (2020). A review on monitoring and advanced control strategies for precision irrigation. *Computers and Electronics in Agriculture*, *173*, 105441. doi:10.1016/j.compag.2020.105441

Adeyemi, O., Grove, I., Peets, S., & Norton, T. (2017). Advanced monitoring and management systems for improving sustainability in precision irrigation. *Sustainability (Basel)*, *9*(3), 353. doi:10.3390/su9030353

Balafoutis, A., Beck, B., Fountas, S., Vangeyte, J., Van der Wal, T., Soto, I., ... Eory, V. (2017). Precision agriculture technologies positively contributing to GHG emissions mitigation, farm productivity and economics. *Sustainability (Basel)*, *9*(8), 1339. doi:10.3390/su9081339

Barragan, J., Bralts, V., & Wu, I. P. (2006). Assessment of emission uniformity for micro-irrigation design. *Biosystems Engineering*, *93*(1), 89–97. doi:10.1016/j.biosystemseng.2005.09.010

Bhatt, R., Arora, S., & Chew, C. C. (2016). Improving irrigation water productivity using Tensiometers. *Journal of Soil and Water Conservation*, *15*(2), 120–124.

Evans, R. G., & Waller, P. M. (2007). 8. Application of chemical materials. In *Developments in Agricultural Engineering* (Vol. 13, pp. 285–327). Elsevier.

García, L., Parra, L., Jimenez, J. M., Lloret, J., & Lorenz, P. (2020). IoT-based smart irrigation systems: An overview on the recent trends on sensors and IoT systems for irrigation in precision agriculture. *Sensors (Basel)*, *20*(4), 1042. doi:10.3390/s20041042 PMID:32075172

Haghnazari, F., Shahgholi, H., & Feizi, M. (2015). Factors affecting the infiltration of agricultural soils. *International Journal of Agronomy and Agricultural Research*, *6*(5), 21–35.

Jamroen, C., Komkum, P., Fongkerd, C., & Krongpha, W. (2020). An intelligent irrigation scheduling system using low-cost wireless sensor network toward sustainable and precision agriculture. *IEEE Access : Practical Innovations, Open Solutions*, *8*, 172756–172769. doi:10.1109/ACCESS.2020.3025590

Jin, J., & Tang, L. (2011). Coverage path planning on three-dimensional terrain for arable farming. *Journal of Field Robotics*, *28*(3), 424–440. doi:10.1002/rob.20388

Jolly, C. M., Nyandat, B., Yang, Z., Ridler, N., Matias, F., Zhang, Z., Murekezi, P., & Menezes, A. (2023). Dynamics of aquaculture governance. *Journal of the World Aquaculture Society*, *54*(2), 427–481. doi:10.1111/jwas.12967

Morrison, M. (2009). Encouraging the adoption of decision support systems by irrigators. *Rural Society*, *19*(1), 17–31. doi:10.5172/rsj.351.19.1.17

Smith, R., & Baillie, J. (2009, January). Defining precision irrigation: A new approach to irrigation management. In *Irrigation Australia 2009: Irrigation Australia Irrigation and Drainage Conference: Proceedings*. University of Southern Queensland.

Sokol, J., Amrose, S., Nangia, V., Talozi, S., Brownell, E., Montanaro, G., Abu Naser, K., Bany Mustafa, K., Bahri, A., Bouazzama, B., Bouizgaren, A., Mazahrih, N., Moussadek, R., Sikaoui, L., & Winter, A. G. (2019). Energy reduction and uniformity of low-pressure online drip irrigation emitters in field tests. *Water (Basel)*, *11*(6), 1195. doi:10.3390/w11061195

Tian, J., Ni, L., Song, T., Olson, J., & Zhao, J. (2018). An overview of operating parameters and conditions in hydrocyclones for enhanced separations. *Separation and Purification Technology*, *206*, 268–285. doi:10.1016/j.seppur.2018.06.015

Varma, S., & Namara, R. E. (2006). Promoting micro irrigation technologies that reduce poverty. *Water Policy Briefing*, *23*, 1–6.

Wen-Yong, W., Yan, H., Hong-Lu, L., Shi-Yang, Y., & Yong, N. (2015). Reclaimed water filtration efficiency and drip irrigation emitter performance with different combinations of sand and disc filters. *Irrigation and Drainage*, *64*(3), 362–369. doi:10.1002/ird.1909

Zhang, J., Guan, K., Peng, B., Jiang, C., Zhou, W., Yang, Y., Pan, M., Franz, T. E., Heeren, D. M., Rudnick, D. R., Abimbola, O., Kimm, H., Caylor, K., Good, S., Khanna, M., Gates, J., & Cai, Y. (2021). Challenges and opportunities in precision irrigation decision-support systems for center pivots. *Environmental Research Letters*, *16*(5), 053003. doi:10.1088/1748-9326/abe436

Chapter 11
Fabrication of Equipment and IoT-Assisted Monitoring for Enhanced Zero-Till Farming

P. V. Rajesh
https://orcid.org/0000-0002-4858-9032
Saranathan College of Engineering, India

J. Ashvin
Saranathan College of Engineering, India

M. Harish
Saranathan College of Engineering, India

G. Kishore
Saranathan College of Engineering, India

S. Nidheesh
Saranathan College of Engineering, India

ABSTRACT

As the global population continues to burgeon, sustainable and efficient agricultural practices are essential to meet the increasing demands for food production while safeguarding the environment. Zero-till farming, characterized by minimal soil disturbance during cultivation, offers a promising approach to achieve these objectives. The present methods of seed sowing are time consuming and inconvenient to handle. To overcome these difficulties, the present research work focuses on the development of specialized equipment and an internet of things (IoT)-based monitoring system to enhance zero-till farming practices. The equipment component of this study encompasses the development of a non-tilling seed sower with solid fertilizer sprayer, alongside field monitoring capabilities. This research contributes to the advancement of precision agriculture and the promotion of sustainable food production systems, ultimately addressing the pressing challenges of global food security and ecological preservation.

DOI: 10.4018/979-8-3693-4864-2.ch011

1. INTRODUCTION

The search for efficient and sustainable farming methods has become critical in the ever-changing agricultural landscape in order to meet the rising demand for food around the world. Zero-till farming has become a highly promising approach that minimizes the environmental impact of traditional tillage practices while simultaneously conserving soil moisture and reducing erosion. Using contemporary technologies more especially, the Internet of Things (IoT) offers a novel approach to the optimization and advancement of zero-till farming. The goal of this project is to close the gap between conventional farming methods and state-of-the-art technology by developing equipment specifically for zero-till farming and putting in place an Internet of Things-assisted monitoring system. By using a multidisciplinary approach, the project aims to make a substantial contribution to the agricultural domain by improving the accuracy and efficiency of zero-till farming. Over time, it has become clear that traditional agricultural practices, which involve a lot of plowing, are not environmentally sustainable (Oukaira, A., et al., 2021). Zero-till farming, which requires very little soil disturbance, has become popular because it can preserve soil structure, use less water, and lessen greenhouse gas emissions. But in order to fully realize the benefits of zero-till farming, specific machinery that can respond to a variety of agricultural environments is required. In order to create equipment specifically for zero-till farming, tools like planters, cultivators, and seed drills must be designed and developed to function with the least amount of disturbance to the soil. These sophisticated instruments are essential to contemporary agriculture since they maximize the sowing process while also conserving resources. Simultaneously, the assimilation of IoT technologies into agricultural processes has accelerated, providing the capacity for data-driven decision-making and real-time monitoring. In order to apply IoT to zero-till farming, sensors and networked devices must be placed throughout the farmland. These gadgets gather and send information to a centralized system on temperature, crop health, soil moisture content, and other important factors. The goal of this project is to build a synergy between manufacturing processes and IoT technologies to improve the overall effectiveness of zero-till farming. Farmers that use real-time monitoring are better equipped to plan ahead, use resources more efficiently, and react quickly to changing environmental circumstances. With the use of predictive analytics and the extensive information created by IoT, farmers can anticipate problems ahead of time and optimize harvests (Balaji, G. N., et al., 2018). This research has two main goals: first, to design, build, and optimize zero-till farming equipment that is compatible with a variety of crop types and agricultural landscapes; second, to deploy an Internet of Things (IoT)-assisted monitoring system that offers real-time data on crucial agricultural parameters, enabling precision farming techniques. These aims, which maximize total output, minimize environmental effect, and save resources, are in line with the larger mission of fostering sustainable agriculture. The study's importance stems from its potential to transform traditional farming methods through the integration of traditional knowledge with modern technology innovations. The production of specialized machinery meets the practical requirements of zero-till farmers by providing them with instruments engineered for maximum efficiency. Concurrently, the incorporation of IoT technology guarantees that these instruments function in accordance with the dynamic and developing character of agriculture. Furthermore, by offering a comprehensive method for precision farming, the work benefits the scholarly and research community. The research's multidisciplinary approach, which incorporates elements of mechanical engineering, electronics, and agronomy, encourages cooperation between several academic disciplines (Rajak, P., et al., 2023). The study methodology includes a methodical approach to the development of devices as well as the application of Internet of Things technology. Specialized implements are conceptualized and designed during the fabrication process, and then they are prototyped, tested, and optimized. Concurrently, the construction of a centralized monitoring system, networking solutions, and the selection and deployment of sensors are all part of the integration of IoT technologies. In order to make sure that the solutions created are workable, sustainable, and flexible enough to meet the wide range of needs of farmers, the implementation will be carried out in cooperation with the nearby farming communities. Because the process is iterative, it is possible to continuously develop it in response to end-user feedback and changes in the agricultural environment (Kumutha, S., et al., 2018). The following are expected results of this research: specialized machinery that improves zero-till farming's efficiency and adaptability; A functional IoT-enabled surveillance system offering up-to-date information for precision farming, enhanced local community acceptance of sustainable farming methods and resource optimization that produces positive effects on the economy and environment. This research has an influence that goes beyond academic circles; it has an impact on grassroots farming practices and adds to the worldwide conversation on sustainable food production. To sum up, the combination of manufacturing methods with Internet of Things-assisted observation has great potential for the advancement of zero-till agriculture (Annapurna, B., & Anusha, B., 2017). This study aims to contribute concretely to the progress of agriculture by tackling the practical issues faced by farmers through the manufacturing of specialized equipment and utilizing the Internet of Things for real-time monitoring. By means of inventiveness, cooperation, and multidisciplinary investigation, the path towards improved zero-till farming is anticipated to produce abundant crops and a sustainable future for worldwide agribusiness. Men and women work equally in agriculture, which is the primary occupation of the rural Indian population. For a very long time, agriculture has been and will continue to be the foundation of the Indian economy. It must provide for nearly 17% of the world's population using 2.3% of the planet's land area and 4.2% of its water resources. India's continuous expansion has always been supported by its agriculture. In light of these fertilizer and seed-sowing-related considerations, an effort is made to develop and build equipment that will enable both tasks to be completed more effectively and affordably. Reduce operating expenses by implementing a novel method (Rayhana, R., et al., 2021).

1.1 Overview of the Current State of Agricultural Technology

The state of agricultural technology today includes a wide variety of developments meant to improve sustainability and production. By using cutting-edge sensors, drones, and data analytics, precision agriculture enables farmers to maximize inputs like water, fertilizer, and pesticides, leading to more productivity and less environmental impact. Agro-forestry incorporates trees into agricultural systems, providing a number of advantages like improved biodiversity, preserved soil, and extra revenue streams. Through encouraging resource conservation, reducing waste, and increasing agricultural yields, these technologies are completely changing conventional farming methods. Modern agriculture technologies are promoting resistance against climate change and guaranteeing food security for future generations by integrating digital tools and environmental practices (Deshmukh, B., & Verma, D.,2018).

1.2 Environmental Sustainability

There is a lot of potential for the agricultural system to have an impact on the environment, both positively and negatively. Extensive monoculture farming is one practice that can worsen soil erosion, which reduces land productivity and causes the loss of fertile topsoil. Moreover, water bodies can become contaminated by fertilizer runoff from agricultural fields, which can damage aquatic ecosystems and cause eutrophication. Furthermore, raising cattle increases greenhouse gas emissions, particularly those of nitrous oxide and methane, which exacerbates climate change. However, by encouraging soil conservation, lowering fertilizer runoff, and sequestering carbon, sustainable farming techniques like conservation tillage, cover crops, and agro-forestry can help ameliorate these environmental challenges as well. Integrated farming systems that put ecosystem health and biodiversity first present viable ways to reduce agriculture's environmental impact while maintaining long-term sustainability (Sengodan, P., & Jbara, Y. H. F., 2017).

1.3 Food Security Aspects

In areas where access to agricultural resources is restricted, advanced agricultural technology has the potential to significantly increase both production and food security. In urban and desert places where traditional farming is difficult, food production is made possible by techniques like hydroponics, aquaponics, and vertical farming. Furthermore, by maximizing resource use, precision agriculture instruments boost yields and increase resistance to environmental stresses like pests and droughts. Food security can be strengthened by biotechnological breakthroughs, such as genetically modified crops, which can enhance agricultural qualities like drought resistance and nutritional value. These technologies guarantee that food security may be improved even in resource-constrained locations, supporting global efforts to reduce hunger by encouraging sustainable behaviours and increasing efficiency (Saravanan, K., et al., 2018).

1.4 Social Aspects of Agricultural Technology

The development of agricultural technology raises important social issues, especially those related to equality and economic effects. Small-scale farmers run the risk of being marginalized since they might find it difficult to acquire or adjust to expensive equipment and software, which could lead to a rise in socioeconomic inequality. Furthermore, a reliance on technology can replace conventional farming methods, causing farming communities to lose their cultural identity. In addition, unequal access to digital literacy and technology could worsen already-existing socioeconomic divides by preventing marginalized farmers from taking advantage of new developments in technology. Agricultural policy must place a high priority on inclusive innovation frameworks that guarantee fair access to technology, assist smallholders, and encourage environmentally friendly and culturally aware farming methods (Tăbăraşu, A. M., et al., 2021).

1.5 Current Trends in Agricultural Technology

The emphasis on efficiency, digitalization, and sustainability is expanding in agricultural technology. Through the use of precision agriculture techniques, such as GPS-guided machinery and sensor technology, farmers may reduce their environmental impact and boost efficiency by optimizing inputs like water, fertilizer, and pesticides. Utilizing IoT devices and farm management software to enable farmers to gather and evaluate data to enhance their operations is another trend in data-driven decision-making. Alternative farming techniques that provide answers for urban agriculture and resource-constrained settings, such as hydroponics, aquaponics, and vertical farming, are also gaining popularity. By fusing digital technologies and precision agriculture principles, the system under development is in line with these developments. It has sensors to track soil moisture and nutrient levels, giving farmers the information they need to make informed decisions about fertilization and irrigation. Moreover, the system makes use of cloud-based software for data administration and analysis, enabling farmers to increase productivity and optimize their agricultural methods. Its modular architecture also permits scalability, which makes it appropriate for a variety of agricultural contexts, including urban and small-scale farming operations. All things considered, the system's combination of sustainability, digitization, and accuracy puts it at the forefront of current agricultural technology developments and presents creative ways to improve farming's sustainability and productivity (Tăbăraşu, A. M., et al., 2021).

2. BACKGROUND TO THE STUDY

2.1 Seed Sowing Machine

A seed sowing machine is a tool that helps farmers save time and money by assisting with the proper seed sowing position. The main goals of the sowing process are to arrange the seeds in rows at the appropriate depth and distance from one another, cover the seeds with soil, and ensure that the soil is properly compacted over the seeds. The discussion of various seed-sowing machine aspects in this project will aid in the agriculture industry's transition to mechanization. India's consistent growth has always been supported by its agriculture sector. The demand for food rises in tandem with India's growing population (Sagar, M., et al., 2020). Therefore, the need for advances in seed sowing devices is increasing. Farms are growing multiple crops, which calls for powerful machinery with large capacities. Because of a lack of expertise and access to sophisticated equipment and tools, mechanization of the agricultural sector in India is still in its infancy. Traditional seed sowing techniques involve broadcasting by hand, using a plough to make furrows, then physically distributing the seeds. India's consistent growth has always been supported by its agriculture sector. The demand for food rises in tandem with India's growing population. As a result, farmers are required to plant multiple crops, which calls for effective and time-saving machinery. The article addresses various seed-planting equipment types that will aid in the agricultural sector's transition to mechanization (Annapurna, B., & Anusha, B., 2017).

2.2 Traditional Sowing Methods

Conventional techniques involve manual broadcasting, using a country plough to create furrows, and manually distributing seeds into the furrow using a bamboo or metal funnel fastened to the plough. Dibbling, or creating holes or slits with a stick or other object and manually dropping seeds, is a common method of sowing in tiny areas. Expert farmers use multi-row traditional seeding equipment with manual seed metering, which are highly popular. It is impossible to ensure consistent seed dispersal when sowing by hand. Even if a farmer sows at the appropriate seed rate, there may still be gaps and bunching in the field due to unequal seed distribution both within and between rows. The following are the drawbacks of traditional sowing techniques are It is impossible to ensure consistent seed dispersal when sowing by hand. Even if a farmer sows at the appropriate seed rate, there may be irregular seed distribution both inside and between rows, which could lead to bunching and gaps in the field. inadequate management of seed planting depth. The labour need is considerable because dumping seed and seed requires two people. When it comes to crops, the impact of inaccurate seed distribution on plant stand is larger. India is expected to be mostly an agricultural nation, with 75% of its people living either directly or indirectly from farming. For centuries, our farmers have been employing the same tools and techniques. For instance, weeding, spraying, and seeding. The majority of Asia's emerging nations struggle with having a higher population density than developed ones and having less productive land. Low levels of farm mechanization and insufficient power supply are two major causes of low productivity. This particularly applies to India. Globally, it is increasingly acknowledged that agriculture must modernize in order to satisfy the demands of an expanding population and accelerating industrialization (Cucho-Padin, G., et al., 2020). It is stated that poor seedbed preparation and postponed planting, harvesting, and threshing cause production issues on many farms. Through accurate metering that ensures better distribution, lowering the quantity required for better response, and preventing losses or wasting of applied inputs, mechanization makes it possible to conserve inputs. Mechanization lowers manufacturing costs per unit by increasing output and conserving inputs. The government's program for agricultural machinery and implements has focused on selective mechanization in order to maximize the utilization of human, animal, and other power sources. The supply of implements, tractors, power tillers, combine harvesters, and other power-operated machinery was increased, as was the production and availability of improved animal-drawn implements, in order to fulfil the requirements. The latter was given particular attention because over 70% of farmers are classified as small and marginal farmers. It's commonly believed that small farms are challenging to mechanize. Japan's typical land holding is much lower than ours, yet with the right mechanization, agriculture there has reached unprecedented heights. To reduce the amount of work that small farmers have to do, boost productivity, and free up time for them to engage in other income-generating activities, it is necessary to appropriately encourage the use of contemporary time-saving devices and machinery of the right size (Swapnil, T.,et al., 2018).

2.3 Fertilizer Spreader Machine

Spreaders of fertilizer, sometimes referred to as "broadcast seeders" or "broadcast spreaders." A farm tool called a fertilizer spreader is typically used to spread sand, fertilizer, lime, ice melt, seed, and, in an alternate application, drop spreaders. By using mass flow controllers to monitor the amount of fertilizer needed for each subplot and calibration systems to regulate fertilizer capacity, it is possible to increase ramp up production and crop yields. The rotary or broadcast, drop, and liquid or spray spreaders are the three main varieties of fertilizer spreaders. Every one of these spreader kinds has benefits, depending on what you need to disseminate. Granular fertilizer and other dry fertilizers with uniform particle sizes are spread using a rotary spreader. In addition to spreading salt on winter ice and reseeding lawns, rotary spreaders balance out soil covering. Fertilizer is discharged from the rotating fertilizer spreader onto a disk that propels itself ahead of the machine in a semicircle. There are spreaders in a variety of sizes, from compact handheld devices for tiny gardens to bigger push-along models for huge yards. Because handheld devices allow for a higher degree of accuracy, they are perfect for compact spaces and places near pathways and patios (Rao, V. V., et al., 2013). Even coverage can only be achieved by steadily turning a hand crank. Since push versions may spread fertilizer in an arc up to three feet around the hopper, they work best in more open spaces. When utilizing this technique, you must walk steadily to prevent spreading fertilizer to different locations unevenly. Granular fertilizer and uneven materials like mulch or compost can be applied with a drop spreader. Given that it needs to be worked in straight, overlapping lines, free space is necessary for optimal outcomes. You cover half of the preceding pass with a drop fertiliser spreader on each pass. Usually, to do this, a marker is matched with the wheel track from the preceding pass. The "drop" spreader gets its name from the way these fertilizers operate fertilizer is forced out of a hole in the hopper bottom and onto the ground below. For drop fertilizer spreaders, push-along and pull-along types are the two most popular sizes. When the grass is being mowed, pull-along spreaders that are attached to riding mowers distribute fertilizer. Because of this, they are the best option for applying mulch and grass clippings from a mower bag evenly. In narrow spaces where a riding mower cannot even pass, push models work well. The way liquid fertilizer spreaders operate is by fitting a garden hose with an adapter nozzle onto a canister. The fertilizer in the canister is dry and meant to dissolve quickly. The adapter nozzle's end is then used to spray the fertilizer. While liquid spreaders are useful for rapid, light fertilization tasks, they fall short of rotary or drop spreaders in terms of accuracy and uniformity of application (Kulkarni, S. R., et al., 2015).

2.4 Need of Fertilizer Spreader

Fertilizer application is necessary to provide a decent yield of sugarcane. Fertilizers give plants the key chemical elements nitrogen, phosphorus, and potassium-that they require to flourish. Solid chemical fertilizers are one of the most significant sources of plant nutrition because they are less expensive than liquid fertilizers. In addition to giving plants the essential nutrients they need to grow throughout their life cycles, solid fertilizers also work to improve the characteristics of the soil, such as the degree of acidity and soil structure. Fertilizer application is highly necessary, particularly in areas with poor soil, to preserve essential ingredients for raising crop yield and quantity. A fertilizer spreader is a technological device that has been upgraded to distribute fertilizer for optimal outcomes. Our current method is less time-consuming, more comfortable, and poses no health risks than the conventional way of distributing fertilizer by hand, which is more tedious, time-consuming, and hazardous (Kumbhare, D. N., et al., 2016).

2.5 Internet of Things (IoT)

The term "Internet of things" (IoT) describes the idea of expanding Internet connectivity into a wide range of conventionally "dumb" or non-internet-enabled physical items and daily objects, in addition to traditional computing platforms like personal computers and mobile devices. These devices may be remotely monitored and controlled, and they can communicate and interact with others over the Internet thanks to their embedded electronics, Internet connectivity, and additional hardware (such sensors). Due to the confluence of several technologies, real-time analytics, machine learning, commodity sensors, and systems, the concept of the Internet of things has changed. The Internet of things is made possible by the traditional domains of embedded systems, wireless sensor networks, control systems, automation (including automation of homes and buildings), and others. Simplified, an IoT system architecture is composed of three tiers: Tier 1 is made up of devices, Tier 2 is the edge gateway, and Tier 3 is the cloud. Devices are networked objects, like the sensors and actuators used in Internet of Things (IoT) equipment, especially those that link to an Edge Gateway via proprietary or Modbus protocols (Sengodan, P., & Jbara, Y. H. F., 2017).The Edge Gateway is made up of systems for aggregating sensor data called Edge Gateways. These systems can do pre-processing, secure cloud communication, use Web Socket, the event hub, and, in some situations, edge analytics or fog computing. The cloud application designed for the Internet of Things (IoT) utilizing the micro services architecture is included in the last layer. Micro services are typically polyglot and intrinsically secure using HTTPS/OAuth. It encompasses different database systems that use backend data storage systems (like Cassandra and Postgres) to store sensor data, including time series databases or asset stores. Most cloud-based IoT systems have a cloud tier with event queuing and messaging systems to manage communication across all tiers. To put it succinctly, an Internet of Things ecosystem is made up of web-enabled smart devices that gather, transmit, and act upon data they obtain from their surroundings through the use of embedded computers, sensors, and communication gear (Solanki, S. N., et al., 2022).

2.6 Usage of IoT in Mechanical Systems

To meet an increasing demand for specifications, the development of sophisticated mechanical systems and production has been continuously pushed in recent years. because there is a need for products with higher and more consistent quality, greater product variety, lower costs, products that are competitive worldwide, and shorter product life cycles. Connecting sensors, robotics, and devices is made possible by the Internet of Things (IoT), which advances data transfer and storage. To put it another way, the rise of connected production floors has reduced the need for humans in the workplace by giving them more authority over choices, manufacturing processes, and even maintenance. More and more value is being placed on the Big Data produced by the manufacturing process than on the hardware it produces. But, in order to utilize it, producers need to separate out the relevant data from Big Data. As a result, in addition to big data, manufacturers also need the capacity to handle their data, transform it into actionable information, and increase their profit margin. Therefore, the goal of our project is to assess air temperature and track the seed and fertilizer in their corresponding hoppers using Internet of Things devices (Liu, W., 2023).

3. RESEARCH QUESTIONS

The real time problems which are being faced by existing seed sowing and fertilizer spraying machines while farming are as follows:

- The Major Problem is tilling the farm land before farming which is a tedious process.
- Various sizes of seeds cannot be sowed seamlessly unless it is equipped with suitable indexing mechanism.
- Complicated IoT systems are used which is bit complicate to rectify any problem occurs.
- Still there is a traditional method of monitoring the lands which is to be eliminated.
- Bigger size seeds cannot be injected at required depth.
- For Integrated farming (seed sowing and fertilizer spraying) there is no equipment is existing.
- The dispersing length of the solid fertilizer cannot be varied.

4. THEORETICAL FRAMEWORK

The theoretical framework is given in the form of flow chart in Fig.1

Figure 1. Theoretical framework

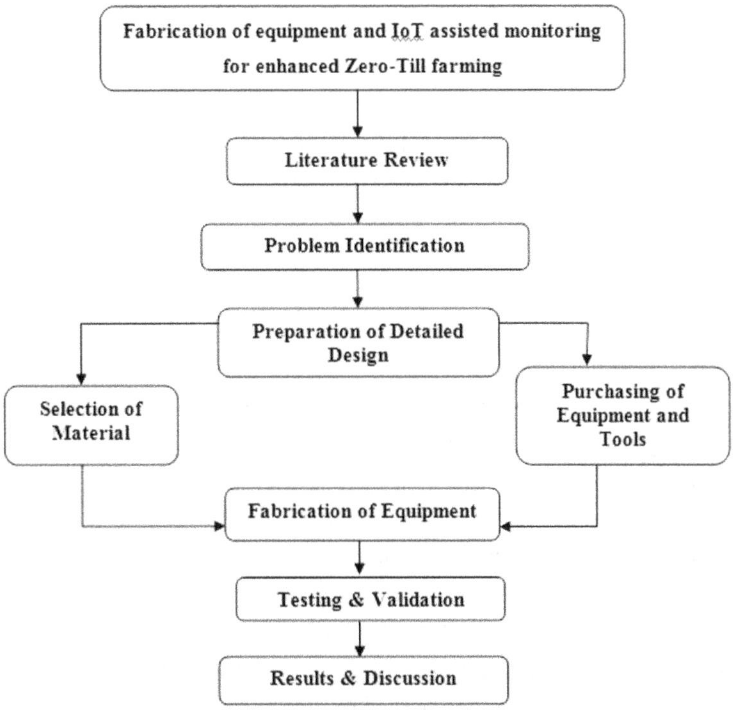

5. IN DEPTH LITERATURE DISCUSSION

Bhushan Deshmukh et al. fabricated a seed sowing machine which has a DC motor as a prime mover. It consists of a seed metering disc with a sprocket mechanism and seed metering mechanism with a rectangular metering disc to control the flow of seeds. A seed pipe is used for transferring the seed to the furrow opener. Furrow opener is a triangular curved shape cutting edge made up of metal. Furrow opener is used for ploughing the soil. Furrow closer employed to cover the seed with soil. 89C51 micro-controller used for controlling the machine. Here they developed an assembly for cultivating, ploughing land automatically. It requires less power. They eliminated the blocking of seed problem with the help of water pressure.

Akshatha et al. developed a Solar Energy based IoT Embedded Multipurpose Agricultural Bot .That system mainly consist of the seeding, ploughing, mud levelling and plant cutting functions. The ultra-sonic sensor device is utilized to detect the obstacle present in the surroundings. LCD monitor is used to display the error message and the corresponding functionality names. The IR sensor is used to sense the seeds in the seed container. Their agricultural bot used towards investigating the utilization of little and lightweight machines in the other agricultural operations.

G.Naveen Balaji et al. developed an IoT Based Smart Crop Monitoring in Farm Land. They used thermal imaging for irrigation in the crop field. Thermal imaging is used to measure stomatal conductance, plant temperature and evapo-transpiration rate by determining stomatal responses. Arduino microcontroller ATmega328 is used for controlling the IoT. They proposed a method for efficient crop monitoring for agricultural field. With the application of IoT the data can be stored and retrieved from anywhere.

Aditya Vishwas Kanade et al. developed an IoT Controlled Agri-Rover for Automatic Seeding. the control system of Agri-Rover, Android Smartphone is connected to the Blynk server. ESP8266-12 Wi-Fi module is used to connect to the server by using unique key provided by the Blynk server. They used ATMEGA 328 as the system MCU. The ATmega328 used by them has 32 KB with 2 KB of SRAM and 1 KB of EEPROM. Here that concluded IoT enabled Agri-Rover with newly developed Flap-servomotor mechanism provides extensive advantage over the existing mechanisms.

K.Saravanan et al. fabricated an automatic seed sowing robot for agricultural field. Chickpeas Seed hopper is used as a Storage device of the system. Four holes in the digging plates are used to dig the land and to spray the seeds. The PIC microcontroller PIC16F877A microcontroller is used in the robot.

B.Mohan et al. fabricated electrical based seed sowing machine. In this machine, for each complete rotation of rotating wheel there is seed falls from this seed drum and the seed plantation process can take place smoothly as well as without wastage of seeds. The end of system machine reached and it create an alarm. This system provides information that the storage tank is empty. the facility helps the farmer sow the seeds very easily. They used Zn-Mn batteries as a power source for dc motor. This project can also be implemented for other kind of crops such as fruits, paddy, sugarcane. It does not need any operator.

B. Annapurna et al. developed an Automated Seed Sowing Machine Using Atmega2560.the The ATMEGA2560 used in seed sowing machine board has 54 digital I/O ports (15 can be used for PWM outputs), 16 analog inputs, 4 UARTs (hardware serial ports), USB connection and 16 MHz crystal oscillator, power jack, ICSP header and reset button. The board also has an IR sensors attached at the front wheels to find obstacles. here they concluded that the seed farmstead instrument present in the system has great probability for expanding the fertility of the plantation.

Thorat Swapnil V et al. developed a Seed Sowing Machine. In that system, the seeds falling from tank is varied according to requirements by varying the disc opening angle. Seed metering mechanism is used for Metering of the seed to be sowed at a required rate. (e.g. kg/ha or seeds/meter of row length). Metering should be accurate as per the requirements. It also ensures that there should not be any damage to the seeds during metering. Here they concluded that their seed plantation machine has great potential for increasing the productivity of the planting. It also cost-effective method.

Varikuti Vasantha Rao et al. fabricated Multiple Power Supplied Fertilizer Sprayer. Its biggest problem is charging time .it needs 16.67 hours to charge fully. But it can be used for 6 acres of land. Which was 575 litres of pesticides. Existing hand driven sprayer needs high man power and fuel operated sprayer pollutes the environment. The system has PIC16F72 microcontroller for effective functioning. Here the researchers concluded that the Multiple Power Supplied Fertilizer Sprayer has an advantage over hand driven sprayer by reducing manpower. It does not need fuel like fuel driven sprayer.

S R Kulkarni et al. developed Foot Operated Agricultural Fertilizers and Pesticides Spraying Pump. It converts kinetic energy to spray fertilizer. it needs manpower to spray. it works when foot piston pump is pressed. Developed The foot walking pesticide spraying machine is fitted on the foot activated by kinetic force being applied on it, when the foot is stamped on the ground, it pushes the assembly to pump air into the tank, the power is very less but sufficient enough to use the pressure for pesticide spraying

Bhushan Deshmukh and Durgesh Verma fabricated a seed sowing machine which has a DC motor as a prime mover. Seed metering disc consists of flat rod. It has 8 holes around the circumference of the circle at equally spaced distance and it is used for sowing beans. The same circular rod having 5 holes equally spaced in its circumference can be used for sowing maize. It consists of a seed metering disc with a sprocket mechanism and seed metering mechanism with a rectangular metering disc to control the flow of seeds. A seed pipe is used for transferring the seed to the furrow opener. Furrow opener is a triangular curved shape cutting edge made up of metal. Furrow opener is used for ploughing the soil. Furrow opener is provided with a sharp cutting edge to cut the soil. The concave curved portion of the plough is forward and the convex part is backward. It is a two way plough like opener. The angle between two wings of the openers is 135°. It makes an angle of 45° with horizontal which facilitates the easy penetration into the soil. Furrow closer employed to cover the seed with soil. Furrow Closer is made of mild steel sheet. It is cut into the form of crescent shape. It is curved. The concave part is backside. The angle between wings is 135°. It facilitates the soil to cover the seed. The D.C. motor is operated on 12V with 100 rpm. 89C51 microcontroller used for controlling the machine. Relay driver and display are interfaced to the various ports of 89C51.theydeveloped an assembly for cultivating, ploughing land automatically. It requires less power. They eliminated the blocking of seed problem with the help of water pressure. The model has been proposed with the aim to establish a 12kg per hour capacity seed processing plant.

Akshatha T and Bharathi Malakreddy A developed a Solar Energy based IoT Embedded multipurpose Agricultural Bot .That system mainly consist of the seeding, ploughing, mud levelling and plant cutting functions. The ultrasonic sensor device is utilized to detect the obstacle present in the surroundings. The robotic system and the remote framework are associated via networking framework.

DC motors are utilized for route of the robotic system. DC motors speed can be managed by utilizing controller. The solenoid is utilized to control the seeding operation and treating.LCD monitor is used to display the error message and the corresponding functionality names. The IR sensor is used to sense the seeds in the seed container. Their agricultural Bot used towards investigating the utilization of little and lightweight machines in the other agricultural operations.

G. Naveen Balaji, V. Nandhini, S. Mithra, N. Priya and R. Naveena developed an IoT Based Smart Crop Monitoring in Farm Land. They used thermal imaging for irrigation in the crop field. They said that there is no need for modifications in the surface temperatures when thermal imaging technique is used which is a noncontact and nonintrusive technique. Water stress, gas exchange, evapo-transpiration rate stomatal conductance. Thermal imaging is used to measure stomatal conductance, plant temperature and evapo-transpiration rate by determining stomatal responses. The advantage of using thermal imaging is that thermal imaging can provide temperature value of all pixels in the field when compared to thermometry which only provides an average value. Image processing and data analytic techniques are used in thermal imaging for reducing crop water stress and to provide scheduled irrigation. The different sensors used In their system are temperature and humidity sensor and soil moisture sensor. Arduino microcontroller ATmega328 is used for controlling the IoT. The information collected by the sensors in their system is sent to the Arduino microcontroller ATmega328.The collected information is displayed in a LCD display. A webpage is created by them and the information collected by the sensors is updated periodically in it through Wi-Fi. A GSM module is connected with the microcontroller through which the message about the farm condition is sent to the authorized person. Soil moisture sensor used in their system is YL-69. The DHT11 digital temperature and humidity sensor was used by them to collect data related to temperature. They proposed a method for efficient crop monitoring for agricultural field. With the application of IoT the data's can be stored and retrieved from anywhere. They concluded in that proposed work, the sensor part is limited only for monitoring of crops hence in future it can be automated for irrigation and the system can be enhanced with security of farmland under video surveillance which prevents it from obtrude intrusion.

Aditya Vishwas Kanade, Arockia Selvakumar A, Dnyanesh Jalamkar, Prabaharan Sengodan and Yousef Hasan F. Jbara developed an IoT Controlled Agri-Rover for Automatic Seeding. They said that

An Agri-Rover is the rover which can accomplish agricultural tasks. Different types of rovers can be used to do functions like spraying, fruit picking etc in the farm. The robotic system developed by them will navigate autonomously into rubbles and to search for living human body heat using its thermal array sensor. The developed system uses a simple and less complex mechanism for the seed sowing. Their mechanism contains a tapered feeder, a servomotor driven by the ARDUINO controller and one flap having opening. Arduino program is developed in such way that it Arduino UNO is used to control angular speed of the motor and ultimately the opening and closing of the container. The time delay for opening and closing of the feeder can be adjusted by varying the Arduino programming. Using different time delay seed falling distance can be varied in the actual environment. Considering variation in diameter of seeds ranging from 2.5 mm to 9 mm, the feeder has given 13 mm hole at the end. the control system of Agri-Rover, Android Smartphone is connected to the Blynk server. ESP8266-12 Wi-Fi module is used to connect to the server by using unique key provided by the Blynk server. They used ATMEGA 328 as the system MCU. The ATmega328 used by them has 32 KB with 2 KB of SRAM and 1 KB of EEPROM. They concluded IoT enabled Agri-Rover with newly developed Flap-servomotor mechanism provides extensive advantage over the existing mechanisms.

K.Saravanan, S.P. Sundar Singh Sivam, S.Rajendra Kumar and K.Sathiya Moorthy fabricated an automatic seed sowing robot for agricultural field. The main purpose of seed sowing disc is to dropping the seeds into the field. From the surveys to calculate the distance between two seeds, they have make conclusion that the approximate distance between two seeds is normally 2.5 to 3 inches. Chickpeas Seed hopper is used as a Storage device of the system. Digging plates are used to dig the land and to spray the seeds. There are four holes in the plate. It is used to mount with rod. Digging plates are made of mild steel. These plates are mounted in front of the seed container and coupled in the rod with particular distance using bolts and nuts. By adjusting the bolts and nuts we can vary the position of plates. The PIC microcontroller PIC16F877A microcontroller is used in the robot. It has a total number of 40 pins and there are 33 pins for input and output. PIC16F877A is used in many PIC microcontroller projects. PIC16F877A also have much application in digital electronics circuits.PIC16F877A finds its applications in a huge number of devices. They concluded that robot can not exactly able to sow in straight line due to high torque and employed bearings.

B.Mohan, M.V.Ramesh and T. Srinivasa Rao fabricated electrical based seed sewing machine. In this machine, for each complete rotation of rotating wheel there is seed falls from this seed drum and the seed plantation process can taken place smoothly as well as without wastage of seeds. The end of system machine reached and it create an alarm. This system provides information that the storage tank is empty. the facility helps the farmer sow the seeds very easily. They used Zn-Mn batteries as a power source for dc motor. Their project can also be implemented for other kind of crops such as fruits, paddy, and sugarcane. It does not need any operator.

B. Annapurna, B. Anusha, S.V.S Prasad and B. Somanaidu developed an Automated Seed Sowing Machine Using Atmega2560.the TheATMEGA2560 used in seed sowing machine board has 54 digital I/O ports (15 can be used for PWM outputs), 16 analog inputs, 4 UARTs (hardware serial ports), USB connection and 16 MHz crystal oscillator, power jack, ICSP header and reset button. They have used motor driver L2930 which drives the shaft motor by coupling with battery bank. As motor starts rotating this vehicle starts moving. Seed storage tank is coupled at finest of module nearby wheels. The board also has an IR sensors attached at the front wheels to find obstacles. They concluded that the seed farmstead instrument present in the system has great probability for expanding the fertility of the plantation.

Varikuti Vasantha Rao, Sharanakumar Mathapati and Dr. Basavaraj Amarapur fabricated Multiple Power Supplied Fertilizer Sprayer. Its biggest problem is charging time .it needs 16.67 hours to charge fully. But it can be used for 6 acres of land. Which was 575 litres of pesticides. Existing hand driven sprayer needs high man power and fuel operated sprayer pollutes the environment. The system has PIC16F72 microcontroller for effective functioning. They concluded that the Multiple Power Supplied Fertilizer Sprayer has an advantage over hand driven sprayer by reducing manpower. It does not need fuel like fuel driven sprayer.

6. DISCUSSION

6.1 Selection of Construction Materials

Basically our model is divided in to two types of components one is Mechanical components and other is Electronic components where these components are selected and fabricated with respect to our requirements and consideration of certain constrains which is to be followed universally. When we consider about our model mostly 90% of the components are mechanical parts and they are fabricated by suitable processes such as cutting, welding, grinding etc.

6.1.1 Mechanical Components

External frame - As a supporting member for our model angle rod (L-bar) is used with required dimensions. The material for that angle frame is mild steel.

Supporting rod - Mild Steel Rod of diameter 15mm is used as a supporting element in between every component in our model as per requirement.

Hopper - It is used for storing of seed and solid fertilizers we have used hopper like structure which is made up of Galvanized iron sheet of 20 gauge thickness. The requirements of plain galvanized steel sheets and strips (coils) and corrugated galvanized sheets shall conform to the IS 277: 2003. Galvanized steel sheets covered by this standard are intended to be used for purposes such as panelling, roofing, lock forming etc.

Pneumatics - 'T joint' is used for distributing of air to respective lines that is seed outlet and fertilizers outlet. The flow control valve is for controlling of flow of air in their respective lines. Air hose(blue hose) of diameter 10 mm is used as a connecting hose in-between compressor and respective outlets.

Compressors - Here in our model compressed air is used to two inputs one is for seed sowing and the other is for fertilizer spraying. Separate controlling valves is provided to controlling of air inlet separately

PVC Components - Ball valve, PVC Connecting Pipes, PVC Tee Joint, PVC Reducer

6.1.2 Electronic Components

In our model there is no relation between mechanical and electronic components because only 10% is been contributed with electronic components such as sensors which is going to monitor certain parameters.

Hardware periphery - The following are the Hardware used in the project such as Simple link Wi-Fi cc3200 launchpad and Ultrasonic sensor.

Simple link Wi-Fi cc3200 launchpad - Start your design with the industry's first Wi-Fi CERTIFIED single-chip microcontroller unit (MCU) with built-in Wi-Fi connectivity. Created for the Internet of Things (IoT), the Simple link CC3200 device is a wireless MCU that integrates a high-performance ARM Cortex-M4 MCU, allowing customers to develop an entire application with a single IC. With on-chip Wi-Fi, Internet, and robust security protocols, no prior Wi- Fi experience is required for faster development. The CC3200 device is a complete platform solution including software, sample applications, tools, user and programming guides, reference designs, and the TI E2E™ support community. The device is available in a QFN package that is easy to layout. The applications MCU subsystem contains an industry-standard ARM Cortex-M4 core running at 80 MHz. The device includes a wide variety of peripherals, including a fast parallel camera interface, I2S, SD/MMC, UART, SPI, I2C, and four-channel ADC. The CC3200 family includes flexible embedded RAM for code and data and ROM with external serial flash bootloader and peripheral drivers. The Wi-Fi network processor subsystem features a Wi-Fi Internet-on-a-Chip and contains an additional dedicated ARM MCU that completely offloads the applications MCU. This subsystem includes an 802.11 b/g/n radio, baseband, and MAC with a powerful crypto engine for fast, secure Internet connections with 256-bit encryption. The CC3200 device supports Station, Access Point, and Wi-Fi Direct modes. The device also supports WPA2 personal and enterprise security and WPS 2.0. The Wi-Fi Internet-on-a-chip includes embedded TCP/IP and TLS/SSL stacks, HTTP server, and multiple Internet protocols. The power-management subsystem includes integrated DC-DC converters supporting a wide range of supply voltages. This subsystem enables low-power consumption modes, such as the hibernate with RTC mode requiring less than 4 μA of current.

Features - CC3200 Wi-Fi wireless MCU. USB interface to PC for CCS/IAR using FTDI USB drivers. Flash update over the USB using Simple Link Programmer 2 20-pin connectors enables compatibility with other TI Launchpads with added functions. Standalone development platform featuring sensors, LEDs and push-buttons. Power from USB for the Launchpad as well as external Booster Pack Operates from 2 AA alkaline batteries. On-board antenna and U.FL connector selectable using a capacitor re-work. Supports 4 wire JTAG and 2 Wire SWD. GNU Debugger (GDB) support over Open On chip debugger (OpenOCD)

Processor - The high-performance ARM Cortex-M4 processor provides a low-cost platform that meets the needs of minimal memory implementation, reduced pin count, and low power consumption, while delivering outstanding computational performance and exceptional system response to interrupts. The ARM Cortex-M4 core has low-latency interrupt processing with the following features, A 32-bit ARM Cortex Thumb® instruction set optimized for embedded applications. Handler and thread modes. Low-latency interrupt handling by automatic processor state saving and restoration during entry and exit. Support for ARMv6 unaligned accesses

Nested vectored interrupt controller (NVIC) closely integrated with the processor core to achieve low latency interrupt processing.

Wake-up interrupt controller (WIC) providing ultra-low power sleep mode support.

Bus interfaces - Three advanced high-performance bus (AHB-Lite) interfaces: ICode, DCode, and system bus.

Low-cost debug solution featuring - Debug access to all memory and registers in the system, including access to memory-mapped devices, access to internal core registers when the core is halted, and access to debug control registers even while SYSRESET is asserted. Serial wire debugs port (SW-DP) or serial wire JTAG debug port (SWJ-DP) debug access. Flash patch and breakpoint (FPB) unit to implement breakpoints and code patches.

External memory requirement -The CC3200 device maintains a proprietary file system on the SFLASH. The CC3200 file system stores the service pack file, system files, configuration files, certificate files, web page files, and user files. By using a format command through the API, users can provide the total size allocated for the file system.

Internal memory - The CC3200 device includes on-chip SRAM to which application programs are downloaded and executed. The application developer must share the SRAM for code and data. To select the appropriate SRAM configuration, see the device variants listed in the orderable addendum at the end of this datasheet. Themicro direct memory access (μDMA) controller can transfer data to and from SRAM and various peripherals.

ROM - The internal zero-wait-state ROM of the CC3200 device is at address 0x0000 0000 of the device memory and programmed with the following components are Bootloader and Peripheral driver library (DriverLib) release for product-specific peripherals and interfaces. The bootloader is used as an initial program loader (when the serial flash memory is empty). The CC3200 DriverLib software library controls on-chip peripherals with a bootloader capability.

Ultrasonic sensor - Ultrasonic Ranging Module HC - SR04 λ.Two sensors were used, one is to detect the amount of filament and other alerts technician on cut off of filament while running through the 3D printer. The specifications are Working Voltage: DC 5V, Working Current: 15mA, Working Frequency: 40Hz, Max Range: 4m, Min Range: 2cm, Measuring Angle: 15 degree, Trigger Input Signal: 10μS TTL pulse, Echo Output Signal Input TTL lever signal and the range in proportion, Dimension 45 * 20 * 15mm.The product features are Ultrasonic ranging module HC - SR04 provides 2cm - 400cm non-contact measurement function, the ranging accuracy can reach to 3mm. The modules include ultrasonic transmitters, receiver and control circuit. The basic principle of work,Using IO trigger for at least 10us high level signal, The Module automatically sends eight 40 kHz and detect whether there is a pulse signal back. IF the signal back, through high level, time of high output IO duration is the time from sending ultrasonic to returning. Test distance = (high level time × velocity of sound (340M/S) / 2.

6.1.3 Software Periphery

Energia IDE - ENERGIA IDE is used for programming and upload the sketch to the interfacing unit (CC3200 launchpad). It is the copy software of Arduino IDE,where all the featured that are in Arduino IDE is presented.

Blynk application - Blynk is a Platform with IOS and Android apps to control Arduino, Raspberry Pi and the likes over the Internet. It's a digital dashboard where you can build a graphic interface for your project by simply dragging and dropping widgets. There are three major components in the platform are Blynk App, Blynk Server, Blynk Libraries.

6.2 Fabrication Process

6.2.1 Mechanical Work

Initially external frame has been fabricated by welding, where 'L-Bar has been used with thickness of 5mm. Then Wheels along with axels has been jointed with required bearings and the axels has been fastened to the frame. Now two various dimensioned hoppers are made with GI sheet for seed and fertilizer respectively, the smaller one for solid fertilizer and the bigger one for seeds. The Fertilizers hopper is connected to a draft tube like structure which act as a nozzle and along with air input where the air is supplied through air compressor. A ball valve is used in the fertilizer hopper to control the flow rate of the fertilizers. The seed hopper is connected to a PVC nozzle along with a ball valve where the flow of seeds is controlled. Now the two hoppers are jointed to the frame with required supporting rod, for jointing welding process is used. Then pneumatic components are connected along with a connection to the air compressor. The pressure set in the compressor is not constant it should be varied with respect to the distance of the seed/fertilizer should sprayed.

6.2.2 Electronic Work

CC3200 launchpad board is used as a CPU here where the ultrasonic sensors are connected and programmed to it. Bread Board is used to serial wire connections. Here for our convenient power source is given by Mobile Power bank. Two ultrasonic sensors are used one for seed hopper and the other for fertilizer hopper. An wooden board has been fixed in between both the hoppers where the CC3200 board has been fixed. Both the ultrasonic sensors have been fixed to it with an supporting element. For monitoring results through mobile "Blynk" mobile application is used which connects the board and mobile an we can monitor the reading through it.

6.3 Experimental Procedure

The seed or fertilizer which has to be farmed is filled in their hoppers, where the ball valve should be closed. Then if fertilizer should be sprayed the air intake from the compressor is opened simultaneously the ball valve also should be opened where the flow of fertilizer will be perpendicular to the flow of air. Now in case of seeds it is very simple that the ball valve is opened and closed to control the flow rate of seeds. For monitoring the level of seeds and fertilizers in their respective hoppers the CPU board is switched ON and then using "Blynk" mobile application the board and mobile is connected. Where the Ultrasonic sensor sends the data to CPU and then we can monitor in our mobile. A temperature sensor is presented in the board where the atmosphere's temperature can also be monitored. In case we need to spray seeds the fertilizers hopper can be used to spray seeds. Using the Blynk mobile application the seed/fertilizer in their respective hoppers can be determined

6.4 Sensor Program

```
#define BLYNK_PRINT Serial // Comment this out to disable prints and save space
#include <SPI.h>
#include <WiFi.h>
#include <BlynkSimpleEnergiaWiFi.h>
#include <Wire.h>
#include "Adafruit_TMP006.h"
#include <BMA222.h>
#define dht_apin 6
// Analog Pin sensor is connected to
// Sensor objects
```

```
BMA222 mySensor;
Adafruit_TMP006 tmp006(0x41);
// You should get Auth Token in the Blynk App.
// Go to the Project Settings (nut icon).
const char* auth = "LUTG64FVgtOT29h8SMExHwm4K6TFAYmZ";
// Your WiFi credentials
char ssid[] = "ARUL";
const char* pass = ""; // Set to "" for open networks
#define st1 7
#define se1 8
#define st2 5
#define se2 6
#define sensor_pin 23
float tm1 =0;
float distance1 =0;
float tm2 =0;
float distance2 =0;
int output_value ;
void setup()
{
//Open a serial terminal with the PC
Serial.begin(9600);
//Set up a blynk connection with your WiFi credentials
Blynk.begin(auth, ssid, pass);
//TMP006 setup
tmp006.begin();
pinMode(st1,OUTPUT);
pinMode(se1,INPUT);
pinMode(st2,OUTPUT);
pinMode(se2,INPUT);
//Setup RED LED to be an output }
// Virtual Pin 1 - Toggles the LED high or low depending on the mobile app in-
put
//Virtual pin 5 - Read the TMP006 value when called by the app
BLYNK_READ(11)
{ digitalWrite(st1,LOW);
delayMicroseconds(2);
digitalWrite(st1,HIGH);
delayMicroseconds(10);
digitalWrite(st1,LOW);
delayMicroseconds(2);
tm1 = pulseIn(se1,HIGH);
distance1 = tm1*340/20000;
Blynk.virtualWrite(V11,((int)distance1-55));
```

```
Serial.println((int)distance1-55);
delay(10);
}
BLYNK_READ(12)
{
digitalWrite(st2,LOW);
delayMicroseconds(2);
digitalWrite(st2,HIGH);
delayMicroseconds(10);
digitalWrite(st2,LOW);
delayMicroseconds(2);
tm2 = pulseIn(st2,HIGH);
distance2 = 25;
Blynk.virtualWrite(V12,((int)distance2));
Serial.println((int)distance2);
delay(10);
}
BLYNK_READ(10)
{
output_value= analogRead(sensor_pin);
output_value = map(output_value,550,0,0,100);
Blynk.virtualWrite(V10,(int)output_value);
Serial.println(analogRead(sensor_pin));
delay(10);
BLYNK_READ(9)
float objt = tmp006.readObjTempC();
Serial.print("Object Temperature: ");
Serial.print(objt);
Serial.println("*C");
Blynk.virtualWrite(V9, (int)objt);
delay(10);
// The main loop listens for commands from the mobile app
void loop()
Blynk.run();
```

7. FINDINGS AND RESULTS

7.1 Mechanical Work

Here for sample we used Solid fertilizer 'Potassium' for testing for both seed and fertilizer potassium has been used. In the below Fig.2 and Fig.3 the flow of seeds (here potassium) through the PVC nozzle has been shown, where there is a flow of potassium (fertilizer) can be seen through.

Figure 2. Seed nozzle

In the below Fig.4 and Fig.5 the fertilizer nozzle are shown where the fertilizers are sprayed.

7.2 Electronic Work

The Fig.6 shows the interface of the Blynk mobile application where the level of two various hoppers and then the atmosphere temperature is displayed. Here there is 3 boxes where smaller one is to display air temperature and the other two is to display the level of seed and fertilizers hopper respectively.

The below Fig.7 and Fig.8 shows the top view and front view of the equipment.

Figure 3. Seed nozzle with seeds

Figure 4. Fertilizer nozzle

8. RECOMMENDATIONS

The Non-Tilling farming equipment (seed sowing and fertilizer spraying gas) has been fabricated with the help of IoT systems. A seed hopper is used for seed sowing, and a nozzle is attached to it to inject the seed. A PVC ball valve is placed in the seed's path to control the flow rate. Then, with the aid of compressed air flow, a specific quantity of solid fertilizer (potassium) is sprayed along it. Here too, a PVC ball valve is positioned in the fertilizer's path to regulate the flow rate. Compressed air from the air compressor is used as one input and two outputs in the pneumatic process: one for the fertilizer nozzle and the other for the seed nozzle. Separate ball valves are also utilized in pneumatic construction to regulate the flow rate. The "CC3200 launchpad" board is used as the CPU in Internet of Things systems, while the Blynk mobile application is used as a software device to view the output readings. The Energia IDE tool facilitates appropriate programming to link the hardware and software components.

Figure 5. Fertilizer nozzle with hopper

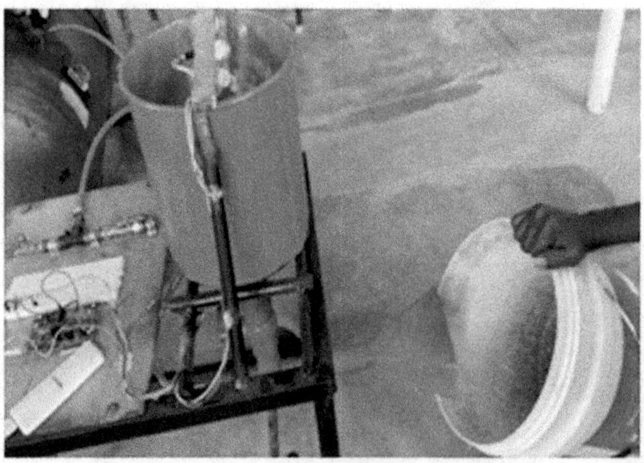

Figure 6. Blynk application's working interface

Figure 7. Top view of final equipment

Figure 8. Front view of final equipment

8.1 Future Implications and Trends

Developments in Artificial Intelligence and Machine Learning: As these fields develop, agriculture will be able to make more accurate and data-driven decisions thanks to the advanced algorithms used in AI and machine learning. With the use of these technologies, crop monitoring, disease detection, yield prediction, and pest management will all be transformed, resulting in increased output and resource efficiency. Extension of IoT and Sensor Networks: As IoT devices and sensor networks proliferate, real-time agricultural operation monitoring and management will be made easier. Farmers will be able to make proactive decisions and maximize resource use thanks to their access to complete data on crop conditions, weather patterns, soil health, and equipment performance. Integration of Robotics and Automation: The simplification of labour-intensive operations like planting, weeding, and harvesting will be greatly aided by robotics and automation. Artificial intelligence (AI)-enabled smart technology, autonomous drones, and robots will improve productivity, lower labour costs, and address the manpower shortage in the agriculture sector. Agriculture Tech Solutions and Digital Farming Platforms: Agriculture Tech solutions and digital farming platforms will proliferate and offer farmers all-inclusive tools for data analysis, decision assistance, and farm management. These platforms will provide adaptable solutions catered to particular farming requirements, encouraging farmers of all sizes to utilize technology more widely. Sustainable and Regenerative Agriculture: Methods of sustainable and regenerative agriculture that protect soil health, preserve water resources, and lessen climate change will become increasingly important. Environmentally friendly farming methods will be greatly aided by technologies like carbon sequestration, cover crops, agro-forestry, and precision irrigation. Vertical farming and controlled environment agriculture (CEA) are becoming more popular as viable methods for producing food in urban regions and resource-constrained settings as a result of population expansion and urbanization. Modern hydroponic, aeroponic, and aquaponic systems will make it possible to cultivate high-value crops year-round with no negative influence on the environment. Collaboration and Data Sharing: Among the various players in the agricultural ecosystem, such as farmers, academics, technology suppliers, and legislators, there will be a growing tendency toward collaboration and data sharing. A constant improvement in agricultural methods will be facilitated by open data platforms, interoperable technologies, and collaborative activities that promote information exchange and innovation diffusion. All things considered, agricultural technology has a great deal of promise for solving global issues like sustainability, climate change, and food security. The agricultural industry may develop to become more resilient, effective, and environmentally sustainable by embracing innovation and utilizing cutting-edge technologies, protecting the interests of both farmers and consumers.

9. LIMITATIONS

Research on IoT and smart sensor-based smart farming conducted globally has yielded positive outcomes. Numerous small-scale agricultural fields have used IoT-based technology. Large-scale implementation of the same is still a ways off. The financial burden associated with the deployment and installation of IoT-tagged sensors and accessories across a sizable acreage of agricultural land is one of the main obstacles. Furthermore, there is a degree of uncertainty surrounding the implementation costs and the profit margin. The cost of purchasing hardware, installing software, and running the system are the main expenses associated with implementing IoT-enabled technology. Energy consumption, system maintenance, service registration, and labour costs related to running the integrated hardware devices and related software will all incur additional expenditures. An extensive global push to improve farmers' technological literacy is necessary for the IoT to be applied globally. Underutilization of the smart system in agriculture may arise from a lack of appropriate information and literacy regarding the scope, method of action, and use of IoT-based technologies in farming. It may be more difficult to deploy IoT-connected smart sensors and accessories in rural farming areas where farmers are less accustomed to cutting-edge technologies.

REFERENCES

Andure, M. W., Rajiv, I., Nikesh, J., Vaibhav, K., & Rahul, K. (2018). Design and fabrication of manual seed sowing machine with fertilizer. *International Journal for Research in Applied Science and Engineering Technology*, *6*(2), 515–519. doi:10.22214/ijraset.2018.2073

Annapurna, B., & Anusha, B. (2019). Automated Seed Sowing Machine Using Atmega2560. *International Journal of Innovative Technology and Exploring Engineering*, *8*, 300–302.

Balaji, G. N., Nandhini, V., Mithra, S., Priya, N., & Naveena, R. (2018). IoT based smart crop monitoring in farm land. *Imperial Journal of Interdisciplinary Research*, *4*(1), 88–92.

Cucho-Padin, G., Loayza, H., Palacios, S., Balcazar, M., Carbajal, M., & Quiroz, R. (2020). Development of low-cost remote sensing tools and methods for supporting smallholder agriculture. *Applied Geomatics*, *12*(3), 247–263. doi:10.1007/s12518-019-00292-5

Deshmukh, B., & Verma, D. (2018). Fabrication and implementation of automatic seed sowing machine. *International Journal of Engineering Sciences & Research Technology*, *3*, 274–281.

Jurišić, M., Plaščak, I., Barač, Ž., Radočaj, D., & Zimmer, D. (2021). Sensors and Their Application in Precision Agriculture. *Tehnički glasnik, 15*(4), 529-533.

Kulkarni, S. R., Nyamagoud, R. V., Naik, H., &Futane, M. (2015). Fabrication of Portable Foot Operated Agricultural Fertilizers and Pesticides Spraying Pump. *International Journal of Engineering Research and Technology*.

Kumbhare, D. N., Singh, V., Waghmare, P., Ansari, A., Tiwari, V., & Gorle, R. D. (2016). Fabrication of automatic pesticides spraying machine. *International Research Journal of Engineering and Technology (IRJET), 3*(4).

Kumutha, S., Gayathri, D. N., & Mohanbabu, S. (2018). IoT Concept for Smart System Monitoring Agricultural Land. *International Journal of Engineering Research & Technology (Ahmedabad), 6*(8), 1–8.

Liu, W. (2023). Smart sensors, sensing mechanisms and platforms of sustainable smart agriculture realized through the big data analysis. *Cluster Computing, 26*(5), 2503–2517. doi:10.1007/s10586-021-03295-3

Oukaira, A., Benelhaouare, A. Z., Kengne, E., & Lakhssassi, A. (2021). FPGA-embedded smart monitoring system for irrigation decisions based on soil moisture and temperature sensors. *Agronomy (Basel), 11*(9), 1881. doi:10.3390/agronomy11091881

Rajak, P., Ganguly, A., Adhikary, S., & Bhattacharya, S. (2023). Internet of Things and smart sensors in agriculture: Scopes and challenges. *Journal of Agriculture and Food Research, 14*, 100776. doi:10.1016/j.jafr.2023.100776

Rao, V. V., Mathapati, S., & Amarapur, B. (2013). Multiple power supplied fertilizer sprayer. *International Journal of Scientific and Research Publications, 3*(8), 1–5.

Rayhana, R., Xiao, G., & Liu, Z. (2021). RFID sensing technologies for smart agriculture. *IEEE Instrumentation & Measurement Magazine, 24*(3), 50–60. doi:10.1109/MIM.2021.9436094

Sagar, M., Hullur, P., Ankush, S., Kannolli, V., & Rayakar, V. (2020). *Design and Fabrication of Semi-Automatic Seed Sowing Vehicle*. Academic Press.

Saravanan, K., Sivam, S. S. S., Kumar, S. R., & Moorthy, K. S. (2018). Design and fabrication of automatic seed sowing robot for agricultural field. *International Journal of Pure and Applied Mathematics, 120*(6), 11749–11766.

Sengodan, P., & Jbara, Y. H. F. (2017). Development of IoT controlled agri-rover for automatic seeding. *International Journal of Pure and Applied Mathematics, 114*(11), 241–251.

Solanki, S. N., Ramteke, R. T., & Dhande, P. S. (2022). Agricultural Automatic Seed Sowing Machine: A Review. *International Journal of Research in Engineering. Science and Management, 5*(2), 114–119.

Swapnil, T., Kasturi, M. L., Girish, P., & Rajkumar, P. (2017). Design and fabrication of seed sowing machine. *International Research Journal of Engineering and Technology, 4*(09), 704–707.

Tăbăraşu, A. M., Anghelache, D., & Dumitru, D. (2021). Benefits Regarding the Implementation of Agriculture 4.0 In the Current Context. Annals of the University of Craiova-Agriculture, Montanology. *Cadastre Series, 50*(2), 544–551.

KEY TERMS AND DEFINITIONS

Agri Rover: Agricultural rover is a multifunctional robot platform that is often used in the field. This robot platform with suitable payload can do soil sampling, mapping, and fruit picking.

Arduino: Arduino is an open-source electronics platform based on easy-to-use hardware and software.

Big Data: Big data is a combination of structured, semistructured and unstructured data collected by organizations that can be mined for information and used in machine learning projects, predictive modeling and other advanced analytics applications.

Blynk: It is an IoT platform for iOS or Android smart phones that is used to control Arduino, Raspberry Pi and Node MCU via the Internet. This application is used to create a graphical interface or human machine interface (HMI) by compiling and providing the appropriate address on the available widgets.

Bot: A bot is an automated software application that performs repetitive tasks over a network. It follows specific instructions to imitate human behavior but is faster and more accurate. A bot can also run independently without human intervention.

Bread Board: A breadboard, solder less breadboard, or protoboard is a construction base used to build semi-permanent prototypes of electronic circuits.

Enhanced Zero-Till Farming: Zero-till farming augmented with technological solutions aimed at improving productivity, efficiency, and sustainability.

IDE: An integrated development environment (IDE) is a software application that helps programmers develop software code efficiently. It increases developer productivity by combining capabilities such as software editing, building, testing, and packaging in an easy-to-use application.

IoT (Internet of Things): A network of interconnected devices embedded with sensors, software, and connectivity capabilities to collect and exchange data, enabling remote monitoring and control.

Microcontroller: A microcontroller is a compact integrated circuit designed to govern a specific operation in an embedded system. A typical microcontroller includes a processor, memory and input/output (I/O) peripherals on a single chip.

Monitoring: The continuous observation and measurement of parameters such as soil moisture, temperature, and crop growth to assess the status and performance of farming operations.

Nozzle: It is often a pipe or tube of varying cross sectional area, and it can be used to direct or modify the flow of a fluid (liquid or gas). Nozzles are frequently used to control the rate of flow, speed, direction, mass, shape, and/or the pressure of the stream that emerges from them.

Sensor: A device that detects and measures physical properties such as moisture, temperature, or light, often used in IoT systems for data collection.

Servomotor: A servomotor (or servo motor or simply servo) is a rotary or linear actuator that allows for precise control of angular or linear position, velocity, and acceleration in a mechanical system.

Zero-Till Farming: Agricultural practice that involves minimal soil disturbance by eliminating traditional ploughing or tilling, promoting soil conservation, water retention, and reduced erosion.

Chapter 12
Review of Terrace Gardening Literature and Suggestions for Increasing Adoption

Chitra Dey
Vellore Institute of Technology, Chennai, India

V. A. Monica
Vellore Institute of Technology, Chennai, India

Amit Karanth Gurpur
Vellore Institute of Technology, Chennai, India

Swetha Venugopal
https://orcid.org/0009-0006-3361-6671
Vellore Institute of Technology, Chennai, India

M. Lavanya
Vellore Institute of Technology, Chennai, India

Sai Choornika Chandana Samudrala
https://orcid.org/0009-0008-7992-4586
Vellore Institute of Technology, Chennai, India

A. Aakash Raja
Vellore Institute of Technology, Chennai, India

J. Aishwar Haris
https://orcid.org/0009-0009-6207-4279
Vellore Institute of Technology, Chennai, India

ABSTRACT

Agricultural lands are swiftly transforming into residential complexes and high-rise buildings, leaving limited space for traditional gardening. A sustainable and creative venture, terrace gardening (TG) focuses on using rooftop areas for the growth of plants, vegetables, and herbs. To give people and communities access to fresh food and a green haven in urban settings, it entails converting underutilized rooftops into gardens. The authors review and present relevant literature on the topic of TG. The review identifies that papers on TG mainly cover themes like motivations to participate in TG, advantages of TG, sustainability and innovation in TG, operations and stakeholder management in TG, and limitations of TG. The review shows that time management, project management, networking, and stakeholder involvement are essential for TG. A SWOC analysis reveals that TG offers potential for business owners seeking to have a positive environmental impact while supplying the rising demand for fresh, locally grown food.

DOI: 10.4018/979-8-3693-4864-2.ch012

INTRODUCTION

The practice of growing plants on terraced surfaces, commonly in urban or suburban settings with constrained space, is known as terrace gardening (TG). It entails building a terrace with several levels, and raised beds that can be used to cultivate a range of plants, such as decorative plants, fruits, herbs, and vegetables (Medha *et al.*, 2022). TG is significant because it has so many positive effects on urban dwellers, sustainability, the environment, and urban living.

Urban farming has been gaining momentum over the years among urban dwellers. The awareness of safe and nutritious food among consumers is driving them to think of ways and means of cultivating fruits and vegetables that they eat. TG can encourage urban green cover and improves the aesthetics of buildings (Winkler *et al.*, 2019). These methods not only contribute to local food security but also promote a healthier and more environmentally conscious lifestyle. Research indicates that terrace gardens might enhance the general and mental health of city dwellers. These also encourage a sense of community engagement and self-sufficiency. By reducing the need for long-distance transportation of food, urban farming and TG contribute to lower carbon footprints and create more resilient, sustainable cities in the face of increasing urbanization (Thomas & Cherian, 2021). TG is a simple but definite step towards increasing our green cover, irrespective of space restrictions. This is an achievable activity for many urban dwellers to cultivate the food crops of their choice.

TG provides environmental benefits, reduces urban heat island effect, helps with waste recycling, energy, and water conservation. By adding plants that can reflect sunlight and enhance air circulation, TG can help lessen the impact of the urban heat island effect. TG can offer food and shelter to insects and birds, fostering biodiversity in urban settings (Deng *et al.*, 2021). Fertile agricultural land in urban areas has been converted into houses and other infrastructures leading to reduction of land for cultivation. The impacts of residential development on the natural environment extend to the realm of agriculture, particularly in the use of chemical pesticides and fertilizers. As urbanization encroaches on agricultural lands, the demand for increased food production often leads to intensive farming practices. So, if one wants to grow vegetables and fruits domestically, a viable alternative is organic TG, which provides the ideal space (Medha *et al.*, 2022). This felt need can become a real movement by sharing and spreading know-how amongst stakeholders.

Despite these advantages, further studies and evaluations of the literature are still needed, especially when it comes to understanding factors important for terrace farming, TG's impact on the environment and innovation in TG. This is because the information that is currently available is dispersed and incomplete, which makes it challenging for practitioners and researchers to find pertinent information. Researchers can summarize current information, identify gaps in the field, and suggest new avenues for future research and implementation by conducting a literature review on TG. This can encourage future researchers to address meaningful research gaps in current literature, and help urban and suburban communities to pursue TG as a sustainable and eco-friendly technique (Deng *et al.*, 2021). In the following section we present the current literature on TG. Following that we present a strength, weakness, threat, and challenge analysis of TG. Finally, we conclude the paper with results and discussion.

LITERATURE REVIEW

We identified journal papers published in peer reviewed journals to collate the literature on TG. The key words used for searching for articles was "terrace gardening" and "roof gardening", and we searched for articles on databases like Web of Science, Scopus and search engine Google Scholar. We ensured the quality of the research by adding articles published in peer reviewed journals.

We classified the articles under seven main themes addressed in the paper: motivations for TG, advantages of TG, sustainability and TG, innovation in TG, operations, stakeholder management, and limitations of TG.

a. Motivations for Terrace Gardening

The main theories used to examine the motivations for people to engaged in TG are the self-determination theory, theory of planned behaviour and theory of motivation. As per the self-determination theory, people engage in TG so that they can contribute to the society and community (Ramirez-Andreotta et al., 2019). Participation in community practices have been found to improve individual and community well-being across different countries (Quested *et al.,* 2018). As communities and families practice this art, TG can also help arouse the interest of children and attract them towards agriculture and gardening activities (Naveena & Sahana, 2017). The second main motivation for individuals to participate in TG is to promote the local flaura and biodiversity through the theory of planned behaviour (Samus *et al.,* 2023). As per this theory, people are more likely to engage in TG as it connects them with nature and promotes a feeling of contributing to biodiversity conservation. The final line of argument that supports participation in TG is for its health benefits for the participants through mental and physical well-being (Chalmin-Pui *et al.,* 2021).

b. Advantages of Terrace Gardening

The adoption rate of TG has seemingly increased over the past few years driven by the trends that are discussed below. As a home business, TG aims to provide fresh and healthy produce by utilizing small spaces on rooftops, balconies, or terraces. Initially, TG was practiced by people as a hobby of growing their plants of choice and obtaining an organically grown plant. But now people have started to practice TG to get profitable benefits by selling their yields to interested customers. This increases the entrepreneurial opportunities not just for the people who do TG but also for the people who help to build this setup. However, one challenge in this area is on determining the adoption rate of TG, as there is no exact figure to get the adoption rate. This is because not many households perform TG for business purposes, so the accuracy of reported data is less. Another significant advantage of TG is its contribution to food security. With the ability to grow a variety of fresh produce right at home, homeowners can reduce their reliance on store-bought items that often travel long distances, resulting in increased carbon emissions. By cultivating seasonal fruits and vegetables on their rooftops or terraces, households can ensure a continuous supply of nutritious, organic food throughout the year.

TG has the potential to drastically reduce the number of steps in the current food supply chain, resulting in numerous advantages. Here are some ways that TG can shorten the conventional food supply chain (Dash & Deole, 2020). With TG, food can be grown directly on rooftops or near urban centers, eliminating the need for long-distance transport from rural farms to urban centers (Chapagain & Raizada, 2017). This reduces the carbon emissions and dependence on fossil fuels associated with transportation. There is no need for extensive warehousing or cold storage facilities when food is grown on rooftops or nearby land (Chatterjee, Debnath & Pal, 2020). Freshly harvested fruits and vegetables can be consumed or distributed immediately, eliminating the need for prolonged storage, refrigeration, and associated energy consumption. TG reduces or eliminates packaging of fruits and vegetables (Akkara & Mohan, 2023). In conventional supply chains, food is frequently packaged multiple times to ensure its freshness during transport. Consumers can harvest produce directly from their terrace gardens, eliminating the need for excessive packaging. This reduces the amount of waste produced and the environmental impact of packaging materials. TG permits harvesting according to demand and hence reduced food waste. Consumers can select only the quantity of produce they require, reducing the likelihood of food waste. This aspect contributes to addressing the significant issue of food waste in the current food supply chain. TG encourages local food production, allowing individuals and communities to have a greater say over their food supply (Vazhacharickal, 2014). It improves food security and resiliency in the face of disruptions or supply chain difficulties by decreasing reliance on external sources. TG improves food quality and nutritional value. With TG, fresh produce can be harvested at its peak ripeness, ensuring greater nutritional value and superior flavor compared to produce that may undergo lengthy transportation and storage in conventional supply chains. This provides consumers with healthier and better food options. TG is an excellent way to access fresh produce while providing numerous benefits, such as low investment costs and improved well-being (Medha *et al.*, 2022). Recycling and managing rubbish at home through composting, specifically vermicomposting, is a practical solution that allows for organic food production at a low cost (Ravi Kumar *et al.*, 2009).

TG also improves biodiversity. TG can become a haven and shelter for birds and insects, especially in urban areas (Patil, Patil & Mail, 2020; Singh, 2017). Experts in this field suggest suitable soil types for different plants, allowing a wide variety of plants to be grown even in large polythene bags. The garden can be diversified by employing a range of containers such as earthen pots, cement pots, and grow bags, or opt for hanging pots available in various shapes, sizes, and colors to enhance the garden's aesthetic appeal (Dash & Deole, 2020). Shade net gardening can also be used which controls the amount of sunlight and yields good productivity in adverse weather conditions.

c. Sustainability and Terrace Gardening

Sustainability in TG refers to the practice of cultivating plants in urban settings, particularly on terraces, with the aim of reducing the environmental impact of food production and consumption by conserving resources such as water, energy, land, and transportation costs.

TG strengthens the economy and society by implementing the circular economy. The circular economy represents a revolutionary approach to resource management, diverging from the linear "take, make, dispose" model to create a regenerative and sustainable system (Corona *et al.*, 2019). Implementing a circular economy approach in TG involves reimagining traditional practices to minimize waste, maximize resource efficiency, and promote sustainability. In this context, circularity begins with the careful selection of materials, opting for reusable and recycled containers, minimizing single-use plastics, and choosing locally sourced, eco-friendly products (Corona *et al.*, 2019). The cultivation of plants focuses on regenerative practices, such as composting kitchen scraps to create nutrient-rich soil and embracing natural pest control methods. Seed saving and plant propagation further contribute to closing the loop, reducing the need for continuous resource inputs. Additionally, terrace gardeners can engage in community-based initiatives, sharing excess produce and knowledge to foster a circular ecosystem. By integrating circular economy principles into TG, individuals not only create flourishing green spaces but also actively participate in a sustainable, closed-loop system that aligns with the broader goal of environmental responsibility (Mohan *et al.*, 2020).

TG also contributes to improving energy consumption in the building by lower ceiling temperatures. It was found that terrace gardens can reduce ceiling temperatures and improve building micro climate in residential buildings when the area of the TG was as small as 15 square metres (Chidambaram *et al.*, 2018). Further use of vertical gardens on a building's façade can also help reduce the building's temperature especially during the summer months (Reddy, Omprasad & Janakiram, 2022).

In addition, TG offers a healthy atmosphere by improving air quality, saving energy for air conditioners, and providing fresh fruits and vegetables free of chemicals (Dash & Deole, 2020). The desire to have organic yields and consume something free of pesticides and insecticides has increased as people have started to keep their health as their priority (Kishore & Patil, 2015). TG is a unique way to grow plants with the use of organic fertilizers such as food waste and cow dung. For a completely organic growth approach, materials like kitchen waste and cow dung can be utilized as fertilizers (Dash & Deole, 2020). It helps to keep the plants healthy and decreasing pollution and increasing oxygen levels.

Waste management in TG involves a thoughtful and sustainable approach to handling various organic and inorganic materials generated within this confined gardening space (Ravi Kumar *et al.*,2009). Several studies (Kumar *et al.,* 2009 is a good example) discusses composting, particularly vermicomposting, as a household-level solution to the environmental, cultural, social, and political factors that impact how municipalities manage household waste. Organic waste, such as kitchen scraps and plant trimmings, can be efficiently managed through composting, creating nutrient-rich compost that enriches the soil. Utilizing composting bins or vermicomposting systems helps minimize the environmental impact of organic waste, turning it into a valuable resource for the garden (Naveena & Sahana, 2017). TG provides an option for managing biodegradable kitchen waste, highlighting the need for efficient management of such waste from an economic and environmental perspective. For inorganic waste, terrace gardeners can adopt practices like upcycling, reusing containers, and properly recycling materials to reduce the overall environmental footprint. By integrating effective waste management strategies, TG not only produces vibrant and green spaces but also serves as a microcosm of sustainable living, showcasing how responsible waste handling contributes to the health and vitality of both the garden and the broader environment (Kishore & Patil, 2015)

The bioeconomy aims to establish a sustainable economy by aligning economic growth with environmental conservation and climate change mitigation through the utilization of natural matter cycles. To attain this goal, it is imperative to develop more sustainable methods of production and consumption (Winkler *et al.,* 2019). TG has indeed offered solutions to the problems in this space. Embracing organic fertilizers and natural pest control methods is another crucial aspect of sustainable TG. By avoiding chemical pesticides and synthetic fertilizers, individuals can reduce their carbon footprint, as these products are often manufactured using fossil fuels. Instead, they can promote biodiversity in their gardens and create a healthier, more balanced ecosystem.

TG helps conserve precious resources and minimizes the environmental impact associated with conventional food production. When it comes to water management, drip irrigation and rainwater harvesting provide solutions for dealing with resource constraints. Drip irrigation systems, for instance, make efficient use of water and reduce the demand for fertilizers. However, we found that there were few papers dealing with issues related to drip irrigation, and that it was practiced by fewer people in the studied markets (Thomas & Cherian, 2021). Implementing drip irrigation systems not only conserves water but also helps deliver nutrients efficiently to the plants. Implementing rainwater harvesting techniques can also significantly decrease the need for municipal water supplies, contributing to water conservation. However, most of the research and practice of rainwater harvesting has been directed towards in-house usage of the harvested water (Amos *et al.*, 2018). Thus, TG is a sustainable and environmentally friendly way to produce fresh and healthy food. By following these eco-conscious methods, individuals can create terrace gardens that benefit both themselves and the planet, contributing to a more sustainable and greener future.

d. Innovation in Terrace Gardening

By adopting eco design principles in TG, individuals contribute to urban biodiversity, mitigate the urban heat island effect, and create harmonious green spaces that are both aesthetically pleasing and environmentally responsible (Mladenović *et al.,*2017). Eco design in TG involves a conscientious approach to creating sustainable and environment friendly green spaces in urban settings. In this context, eco design encompasses thoughtful choices in plant selection, resource utilization, and overall garden layout. Opting for native and drought-resistant plants not only enhances biodiversity but also reduces the need for excessive water consumption (Åstebro & Serrano, 2015). It is important to select plants that thrive in the local climate and conditions, ensuring that the garden remains resilient and productive. This practice offers benefits ranging from a decrease in the amount of pesticides used to sustaining the biodiversity of the planet by safeguarding local plants (Shelef, Weisberg, & Provenza, 2017).

Utilizing recycled or repurposed materials for containers and garden structures aligns with eco-friendly principles, minimizing the environmental impact of TG. Additionally, incorporating composting systems and organic fertilizers nurtures a closed-loop system, where kitchen waste becomes a valuable resource for plant nourishment (Naveena & Sahana, 2017). While proposing a methodology for assessing the performance of various compost models, RaviKumar *et al.* discuss the effectiveness of composting methods and resolving waste separation issues (2009).

Another innovative trend that was observed is improving the water management technology used in TG, like hydroponic systems and aquaponics. The usage of hydroponics makes gardening possible without soil and the yield produced is also increased than the traditional method thereby increasing the profit of the business (Rondhi *et al.*, 2018). Innovative techniques such as automated watering systems using various sensors can be implemented in TG to boost the yield and minimize water usage. Aquaponics, which is cultivation in water, has been found to have a better production and growth rate than other farming methods and economically this method is a viable solution (Palm *et al.,*2018). Technology based on sensors can also be deployed to assess weather, soil, plant growth and irrigation to optimise usage of resources (Revathy, Balamurali & Murali, 2021).

Renewable energy sources, such as solar panels, can be employed to power equipment and lighting on the terrace. In addition, space constraints can be handled by utilizing vertical space with trellises, wall-mounted planters, or vertical gardens. Factors such as sunlight exposure, plant height, and container size should also be considered. For example, choosing plants with a compact growth habit, using containers that fit the available space efficiently, and carefully planning the proper garden layout also maximize the space.

Terrace gardens face exposure to various environmental conditions, including sunlight, wind, and temperature fluctuations. To cope with these conditions, it is important to select resilient plants that can thrive in the specific microclimate of your terrace. Adopting a flexible mindset is essential for learning and experimenting with different gardening techniques (Poornima & Suryawanshi, 2020). Stay open to novel ideas, seek guidance from experienced gardeners, and be willing to try innovative approaches to enhance your terrace garden. If a particular plant is not thriving or suitable for the conditions, be flexible enough to replace it with a more appropriate alternative. TG is a dynamic pursuit that requires adaptability and flexibility. As a terrace gardener, it is crucial to adjust gardening practices based on changing seasons, plant requirements, and environmental factors. A key element in achieving success in TG is maintaining an open mind to new ideas and being receptive to learning from personal experiences.

e. Operations

For successful management of a terrace garden, the stakeholders need to pay attention to different aspects of operations. To take an analogy, organizational research has shown that total quality management and culture are important for green practice implementation in organizations (Khalil & Muneenam, 2021). We identified the following themes related to operations: planning and organizing, time management, financial management, project management, networking, problem solving, and decision making.

Planning and Organizing: A terrace garden requires careful planning and organization. All stakeholders should be included in giving inputs from the phase of design of the building to ensure that the provision for TG is built into the blueprint.

It is important to determine the dimensions and limitations of the area to plan effectively. It is crucial to consider factors like sunlight availability, wind exposure, drainage systems, structural integrity, and space utilization (Weiler & Scholz-Barth, 2009). Developing a clear plan and organizing the gardening activities will help in maximizing the potential of terrace gardens. It is important to develop a layout plan for the terrace garden, considering factors such as plant placement, pathways, seating areas, and water sources. The garden should be planned in such a way to optimize the utilization of vertical space by incorporating hanging pots, trellises, or vertical gardens (Dash *et al.*, 2020).

Time Management: Effective time management is crucial in TG. Plants need regular care and attention, including watering, fertilizing, pruning, harvesting and pest control. It is essential for management to consider creating a schedule for these tasks and allocating appropriate time to ensure the thriving of plants. Based on the urgency and importance of the activities stakeholders should identify and prioritize the essential tasks. Opportunities to optimize time through the grouping or batching of similar tasks should be sought. (Dash *et al.* in 2020). For instance, when plants are being watered, consideration can be given to the simultaneous inspection for pests, weed removal, and evaluation of the overall health of the garden. This approach not only reduces repetitive movements but also ensures the streamlining of the workflow, resulting in time savings.

Budgeting and Financial Management: Proper financial management is vital when it comes to terrace gardens. It is necessary to allocate a budget for supplies, plants, equipment, and maintenance costs. Optimizing the utilization of available resources for a terrace garden project can be achieved by diligently monitoring expenses and identifying cost-effective solutions. This approach enables us to make the most out of the resources at hand, ensuring efficient allocation and management throughout the project. Consideration of factors such as the cost of plants, seeds, soil, containers, tools, fertilizers, and any additional equipment or infrastructure is crucial for finance management (Mladenović *et al.*,

2017). It is important to maintain a record of purchases, receipts, and any unexpected costs that arise, and regularly review expenses to identify areas to adjust or optimize the budget.

Project Management: When undertaking a larger-scale TG project, like designing and implementing a garden for a community, it becomes crucial to possess project management skills. These skills encompass setting clear goals, establishing a well-defined timeline, coordinating various tasks, and overseeing the entire project from its initiation to completion. It is of utmost importance to clearly and appropriately define the objectives of the TG project such as growing vegetables, creating a relaxing space, or enhancing terrace aesthetics. This will help recognize potential risks and challenges that may emerge, such as pest infestation, extreme weather conditions, or plant diseases (Mladenović et al., 2017).

Networking: Effective communication and collaboration skills are essential when working with a team or seeking advice from experts. Collaborating with others not only allows for knowledge sharing but also facilitates resource sharing, ultimately contributing to the overall success of the terrace garden project. Establishing extensive networks and affiliations with diverse organizations, colleges, and universities can significantly contribute to the growth and advancement of our organization. Furthermore, such collaborations serve as prominent platforms to showcase our work, enhancing the visibility and promotion of our organization.

Problem-solving and Decision-making: In TG, there are various challenges that need to be faced, including plant diseases, pest infestations, water management issues, or weather-related problems. Overcoming these challenges requires the ability to analyse situations, think critically, and make well-informed decisions. Developing these skills will enable one to effectively tackle the obstacles encountered in the TG endeavours.

f. Stakeholder Management

Stakeholders or key partners are the entities or people who will contribute to the success of a business, provide business necessities, and supply the knowledge and requirements along with raw materials. Stakeholders are critical in a terrace garden venture, irrespective of whether it is an individual or community effort. The partners in TG include the suppliers, farmers, consultants, collaborators, knowledge contributors, landowners and other stakeholders like regulators and investors. Maintaining a terrace garden necessitates the involvement of individuals with diverse skill sets. The absence of any required skills or the key partners can result in significant setbacks for the garden. Here are several crucial skill sets necessary for its successful maintenance:

Garden need suppliers and nurseries: The need includes the gardening suppliers who provide the tools essential for TG, including crop cutters, spades, pruning shears, and gloves. Seed suppliers provide a wide variety of seeds based on the need according to the different plants, climates, and gardening preferences. They must ensure the consistency and quality of the seeds, and encourage more robust plant development and higher yields. Insecticide and pesticide suppliers' needs is to provide quality assurance, which ensures that the products are safe, effective, and eco-friendly, in line with the principles of organic and sustainable TG. Reliable suppliers guarantee timely and consistent delivery of pesticides and insecticides, which will enable prompt action to address pest issues, particularly during critical plant growth phases or unexpected infestations.

Nurseries can be suppliers providing good quality seeds, and plant lings, as well as knowledge partners, inspiration sources and marketing partners for TG produce (Hanson *et al.*, 2021). Nursery workers can also step-in when there is additional work to be done, such as pest control, irrigation, soil maintenance etc.

Farmers and experienced gardeners: Experienced farmers act as a knowledge contributor as they are aware of the overall plant growth process, and local climate, soil conditions, and specific challenges that might impact TG like irrigation methods and pest control (Harvey *et al.*, 2017). They should be included in planning, implementation, and maintenance of a successful terrace garden. They will also help novices identify locally grown plants.

Experienced gardeners can be hired to help novice gardeners with setting up as well as maintaining good health in the TG. Individual TG enthusiasts as well as communities can hire them for training and developing knowledge in specialised areas.

Transportation partners: The transport company ensures the proper coordination to reduce the delays and regulate the proper supply chain logistics of TG business and a reliable transport company makes the smooth import and export of gardening materials and plants and maintains proper schedules efficiently.

Institutions: Several institutions like government agencies, educational and financial institutions can help spread awareness and information about the need and importance of TG, and wellness of it by hosting educational field trips; or through conducting various programs to promote hands-on learning in TG. Government agencies and representatives can provide help in procuring seeds, saplings and setting up composting facilities (Ravikumar *et al.*, 2009). They can also help in finding financing options for people who want to start their own terrace gardens.

Proficient specialists from these institutions, specializing in farming or cultivation or water management can give profitable direction and counsel. Their ability in planting strategies, irrigation, soil richness, pest control and administration can enormously contribute to the TG. For example, water management expert will have an awareness of plant water requirements, irrigation technologies and water conservation which will help optimize water resources effectively. This will help prevent waste and ensure that plants receive proper hydration. Additionally, the knowledge of organic mineral solution mix ratios with water allows for the judicious application of fertilizers, promoting optimal plant growth and health.

Media Partners and Wellness Organizations: Social media is a platform to promote various types of ideas and businesses that would motivate people to try new things. Social media channels for TG projects can help identify fellow enthusiast, business partners, and knowledge partners, and customers (Oktarina *et al.*, 2020). TG enthusiasts can work together to produce interesting content, like blog, gardening guides, or campaigns through social media. It can help focus attention on how TG is good for one's physical and mental health. Broadcasting activities associated with TG can help cultivate a positive mindset, and help share knowledge related to TG (Camille, 2018). Coordinated efforts across media can spread awareness of how small outdoor spaces and gardening encourage healthy habits of body and mind.

Landowners: Individuals or organizations who own the land where the terrace garden is situated provide the necessary space and access for its establishment. Their cooperation and support are crucial for the smooth operation of the garden.

Designers and Architect: Architects and land owners should work together during the design phase of the house or apartment to make provisions for TG taking into consideration space constraints (Akkara & Mohan, 2023). Employing creativity, an awareness of plant aesthetics, and a grasp of spatial design becomes instrumental in planning and designing the layout of the terrace garden. This approach aims to enhance visual appeal and optimize the utilization of available space, ultimately elevating the garden's aesthetic charm. The thoughtful integration of these elements contributes to the creation of a more comfortable and harmonious environment, fostering a space that is not only visually pleasing but also functionally efficient.

Government: The government plays a crucial role in increasing adoption of such sustainable practices by the community. For example, the Bihar Agriculture Department in India has put forth a proposal to incentivize TG on residential rooftops to enhance climatic conditions within the state. This initiative targeted five smart cities—Patna, Muzaffarpur, Gaya, Bhagalpur, and Bihar Sharif. Under this policy, the government plans to provide a 50% subsidy on the necessary inputs' cost for implementing TG in a 300 square feet area for interested individuals (Kumar *et al.,*2022).

g. Limitations of Terrace Gardening

TG also poses several limitations and challenges. Customer perceptions related to TG prove to be a considerable limitation. Additionally, long-term terrace aging and the environmental issues that follow it are reasons people do not prefer TG (Rao *et al.,*2022). The primary constraint lies in the confined space typical of terraces, which restricts the variety and quantity of plants that can be cultivated. Weight limits on the terrace structure require careful consideration to ensure it can bear the combined load of soil, containers, and plants.

Another challenge is sunlight availability, influenced by the orientation of the terrace and surrounding structures. Wind exposure and potential drainage issues demand meticulous planning. Temperature extremes, particularly in urban areas, can impact plant growth. Container gardening on terraces requires regular attention to nutrient levels, and the proximity of plants makes terrace gardens more susceptible to pests and diseases. Transporting materials and adherence to building regulations further add to the complexities.

Using kitchen wastes for TG presents several downsides. The practice is hindered by a limited range of input materials, resulting in a slow overall process. Additionally, it may contribute to uncontrolled greenhouse gas emissions and attract pests and rodents, among other challenges. These challenges must be overcome to have proper manure for TG and be sustainable ecologically and economically in the long run (Blonskaja *et al.,*2014). TG, has also been said to contribute to social inequality, as they are predominantly accessible to richer segments of society who own individual flats (Kumari et al.,2020). In our paper we found that, despite these challenges, successful TG is attainable through strategic planning, suitable plant choices, and diligent maintenance.

SWOC Analysis

Sustainability and innovation are important topics in today's world. A SWOC analysis can be used to understand the strengths, weaknesses, opportunities, and challenges of this innovative and challenging business.

TG affords advantages to all stakeholders. For people engaged in TG, it provides an opportunity for gardening, community work, physical and mental well-being, and an opportunity for business. When communities engaged in TG it results in economies of scale, and benefits from a robust distribution network. The practice maintains a good reputation and retains loyal stakeholders, offering a cost advantage and leveraging assets effectively to provide customers with original and healthy local products. Cultivating plants on balconies and terraces serves as an effective means of mitigating noise, capturing dust from nearby streets, retaining heat, and providing shade, particularly in areas with substantial sunlight exposure during the summer. These combined benefits contribute to the creation of a pleasant environment and enhance the aesthetic appeal of the neighbourhood. The presence of a variety of plants and flowers further yields positive ecological and aesthetic impacts (Mladenović *et al.,*2017).

TG faces weaknesses, including high labor costs and a substantial initial investment. The ongoing expense of maintaining plants poses a financial challenge, while the market's relative newness may result in a lack of interest among customers. Seasonal variations in sales can impact the business, and transportation costs for imported plants add to the overall challenges. TG also poses challenges to the building and will require inputs to strengthen the structure.

TG presents various opportunities for individuals and businesses to explore. With abundant sunlight, it offers an ideal environment for cultivating crops, fostering a hands-on farming experience. Entrepreneurs can turn it into a small business, tapping into the market for locally grown produce. Forming alliances with nearby retail shops enhances distribution channels. Expanding into online shopping caters to the e-commerce trend and donations towards TG initiatives may provide tax benefits.

TG's main challenge could come from the inexperience of novice farmers or gardeners. These new gardeners will require the support of all stakeholders, particularly experienced farmers, government and academic agriculture institutions to be able to manage their new projects. Training and education of these individuals will go a long way in making TG successful and building a strong community of support. TG comes with various challenges, including the issue of seeds not sprouting, which can hinder the initial stages of cultivation. The presence of pests and diseases poses a constant threat to the health of the plants, requiring vigilant pest control measures. Weather changes, such as sudden fluctuations in temperature or unexpected weather events, can adversely impact the growth and well-being of the crops. The scarcity of water and nutrients in a terrace garden can be a limiting factor, affecting the overall productivity. In contrast to ground level, the terrace experiences elevated wind speeds, necessitating adequate protection for plants to prevent potential damage (Attanasio *et al.,*2022). To overcome these challenges, it requires sufficient and proper planning, research, and implementation of appropriate measures. Thus, TG has vast potential for sustainable and innovative business growth, with the correct approach and management.

CONCLUSION

TG is a sustainable solution for urbanization issues, utilizing rooftop spaces to produce local food and enhance greenery while reducing the carbon footprint of food transportation. Despite its benefits, further research is necessary to understand its environmental impact and identify innovate farming techniques and understanding factors influencing its success to aid policymakers in promoting its adoption as an eco-friendly urban agriculture practice, fostering resilient and sustainable cities. Motivated by its potential as a home business opportunity, TG offers urban residents a chance to reconnect with food production, fostering entrepreneurship and community involvement. Its advantages, like shortening the food supply chain and promoting waste management.

Sustainability in TG relies on eco-conscious practices and innovative techniques, supported by meticulous planning, efficient time management, and strategic stakeholder engagement. Despite challenges such as space constraints and customer perceptions, proactive measures and careful management can help TG thrive. A SWOC analysis provides valuable insights into its strengths, weaknesses, opportunities, and challenges, guiding practitioners and policymakers in maximizing its potential. TG represents a transformative force in urban agriculture, offering a pathway towards greener, healthier cities. Continued research, innovation, and collaboration are crucial for creating thriving urban ecosystems in the future.

REFERENCES

Amos, C. C., Rahman, A., Karim, F., & Gathenya, J. M. (2018). A scoping review of roof harvested rainwater usage in urban agriculture: Australia and Kenya in focus. *Journal of Cleaner Production, 202*, 174–190. doi:10.1016/j.jclepro.2018.08.108

Åstebro, T., & Serrano, C. J. (2015). Business partners: Complementary assets, financing, and invention commercialization. *Journal of Economics & Management Strategy, 24*(2), 228–252. doi:10.1111/jems.12095

Attanasio, G., Preghenella, N., De Toni, A. F., & Battistella, C. (2022). Stakeholder engagement in business models for sustainability: The stakeholder value flow model for sustainable development. *Business Strategy and the Environment, 31*(3), 860–874. doi:10.1002/bse.2922

Bernardi, E. (2014). Joining Outside and Inside: Planted Terraces in Multi-Level Buildings. In New Urban Configurations (pp. 944-949). IOS Press.

Blonskaja, V., Põldnurk, J., & Loigu, E. (2014). Utilisation options for biodegradable kitchen waste in Estonia. SWOT analysis. *Proceedings of the International Conference on Environmental Engineering. ICEE, 9*(1).

Camille, F. (2018). Grow what you eat eat what you grow: Urban agriculture as middle class intervention in India. *Journal of Political Ecology, 25*, 222–249.

Chalmin-Pui, L. S., Griffiths, A., Roe, J., Heaton, T., & Cameron, R. (2021). Why garden?–Attitudes and the perceived health benefits of home gardening. *Cities (London, England), 112*, 103118. doi:10.1016/j.cities.2021.103118

Chapagain, T., & Raizada, M. N. (2017). Agronomic challenges and opportunities for smallholder terrace agriculture in developing countries. *Frontiers in Plant Science, 8*, 331. doi:10.3389/fpls.2017.00331 PMID:28367150

Chatterjee, A., Debnath, S., & Pal, H. (2020). Implication of urban agriculture and vertical farming for future sustainability. In *Urban Horticulture - Necessity of the Future*. Intech Open. doi:10.5772/intechopen.91133

Chidambaram, C., Nath, S. S., Varshney, P., & Kumar, S. (2022). Assessment of terrace gardens as modifiers of building microclimate. *Energy and Built Environment, 3*(1), 105–112. doi:10.1016/j.enbenv.2020.11.003

Corona, B., Shen, L., Reike, D., Carreón, J. R., & Worrell, E. (2019). Towards sustainable development through the circular economy—A review and critical assessment on current circularity metrics. *Resources, Conservation and Recycling, 151*, 104498. doi:10.1016/j.resconrec.2019.104498

Dash, D., & Deole, S. (2020). Terrace gardening: An amazing step towards growing vegetables on roof. *Vigyan Varta, 1*(7), 20–24.

Deng, C., Guangye, Z., Yaojun, L., Xiaodong, N., Zhongwu, L., Junyu, L., & Damei, Z. (2021). Advantages and disadvantages of terracing: A comprehensive review. *International Soil and Water Conservation Research, 9*(3), 344–359. doi:10.1016/j.iswcr.2021.03.002

Hanson, H. I., Eckberg, E., Widenberg, M., & Olsson, J. A. (2021). Gardens' contribution to people and urban green space. *Urban Forestry & Urban Greening*, *63*, 127198. doi:10.1016/j.ufug.2021.127198

Harvey, C. A., Martínez-Rodríguez, M. R., Cárdenas, J. M., Avelino, J., Rapidel, B., Vignola, R., Donatti, C. I., & Vilchez-Mendoza, S. (2017). The use of Ecosystem-based Adaptation practices by smallholder farmers in Central America. *Agriculture, Ecosystems & Environment*, *246*, 279–290. doi:10.1016/j.agee.2017.04.018

Josson Akkara, P., & Mohan, R. (2023). Perception and Acceptance of Terrace and Balcony Gardening for Sustainable Future of Bengaluru. In *AI and Business, and Innovation Research: Understanding the Potential and Risks of AI for Modern Enterprises* (pp. 787–796). Springer Nature Switzerland.

Khalil, M. K., & Muneenam, U. (2021). Total quality management practices and corporate green performance: Does organizational culture matter? *Sustainability (Basel)*, *13*(19), 11021. doi:10.3390/su131911021

Kishore, M. V., & Patil, M. S. (2015). a terrace garden from biodegradable waste. *Bionano Frontier*, *8*(3), 1–5.

Kumar, A., Yadav, S. K., Kharga, S., & Jangid, R. (2022). Rooftop Gardening: A New Era of Gardening. *Agriculture & Food: E-Newsletter*, *4*(7), 599–603.

Kumari, S., Lal, K., & Lata, R. (2020). Assessing sustainability of urban and Peri-Urban agricultural development in Mumbai. *Journal of Pharmacognosy and Phytochemistry*, *9*(6S), 114–118.

Medha, M., Lavanya, T., Sunandini, G. P., & Meena, A. (2022). Terrace Gardening: A Sustainable yet Economic Approach. *International Journal of Environment and Climate Change*, *12*(10), 925–932. doi:10.9734/ijecc/2022/v12i1030881

Mladenović, E., Lakićević, M., Pavlović, L., Hiel, K., & Padejčev, J. (2017). Opportunities and benefits of green balconies and terraces in urban conditions. *Contemporary Agriculture*, *66*(3-4), 38–45. doi:10.1515/contagri-2017-0017

Mohan, S. V., Hemalatha, M., Amulya, K., Velvizhi, G., Chiranjeevi, P., Sarkar, O., Kumar, A. N., Krishna, K. V., Modestra, J. A., Dahiya, S., Yeruva, D. K., Butti, S. K., Sravan, J. S., Chatterjee, S., & Kona, R. (2020). Decentralized urban farming through keyhole garden: A case study with circular economy and regenerative perspective. *Materials Circular Economy*, *2*(1), 1–12. doi:10.1007/s42824-020-00011-1

Naveena, K. P., & Sahana, R. T. (2017). Organic terrace gardening, a new vista in bringing back to traditional food system: Economic issues. *The Journal of Rural and Agricultural Research*, *17*(2), 11–14.

Oktarina, S., Sumardjo, S., Purnaningsih, N., & Hapsari, D. R. (2020). Activities of Farmer Women Groups in Utilizing Digital Communication Media in Urban Farming Activities in Bogor City. *International Journal of Progressive Sciences and Technologies*, *19*(1), 241–249.

Palm, H. W., Knaus, U., Appelbaum, S., Goddek, S., Strauch, S. M., Vermeulen, T., Haïssam Jijakli, M., & Kotzen, B. (2018). Towards commercial aquaponics: A review of systems, designs, scales and nomenclature. *Aquaculture International*, *26*(3), 813–842. doi:10.1007/s10499-018-0249-z

Patil, A. R., Patil, A., & Mail, I. D. (2020). The role of the food and fertilizers in antimicrobial resistance in human and its preventive measures. *International Journal of Innovative Science, Engineering and Technology, 1*, 31-37.

Poornima, C. P., & Suryawanshi, P. D. (2020). Vegetable Consumption Pattern and Health Consciousness of Urban Terrace Garden Farmers. *Journal of Extension Education, 32*(1), 6471. doi:10.26725/JEE.2020.1.32.6471-6474

Quested, E., Thøgersen-Ntoumani, C., Uren, H., Hardcastle, S. J., & Ryan, R. M. (2018). Community gardening: Basic psychological needs as mechanisms to enhance individual and community well-being. *Ecopsychology, 10*(3), 173–180. doi:10.1089/eco.2018.0002

Ramirez-Andreotta, M. D., Tapper, A., Clough, D., Carrera, J. S., & Sandhaus, S. (2019). Understanding the intrinsic and extrinsic motivations associated with community gardening to improve environmental public health prevention and intervention. *International Journal of Environmental Research and Public Health, 16*(3), 494–504. doi:10.3390/ijerph16030494 PMID:30754611

Rao, P. B., Inampudi, M. P., & Srikumar, M. S. (2022). Consumer Attitude towards Organic Terrace Gardening-A Study in Bangalore City. *International Journal of All Research Education and Scientific Methods, 10*(9), 1–6.

Ravi Kumar, P., Jayaram, A., & Somashekar, R. K. (2009). Assessment of the performance of different compost models to manage urban household organic solid wastes. *Clean Technologies and Environmental Policy, 11*(4), 473–484. doi:10.1007/s10098-009-0204-9

Reddy, R. V. S. K., Omprasad, J., & Janakiram, T. (2022). Technological innovations in commercial high-tech horticulture, vertical farming and landscaping. *International Journal of Innovative Horticulture, 11*(1), 78–91. doi:10.5958/2582-2527.2022.00008.2

Rondhi, M., Pratiwi, P. A., Handini, V. T., Sunartomo, A. F., & Budiman, S. A. (2018). Agricultural land conversion, land economic value, and sustainable agriculture: A case study in East Java, Indonesia. *Land (Basel), 7*(4), 148. doi:10.3390/land7040148

Samus, A., Freeman, C., Dickinson, K. J., & van Heezik, Y. (2023). An examination of the factors influencing engagement in gardening practices that support biodiversity using the theory of planned behavior. *Biological Conservation, 286*, 110252. doi:10.1016/j.biocon.2023.110252

Shelef, O., Weisberg, P. J., & Provenza, F. D. (2017). The value of native plants and local production in an era of global agriculture. *Frontiers in Plant Science, 8*, 2069. doi:10.3389/fpls.2017.02069 PMID:29259614

Vazhacharickal, P. J. (2014). Balcony and terrace gardens in urban greening and local food production, scenarios from Mumbai Metropolitan Region, India. *International Journal of Food. Agriculture and Veterinary Sciences, 4*(2), 149–162.

Winkler, B., Maier, A., & Lewandowski, I. (2019). Urban gardening in Germany: Cultivating a sustainable lifestyle for the societal transition to a bioeconomy. *Sustainability (Basel), 11*(3), 801. doi:10.3390/su11030801

Chapter 13
Enhancing Farmer-Consumer Linkages Through Technology Adoption and Sustainable Marketing in Zimbabwe's Agricultural Sector

Option Takunda Chiwaridzo
https://orcid.org/0000-0001-9755-9123
University of Science and Technology Beijing, Beijing, China

Tendai Kaponda
https://orcid.org/0009-0003-3287-4335
School of Public Policy and Administration, Central China Normal University, Wuhan, China

Reason Masengu
Middle East College, Oman

ABSTRACT

This quantitative study explores the relationships between precision agriculture, vertical farming, and bio-genetic technologies among 402 smallholder farmers in Zimbabwe. Results highlight the positive impact of these technologies on sustainable farming practices and marketing, with sustainable practices playing a mediating role. The integrated approach emphasizes the importance of combining high-tech solutions with agroecological practices to realize full benefits and meet emerging consumer demands for responsible produce. Recommendations are provided for policy interventions to promote synergistic development, ultimately enhancing productivity, sustainability, and market access for smallholder farmers in Zimbabwe.

DOI: 10.4018/979-8-3693-4864-2.ch013

INTRODUCTION

The agricultural sector in Zimbabwe plays a critical role in the economy, contributing significantly to food security, employment, and rural livelihoods. Agriculture accounts for about 12-15% of Zimbabwe's GDP and over 60% of raw materials used in the manufacturing and industrial sectors (Runganga and Mhaka, 2021). Smallholder farmers make up majority of the agricultural producers, highlighting the sector's importance for rural livelihoods and poverty reduction. However, the sector faces persistent challenges like inefficient traditional marketing channels, poor transport and storage infrastructure, limited access to credit, and low technology adoption (Matandare, 2017, Gwanongodza, 2020, GODFREY, 2016). These limitations hamper productivity, farmer incomes, food security, and sustainability within the sector.

In particular, the lack of adoption of innovative technologies like precision agriculture, vertical farming, and bio-genetic techniques constrains potential efficiency and sustainability gains. Precision agriculture harnesses tools like GPS, satellite imagery, sensors, and variable rate technologies to optimize resource usage, reduce waste, and enhance productivity and traceability across agricultural operations (Zengeya et al., 2021, Musungwini et al., 2023). Vertical farming enables optimized indoor food production through hydroponics, LED lighting, climate control, and multi-level crop cultivation (Ndhlovu and Mhlanga, 2023). It allows for high yields with reduced water and land usage. Bio-genetic technologies like genetic engineering, gene editing, and marker-assisted breeding facilitate rapid development of climate resilient, nutritionally improved, and higher yielding seed varieties (Musara et al., 2012). Wider adoption of such technologies in Zimbabwe, combined with enabling sustainable farming practices, can strengthen farmer-consumer linkages and sustainable marketing.

Sustainable farming practices like integrated pest management, conservation agriculture, agroforestry, and soil conservation have proven environmental and economic benefits but see limited uptake among smallholder farmers in Zimbabwe (Mujeyi et al., 2020). However, greater adoption of precision, vertical, and bio-genetic technologies may directly enable and incentivize such sustainable practices by enhancing efficiency, traceability, climate resilience, and adaptation support. In turn, widespread sustainable practices can facilitate farmer access to premium markets, higher prices, and direct consumer linkages through sustainability certifications and blockchain-enabled transparent food tracing systems.

The aim of this quantitative study is to analyze the relationships between adoption of precision agriculture technology, vertical farming technology, and bio-genetic technology as independent variables, sustainable farming practices as a mediating variable, and sustainable marketing of produce as the dependent variable in Zimbabwe's agricultural sector. Specifically, the study seeks to assess the current level of technology adoption, examine how these technologies influence uptake of sustainable practices, evaluate the role of sustainable practices in enabling access to sustainable agriculture marketing platforms and price premiums, and identify challenges influencing technology adoption and sustainable farming. The findings will provide recommendations for targeted policies and interventions to promote technology adoption, sustainable practices, and direct farmer-consumer market linkages to strengthen agricultural sustainability, productivity, and rural livelihoods in Zimbabwe. Surveys and statistical modeling will be used to analyze the relationships between the variables

LITERATURE AND HYPOTHESIS DEVELOPMENT

Precision Agriculture Technology, Sustainable Farming Practices, and Sustainable Marketing in Agriculture

Precision agriculture technologies like GPS, sensors, robotics, and variable rate application are rapidly transforming farming practices worldwide (Cisternas et al., 2020). Studies show precision agriculture can enable more efficient use of inputs like water, fertilizers and pesticides, leading to both economic and environmental benefits (Sishodia et al., 2020). Key advantages include reduced operational costs through optimized input usage and higher yields from improved crop health and productivity (Erickson and Fausti, 2021, Mendes et al., 2020). Variable rate irrigation and fertilization tailored to within-field soil variations can lower resource waste significantly (Bhat and Huang, 2021).

When combined with monitoring and mapping systems, precision technology allows better targeting of resources, reducing waste (Sanjeevi et al., 2020). Documenting precise field usage can also verify sustainable practices. For example, variable rate irrigation matched to soil conditions and crop needs can lower water usage by 25-60% versus uniform irrigation (Yin et al., 2021). Such water savings are crucial amidst water scarcity concerns globally.

The data and insights from precision technology can promote sustainable farming practices like conservation tillage, integrated pest management (IPM) and optimized nutrient application. (Singh et al., 2020) found that yield monitor data encouraged adoption of no-till practices and IPM among farmers by demonstrating their yield benefits. Reduced tillage and IPM are key sustainability goals given their soil and environmental advantages. Optimized input application from precision tools also supports sustainability. In Italy, precision practices reduced fertilizer use in corn by over 30% without yield declines (Blasch et al., 2022). Similar reductions have been achieved in China, Brazil and other major agricultural countries (Carrer et al., 2022, Lee et al., 2021). This improves farm economics and lowers agriculture's environmental footprint.

However, disadvantages like high upfront costs of equipment and need for technical skills have slowed adoption, especially for smallholder farmers (da Silveira et al., 2021). Lack of technical support and training on utilizing precision data further limits its impact. Proper use of technologies like yield monitors, soil sensors, and VRT applicators is essential to realizing the full benefits (Raj et al., 2022).

Finally, sustainable marketing programs like organic certification, fair trade, and eco-labels rely on documentation of on-farm practices. The detailed field data from precision agriculture systems can verify sustainable practices for market access and price premiums (Karunathilake et al., 2023). For smallholders, obtaining certifications to sell to premium export markets may not be feasible without precision technology. In summary, precision technology shows promise for enabling sustainable farming practices and marketing opportunities globally, but focused efforts are needed to spread these benefits more inclusively. Hence, author proposed the following hypotheses:

H_1: *There is a positive significant relationship between precision agriculture technology and sustainable farming practices.*

H_2: *There is a positive significant relationship between precision agriculture technology and sustainable marketing in agriculture.*

H_3: *There is a positive significant relationship between sustainable farming practices and sustainable marketing in agriculture.*

H_4: *Sustainable farming practices mediates relationship between precision agriculture technology and sustainable marketing in agriculture*

Vertical Farming Technology, Sustainable Farming Practices, and Sustainable Marketing in Agriculture

Vertical farming has emerged as an innovative approach to increase food production in space-constrained urban environments (Van Delden et al., 2021). These technologies utilize multi-level growing platforms, precision irrigation, LED lighting, and climate control to optimize crop yields (Avgoustaki and Xydis, 2020b). Studies have found vertical farming can promote sustainability through reduced water usage, elimination of transport needs, and recycling of nutrients (Roberts et al., 2020, Chatterjee et al., 2020). For example, aeroponic and hydroponic techniques in vertical farms can lower water usage by up to 90% compared to conventional farming (Avgoustaki and Xydis, 2020a, Asseng et al., 2020).

The closed-loop and highly managed growing environment of vertical farms enables sustainable practices like integrated pest management, zero pesticide use, and recycled water usage. (Heo and Lee, 2021, Maheshwari, 2021) found vertical farms in Korea and Japan successfully grew pesticide-free produce using IPM. On-site water recycling systems are also commonly employed, with some vertical farms reusing up to 80% of drain water (Mir et al., 2022). Such capabilities can obtain organic certifications and appeal to health-conscious urban consumers.

Vertical farming's-controlled system facilitates documentation of growing practices to support sustainable marketing claims. In the US, vertical produce labeled as pesticide-free or locally-grown have shown price premiums and consumer preference (Butturini and Marcelis, 2020). As technology costs decrease, vertical farming can supply cities with affordable sustainable produce while reducing the environmental impacts of long supply chains (Van Gerrewey et al., 2021). More policies and incentives can further encourage adoption of vertical technology. However, high capital and operating costs of artificial lighting, cooling systems, and real estate remain a key disadvantage to mainstream adoption (Chatterjee et al., 2020). Technical expertise required to manage integrated systems from lighting to hydroponics also poses a barrier. Government subsidies and renewable energy integration can help improve vertical farming's economic viability (Martinez, 2023). From literature the author therefore proposes the next hypotheses:

H_5: *There is a positive significant relationship between vertical farming technology and sustainable farming practices.*

H_6: *There is a positive significant relationship between vertical farming technology and sustainable marketing in agriculture.*

H_7: *Sustainable farming practices mediates relationship between vertical farming technology and sustainable marketing in agriculture.*

Bio-Genetic Technology, Sustainable Farming Practices, and Sustainable Marketing in Agriculture

Advances in bio-genetic technologies like genetic engineering, gene editing, and marker-assisted breeding are transforming crop improvement globally (Price, 2022, Lin et al., 2021). Key advantages of these technologies include the ability to precisely develop higher yielding, nutrient-dense, and stress resilient crop varieties in relatively shorter timeframes compared to conventional breeding (Rose et al., 2022,

Long et al., 2018). This has major implications for improving food and nutrition security amidst climate change pressures.

While impacts vary case-by-case, certain biotech applications align with sustainability goals. Herbicide tolerant GM crops have enabled greater adoption of no-till agriculture, reducing fuel usage, carbon emissions, and soil erosion (Mall et al., 2019, MEYER, 2019). Such reductions in mechanical tillage are substantial sustainability gains. Biotech-derived pest resistant crops like Bt cotton and Bt brinjal have lowered broad spectrum pesticide applications in India while increasing smallholder farmer incomes (Tawfik et al., 2022)

Lastly, biotech innovations can provide verifiable sustainability traits to support marketing claims. GM drought tolerant corn in South Africa has climate-smart eco-labels for market differentiation (Ugwu, 2021). DNA fingerprinting verifies seed purity and origins for sustainability certifications. While debates persist around genetic engineering, improved regulations and sustainability assessments can aid wider adoption of bio-genetic technologies for sustainable farming worldwide.

Overall, the literature shows bio-genetic technologies have the potential to enable sustainable agricultural practices and marketing opportunities (Escobar-Pemberthy and Calle Saldarriaga, 2020). However, context-specific policies and safeguards are critical to balancing benefits and risks, because there are also risks and uncertainties associated with genetic modification that require careful assessment. Potential for mono-cropping, loss of biodiversity, and equitable access to technology remain key challenges (Glaze-Corcoran et al., 2020). Strict regulatory frameworks for testing and monitoring effects on health, environment and social outcomes are critical. Building from literature review, author proposes the subsequent hypotheses:

H_8: *There is a positive significant relationship between bio-genetic technology and sustainable farming practices.*

H_9: *There is a positive significant relationship between bio-genetic technology and sustainable marketing in agriculture.*

H_{10}: *Sustainable farming practices mediates relationship between bio-genetic technology and sustainable marketing in agriculture.*

RESEARCH METHODOLOGY

This study utilized a quantitative methodology to explore relationships between variables using statistical analysis. As suggested by (Hermida, 2015), a quantitative approach is appropriate when testing hypotheses and models by examining correlations and statistical significance. Quantitative methods allow for precise measurement using scales and surveys. Smart PLS-SEM was used to analyze the data and determine cause-and-effect relationships between variables. Smart PLS-SEM is known for its predictive capabilities (Akter et al., 2017). Bootstrapping, recommended by (Pek et al., 2019), assessed the sample's distribution to make inferences about its characteristics. The standardized quantitative methodology allows for replication in other contexts, enhancing the credibility of the findings. Smart PLS 3 was preferred over CBSEM for its advantages, including the ability to handle multivariate data, sample sizes, complex models, and testing of factor loadings.

Sampling and Questionnaire Process

The study utilized a survey methodology with a structured questionnaire using a 5-point Likert scale, administered electronically via Google Forms. The survey was distributed to a targeted sample of small holder farmers. A simple random sampling method was chosen to reduce selection bias and allow for generalizable findings (Fox et al., 2022). The sample size of 900 was determined based on (Krejcie and Morgan, 1970) table, for large populations, to provide 95% confidence and 5% margin of error. This approach is valued for its simplicity, flexibility, and standardization. Extensive ethical protocols were implemented including informed consent and confidential data collection/analysis. Of the 825 surveys distributed, 402 complete responses were received, a 48.7% response rate attributed to anonymity assurances and repeated reminders.

Measurement and Variables Explanation

This section outlines the process of determining the dependent and independent variables. The paper has three independent variables: The independent variables of precision agriculture technology, vertical farming technology, and bio-genetic technology were selected based on evidence from prior research showing their potential to improve agricultural sustainability and productivity (Pierpaoli et al., 2013; Al-Kodmany, 2018; Qaim & Kouser, 2013). However, adoption of these innovative technologies remains low in Zimbabwe, highlighting the need to examine influencing factors to inform targeted policies and interventions. Sustainable farming practices was included as a mediating variable because despite the known benefits of techniques like conservation agriculture and integrated pest management, their uptake among Zimbabwean smallholders is limited (Makate et al., 2019). This study hypothesizes the adoption of the studied technologies can enable sustainable practices. Finally, sustainable marketing of agricultural produce was chosen as the dependent variable because strengthening farmer linkages to markets through direct sales and sustainability certifications has shown income growth potential (Masanganise et al., 2021). It is hypothesized that widespread sustainable practices will facilitate farmer access to such marketing platforms. Each of the independent, mediating, and dependent variables was represented by five questionnaire items to quantitatively establish their hypothesized relationships. In summary, the selection of these variables was a deliberate process, considering their contextual relevance in Zimbabwe, alignment with the research objectives, and feasibility for data collection and analysis using surveys and statistical modeling. These five variables are visually represented in Figure 1 below:

Key

Direct relationship Indirect relationship

Results

This study used Smart PLS-SEM for data analysis and hypothesis testing. It involved a careful three-step process. First, method bias was thoroughly assessed to ensure the data was accurate and reliable. Next, the model and constructs were evaluated in-depth to validate the integrity of the data. Finally, the hypotheses were rigorously tested using PLS-SEM. This allowed for a comprehensive exploration

Figure 1. Conceptual model

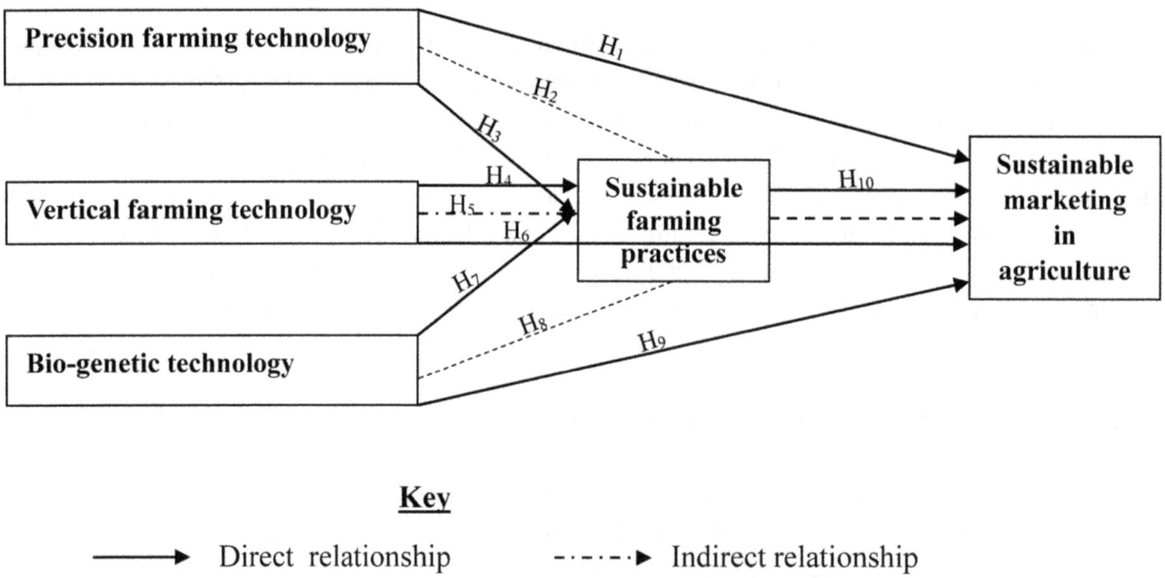

of the relationships between variables and their statistical significance. This step-by-step process was important for strengthening the robustness and credibility of the study's findings. The careful evaluation and testing procedures helped ensure the analysis was thorough and the results were sound.

STEP 1: METHOD BIAS

Non-Response Partiality

When looking at the factors being measured, it's important to think about the possible distortion caused by non-response bias (Khan, 2020). To check for non-response bias, the researcher did a T-test. They compared the scores on the variables between the first people who responded and those who responded at the end. The results showed no big differences between these groups. This means that non-response was not a major concern that affected the results of this study.

Common Method Bias

The researcher in this study followed the methodology suggested by (Aguirre-Urreta and Hu, 2019) to reduce possible biases in their analysis. They used PLS-SEM as their statistical method. This helped deal with issues like outliers and collinearity, improving the accuracy of the study results. Also, to lessen the effect of outliers and prevent overfitting, they used techniques like bootstrapping and cross-validation. Using these recommended methods helped the researcher do a more precise analysis for their study.

Figure 2. Pictorial evidence of factor loadings after PLS algorithm calculations

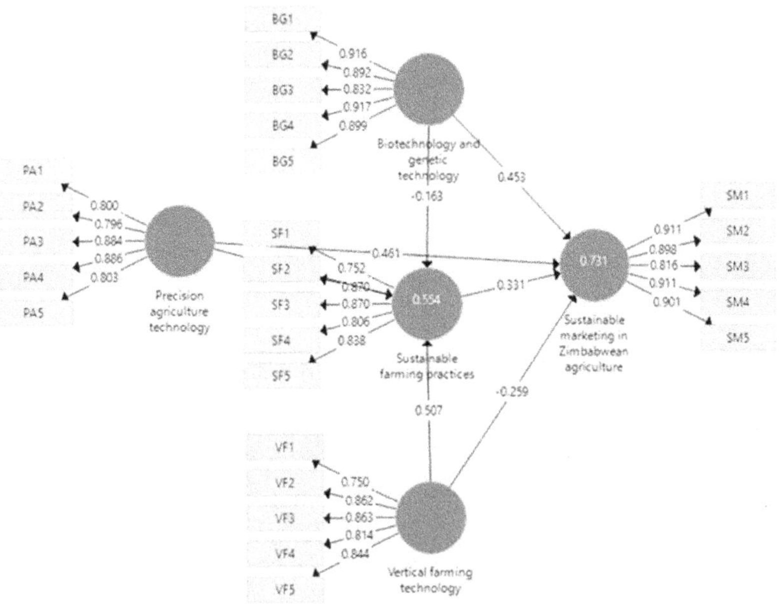

Step 2: Model Valuation and the Assenting of Factor Analysis

The researcher used Smart PLS-SEM to assess, refine, and test hypotheses. The evaluation and factor analysis had two steps: First, they looked at the measurement model. They checked the reliability and validity, using things like outer loadings, Cronbach's Alpha, Average Variance Extracted (AVE), and Composite Reliability. This is shown in Figure 2 and Table 1. Next, they checked for discriminant validity in two ways: the Fornell-Larcker method and the Heterotrait-Monotrait (HTMT) method. These helped establish that the constructs in the measurement model were distinct.Finally, hypothesis testing was done using bootstrapping. This allowed examining the direct relationships between the constructs. Using this step-by-step process helped evaluate the measurements and test the hypotheses rigorously. The methods ensured reliability, validity, and discriminant validity, strengthening the analysis.

Reliability and Scale Assessment

Figure 2 and Table 1 show the results of the reliability and validity assessments done in this study. The researcher used several criteria to check convergent validity: First, they looked at the outer loadings. All were over 0.5, meaning the indicators contributed well to their constructs. This aligns with (Gio, 2022)'s recommendations Second, composite reliability was calculated. All constructs had values over 0.7, indicating consistency based on (Aburumman et al., 2022)'s guidelines. Third, the Average Variance Extracted (AVE) for each construct was assessed. The AVE values were all above 0.5, following (Ardani, 2021)'s suggestions. For reliability, Cronbach's alpha and Rho A were used for each variable. As shown in Table 1, all were over 0.7, demonstrating satisfactory reliability based on (Cho, 2021)'s guidelines. Using these multiple metrics allowed thorough evaluation of reliability and validity.

Table 1. Factor loadings, Cronbach's alpha, AVE, CR, Rho_A, and CA

Description	Constructs	Factor Loadings	Cronbach Alpha	AVE	Composite Reliability	Rho_A
Precision Agriculture Technology	PA1 PA2 PA3 PA4 PA5	0.800 0.796 0.884 0.886 0.803	0.891	0.697	0.951	0.938
Vertical Farming Technology	VF1 VF2 VF3 VF4 VF5	0.751 0.862 0.863 0.814 0.844	0.884	0.685	0.916	0.887
Bio-genetic Technology	BG1 BG2 BG3 BG4 BG5	0.916 0.892 0.832 0.917 0.899	0.936	0.796	0.951	0.938
Sustainable Farming Practices	SF1 SF2 SF3 SF4 SF5	0.752 0.870 0.870 0.806 0.838	0.885	0.686	0.916	0.890
Sustainable Marketing in Zimbabwean Agriculture	SM1 SM2 SM3 SM4 SM5	0.911 0.898 0.816 0.911 0.901	0.933	0.789	0.949	0.934

Note(s): a) All item loadings >0.7 indicates indicator reliability (Hair Jr et al., 2021)
b) All average variance extracted (AVE) > 0.5 indicates convergent reliability (Lleo et al., 2021, Fornell and Larcker, 1981)
c) All composite reliability (CR) > 07 indicates internal consistency (Hair Jr et al., 2021)
d) All Cronbach's alpha >0.7 indicates indicator reliability (Hair Jr et al., 2021)

Discriminant Validity and Assessment

Two methods were used to evaluate the discriminant validity. Firstly, author employed the popular technique proposed by (Fornell and Larcker, 1981) and the outcomes are shown in Table 2. The second method was Heterotrait-Monotrait (HTMT) ratios shown in Table 3. According to the criteria suggested by (Sanmukhiya, 2020), the HTMT ratio must be lower than 0.85 for theoretically diverse concepts and below 0.90 for theoretically comparable concepts.

Table 2 offers a comprehensive view of the correlation coefficients among the seven variables under investigation, a widely accepted criterion for assessing validity, in line with the endorsement of (Fornell and Larcker, 1981). The most prominent correlation emerges between precision agriculture technology and sustainable marketing in Zimbabwean agriculture, revealing a robust positive relationship with a coefficient of 0.759. Likewise, the correlation between sustainable marketing in Zimbabwean agriculture and sustainable farming practices demonstrates a significant correlation effect of 0.624. Individually, the variables exhibit positive correlations as follows: Bio-genetic technology with a coefficient of 0.892, precision agriculture technology registering a coefficient value of 0.835, sustainable farming practices standing at 0.828, sustainable marketing in Zimbabwean agriculture with a coefficient of 0.857 and

Table 2. Fornell-Larker CRITERION

	Bio-Genetic Technology	Precision Agriculture Technology	Sustainable Farming Practices	Sustainable Marketing in Zimbabwean Agriculture	Vertical Farming Technology
Bio-genetic Technology	0.892				
Precision Agriculture Technology	0.402	0.835			
Sustainable Farming Practices	0.349	0.619	0.828		
Sustainable Marketing in Zimbabwean Agriculture	0.597	0.759	0.624	0.857	
Vertical Farming Technology	0.607	0.347	0.584	0.369	0.819

vertical farming technology recording a coefficient value of 0.819. These values collectively affirm the presence of discriminant validity. The subsequent method, as presented in Table 3, employs the HTMT approach for further data scrutiny.

The Heterotrait-Monotrait Ratio (HTMT) analysis presented in Table 3 demonstrates strong evidence of discriminant validity among the constructs in the model, with all HTMT values below the threshold of 0.85, supporting the reliability of the study's findings as supported by (Sanmukhiya, 2020). This indicates that the measurement model effectively distinguishes between Bio-genetic Technology, Precision Agriculture Technology, Sustainable Farming Practices, Sustainable Marketing in Zimbabwean Agriculture, and Vertical Farming Technology. Each construct captures a distinct aspect of the phenomena under investigation without significant overlap with other constructs. These findings enhance the reliability of the study's results, affirming the distinctiveness of the variables and supporting the validity of the research framework.

Table 3. Heterotrait-Monotrait ratio (HTMT)

	Bio-Genetic Technology	Precision Agriculture Technology	Sustainable Farming Practices	Sustainable Marketing in Zimbabwean Agriculture	Vertical Farming Technology
Bio-genetic Technology					
Precision Agriculture Technology	0.435				
Sustainable Farming Practices	0.379	0.689			
Sustainable Marketing in Zimbabwean Agriculture	0.639	0.823	0.685		
Vertical Farming Technology	0.663	0.387	0.659	0.408	

Figure 3. Direct relationships and hypothesis testing

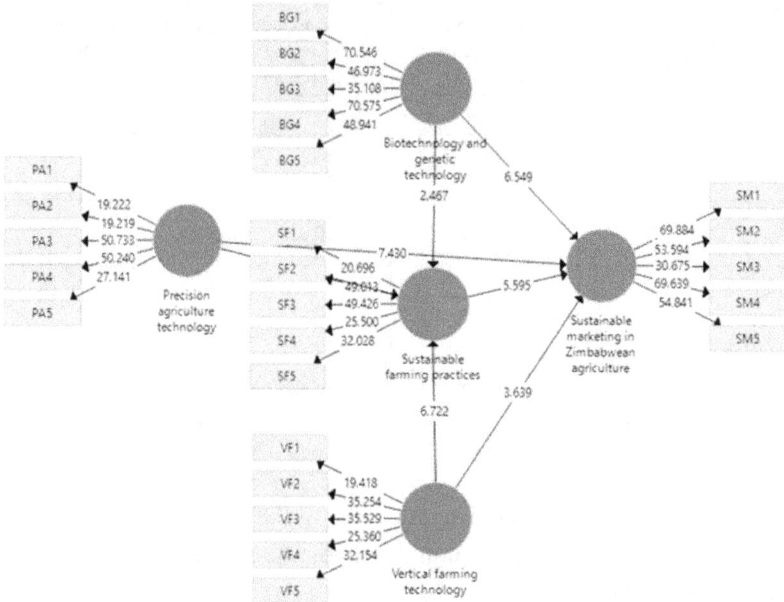

Step 3: Hypothesis Testing

In this study, the researchers followed the guidance of (Hair Jr and Sarstedt, 2021) by employing PLS-SEM to establish predictive relationships and statistical models that elucidate causal correlations. To gauge the significance of structural paths, the author utilized bootstrapping, a statistical method involving the creation of multiple smaller samples from the dataset. Bootstrapping is employed to approximate data normality and provides estimates of the distribution's characteristics within the population, as emphasized by (Cheah et al., 2020). This technique enhances the research by offering more robust estimates of the model's significance and inferential statistics. The summarized findings, as presented in Table 4 and Figure 3, furnish valuable insights into the connections between variables, thereby enhancing the study's statistical rigor and overall credibility.

Direct Relationships and Hypothesis Testing

Direct relationship is the direct effect amongst two constructs in a model, the magnitude and significance of the direct relationship can be interpreted using path coefficients or path weights obtained from the analysis and the research has eleven direct relationships: $\mathbf{H_1}$ bio-genetic technology (BG) to sustainable farming practices (SF), $\mathbf{H_2}$ bio-genetic technology (BG) to sustainable marketing in Zimbabwean agriculture (SM), $\mathbf{H_3}$ precision agriculture technology (PA) to sustainable farming practices (SF), $\mathbf{H_4}$ precision agriculture technology (PA) to sustainable marketing in Zimbabwean agriculture (SM), $\mathbf{H_5}$ sustainable farming practices (SF) to sustainable marketing in Zimbabwean agriculture (SM), $\mathbf{H_6}$ vertical farming technology (VF) to sustainable farming practices (SF), $\mathbf{H_7}$ vertical farming technology (VF) to sustainable marketing in Zimbabwean agriculture (SM) as shown in Table 4 below:

Table 4. Direct relationship and hypothesis testing

Hypothesis	Relationships	Std Beta	Std Value	(t-Value) ^	Decision	(2.5%) CI UL	(97.5%) CI LL
H_1	BG SF	0.258	0.079	2.467	accepted	0.133	0.227
H_2	BGSM	0.646	0.045	6.549	accepted	0.552	0.422
H_3	PA SF	0.274	0.046	9.472	accepted	0.185	0.363
H_4	PASM	0.337	0.053	7.430	accepted	0.089	0.430
H_5	SF SM	0.229	0.083	5.595	accepted	0.076	0.405
H_6	VF SF	0.218	0.075	6.922	accepted	0.078	0.383
H_7	VF SM	0.216	0.050	3.639	accepted	0.115	0.225

The statistical analysis and findings presented in Table 4 demonstrate that all seven direct hypotheses are statistically significant, as indicated by their respective t-values, which exceed the recommended threshold of 1.96 (Bakhar et al., 2023). The standardized beta coefficients (Std Beta) provide insights into the magnitude of the relationships between the constructs. **H1** (Bio-genetic Technology to Sustainable Farming Practices): Std Beta = 0.258, t-value = 2.467, supporting a positive relationship between bio-genetic technology and sustainable farming practices. **H2** (Bio-genetic Technology to Sustainable Marketing in Zimbabwean Agriculture): Std Beta = 0.646, t-value = 6.549, indicating a strong positive relationship between bio-genetic technology and sustainable marketing in Zimbabwean agriculture. **H3** (Precision Agriculture Technology to Sustainable Farming Practices): Std Beta = 0.274, t-value = 9.472, supporting a positive relationship between precision agriculture technology and sustainable farming practices.**H4** (Precision Agriculture Technology to Sustainable Marketing in Zimbabwean Agriculture): Std Beta = 0.337, t-value = 7.430, indicating a positive relationship between precision agriculture technology and sustainable marketing in Zimbabwean agriculture. **H5** (Sustainable Farming Practices to Sustainable Marketing in Zimbabwean Agriculture): Std Beta = 0.229, t-value = 5.595, supporting a positive relationship between sustainable farming practices and sustainable marketing in Zimbabwean agriculture. **H6** (Vertical Farming Technology to Sustainable Farming Practices): Std Beta = 0.218, t-value = 6.922, indicating a positive relationship between vertical farming technology and sustainable farming practices. **H7** (Vertical Farming Technology to Sustainable Marketing in Zimbabwean Agriculture): Std Beta = 0.216, t-value = 3.639, supporting a positive relationship between vertical farming technology and sustainable marketing in Zimbabwean agriculture. These results underscore the significance of bio-genetic technology, precision agriculture technology, and vertical farming technology in promoting sustainable farming practices and marketing in Zimbabwe's agricultural sector.

Mediation Relationships and Hypothesis Testing

A mediation relationship refers to a relationship between three variables: the independent variable (IV), the mediator variable (MV), and the dependent variable (DV). It represents a process in which the IV influences the DV indirectly through the MV. The MV acts as an intermediate variable that mediates the relationship between the IV and the DV. In this study we had three mediation relationships: $\mathbf{H_8}$ bio-genetic technology (IV) sustainable farming practices (MV) sustainable marketing in Zimbabwean agriculture (DV), $\mathbf{H_9}$ precision agriculture technology (IV), sustainable farming practices (MV), sustainable mar-

Table 5. Mediation relationships and hypothesis testing

Total Effects	Direct Effects	Indirect Effects					
β P-value	β P-Value	Hypothesis	β	SD	T-Value	P-Value	Decision
0. 407 0	0.258 0	(H_8) BG SF SM	0.149	0.057	2.160	0.000	Supported
0.283 0	0.170 0	(H_9) PA SF SM	0.113	0.039	4.428	0.000	Supported
0.268 0	0.219 0	(H_{10}) VF SF SM	0.049	0.021	4.186	0.000	Supported

keting in Zimbabwean agriculture (DV), and $\mathbf{H_{10}}$ vertical farming technology (IV), sustainable farming practices (MV) sustainable marketing in Zimbabwean agriculture (DV) as shown in Table 5 below.

he statistical analysis and results presented in Table 5 demonstrate significant mediation effects for hypotheses **H8, H9**, and **H10**, as indicated by their respective t-values, which exceed the recommended threshold of 1.96 (Singleton et al., 2023). **H8** (Bio-genetic Technology to Sustainable Farming Practices to Sustainable Marketing in Zimbabwean Agriculture): The mediation effect is supported (β = 0.149, t-value = 2.160, $p < 0.05$), suggesting that sustainable farming practices partially mediate the relationship between bio-genetic technology and sustainable marketing in Zimbabwean agriculture. **H9** (Precision Agriculture Technology to Sustainable Farming Practices to Sustainable Marketing in Zimbabwean Agriculture): The mediation effect is supported (β = 0.113, t-value = 4.428, $p < 0.05$), indicating that sustainable farming practices partially mediate the relationship between precision agriculture technology and sustainable marketing in Zimbabwean agriculture. **H10** (Vertical Farming Technology to Sustainable Farming Practices to Sustainable Marketing in Zimbabwean Agriculture): The mediation effect is supported (β = 0.049, t-value = 4.186, $p < 0.05$), revealing that sustainable farming practices fully mediate the relationship between vertical farming technology and sustainable marketing in Zimbabwean agriculture. These findings suggest that sustainable farming practices play a significant role in mediating the relationships between bio-genetic technology, precision agriculture technology, vertical farming technology, and sustainable marketing in Zimbabwean agriculture. The results provide valuable insights into the mechanisms through which technology adoption influences sustainable marketing outcomes in the agricultural sector.

DISCUSSION AND IMPLICATIONS

The results for **H1** indicate a positive relationship between bio-genetic technology and sustainable farming practices. This suggests that investing in bio-genetic technology like genetically modified seeds or animals can help farmers in Zimbabwe adopt more sustainable practices. The technology allows crops and livestock to be more resistant to diseases, pests, and environmental stresses. For example, genetic modifications can make maize more drought-tolerant or allow cows to be resistant to certain diseases. This enables farmers to reduce their use of pesticides, herbicides, antibiotics, and other chemicals while still increasing yields. With bio-engineered plants and animals, farmers can achieve higher productivity with fewer inputs. Policymakers should consider incentives like subsidies and improved access to credit to increase adoption of beneficial bio-genetic technologies. The technologies can uniquely help smallholder farmers in Zimbabwe improve productivity, profitability, and environmental sustainability of their operations. Promoting bio-technology can be a key strategy for sustainably enhancing food security and farmer livelihoods.

The strong positive relationship found in **H2** between bio-genetic technology and sustainable marketing is important for connecting farmers to domestic and international value chains. The technology can help farmers grow products that meet market demands for sustainable produce. For example, genetic modifications can produce maize with preferred nutritional traits or aesthetic qualities desired by consumers. Marketers and policymakers should collaborate to promote crops and livestock enhanced by bio-technology to high-value domestic markets like supermarkets and restaurants. They can also open up export opportunities to international markets interested in sustainably produced goods. Marketing bio-genetically engineered products as environmentally sustainable and socially responsible can also appeal to eco-conscious consumers. Strategic branding and advertising campaigns can emphasize the benefits of bio-technology for sustainable farming in Zimbabwe.

For **H3,** precision agriculture technology was also found to have a positive relationship with sustainable farming practices. The use of GPS, sensors, satellite imagery and advanced data analytics can optimize inputs and practices throughout farms in Zimbabwe. For instance, variable rate irrigation and fertilizer application allows precision targeting to reduce waste. Soil and crop monitoring allows early disease and pest detection to guide responsible treatments. This can significantly reduce use of water, fertilizers, pesticides, and energy while lowering costs and improving environmental sustainability. Government extension services and agricultural organizations should promote precision technology through training programs for farmers and increased access through machine sharing models. The technologies can uniquely help Zimbabwe's smallholder farmers practice site-specific management to minimize environmental impact while sustainably increasing productivity.

he positive link in **H4** between precision agriculture technology and sustainable marketing highlights its benefits for connecting farmers to niche markets. Precision technology like GPS tracking and farm management information systems allows for reliable traceability and certification of sustainably farmed products. Marketers can leverage the granular data from precision farming operations to demonstrate sustainability benefits to consumers. This can open up premium domestic and export markets willing to pay a premium for verified sustainably produced goods. Smallholder farmers can especially benefit from the ability to validate their sustainable practices. This allows them to command higher prices by tapping into growing consumer demand for responsibly farmed produce.

For **H5**, sustainable farming practices were found to also positively influence access to sustainable high-value markets. This shows the importance of improved agroecological practices on farms for accessing lucrative sustainable markets. Farmers should prioritize building expertise in techniques like conservation agriculture, integrated pest management, and agroforestry systems. Marketers can then develop distinctive brands and marketing campaigns around sustainably certified produce from farms with verified environmentally and socially responsible practices. Unique and targeted branding can connect consumers willing to pay more for produce grown using specific sustainable farming methods that align with their values.

The **H6** revealed a positive relationship between vertical farming technology and sustainable farming practices. Vertical farming methods like hydroponics, aquaponics, and aeroponics allow high-density production in controlled indoor environments, especially in urban areas. Policymakers should consider incentives like tax breaks and subsidized electricity rates to facilitate investment in vertical farming operations. The technology can promote sustainability by allowing high yields while drastically reducing land, water, and chemical use compared to conventional farming. Vertical farms integrated into cities can also provide urban populations with access to fresher, more local produce while creating employment opportunities.

The **H7** found that vertical farming technology also has a positive link to sustainable marketing opportunities. Urban vertical farming operations can take advantage of their proximity to large consumer bases to market their sustainably grown produce. Unique branding and targeted marketing campaigns promoting produce from vertical farms can appeal to modern urban consumers looking to buy responsibly grown local food. Vertical farming can be promoted as an innovative technological solution for growing food sustainably within urban environments. Marketers can highlight the environmental benefits of vertical farms compared to traditional farming to charge premium prices. Partnerships with urban restaurants and retailers can also be lucrative outlets for vertical farms to market their sustainably grown greens and vegetables.

The mediation effect supported in **H8** suggests sustainable farming practices partially explain the positive relationship between bio-genetic technology and sustainable marketing. This highlights that adopting bio-genetic technology like genetically modified seeds leads to more sustainable farming practices like reduced pesticide use, which in turn enables access to premium sustainable markets. However, bio-technology may also directly link to lucrative marketing opportunities by producing crop traits like enhanced nutrition or aesthetic qualities demanded by eco and health-conscious consumers. Still, improving overall farming sustainability through practices like integrated pest management is an important mechanism for leveraging bio-tech's full potential to connect farmers to environmentally and socially conscious domestic and international markets. Agricultural extension services and farmer organizations should promote sustainable farming methods alongside bio-technologies through training programs and demonstrations. A holistic approach is needed to realize the full marketing benefits from bio-technology adoption.

For **H9**, the mediation effect indicates sustainable farming practices also partially mediate between precision agriculture technology and sustainable marketing. The use of precision technology like satellite imaging facilitates more sustainable farm management through optimized input application, which then provides entry to high-value sustainable produce markets. But precision technology may also directly enable the traceability and data needed for sustainability certifications and marketing claims. Policymakers should encourage both precision technology adoption through subsidies and sustainable farming education through technical institutes. A dual focus is needed to utilize precision agriculture's full potential as a unique selling point to link Zimbabwe's smallholder farmers to lucrative sustainable export and domestic markets.

The full mediation found in **H10** reveals sustainable farming practices fully explain the relationship between vertical farming technology and sustainable marketing. While vertical farming infrastructure allows growing produce sustainably year-round, the technology itself does not directly influence marketing. Instead, vertical farms must combine high-tech facilities like hydroponics with sustainable techniques like integrated pest management in order to appeal to and tap into marketing opportunities from green urban consumers. Policy incentives for vertical farms from local governments should require demonstrating sustainable practices through transparent certification schemes in order to qualify for financial support and tax breaks. This can ensure urban vertical farming realizes its full potential for sustainably feeding cities while creating job opportunities.

Overall, the results reveal positive relationships between the three agricultural technologies (bio-genetic, precision, and vertical farming) and both sustainable farming practices and marketing. This highlights the potential of these technologies to enable more sustainable production and market access when adopted in Zimbabwe's agricultural sector. However, sustainable farming practices were also found to mediate the links between the technologies and sustainable marketing opportunities. This underscores the importance of combining the technologies with agroecological practices like integrated pest management and conservation agriculture in order to leverage the full marketing benefits. The findings suggest an integrated approach that promotes both agricultural technologies and sustainable techniques is imperative. This dual focus can allow Zimbabwe's farmers to capitalize on consumer demand for sustainably produced goods. The synergies between high-tech solutions and ecological practices are key to connecting farmers to premium domestic and international markets for environmentally and socially responsible produce.

Theoretical Contributions

This study makes several theoretical contributions to the literature on technology adoption and sustainability in agriculture. Firstly, it expands adoption models like the Technology Acceptance Model by incorporating sustainability factors. The research confirms that perceived sustainability benefits influence farmers' adoption of new technologies, in addition to factors like ease of use. Secondly, the findings expand on diffusion of innovations theory in agriculture by highlighting the role of sustainable farming practices. The results show practices diffuse alongside technologies to drive sustainable marketing. This contributes a more nuanced perspective on agricultural innovation diffusion. Thirdly, the mediation effects found contribute to technology-organization-environment theory. The results demonstrate sustainable farming practices mediate the marketability gains from new technologies. This highlights the importance of an integrated socio-technical perspective when applying new agricultural technologies.

Practical Contributions

The findings provide several practical implications for promoting technology adoption and sustainable agriculture in Zimbabwe. Firstly, the results underscore the need for policies and programs to develop both technical capabilities and sustainable farming expertise. A holistic approach is key to maximize the benefits from new technologies. Secondly, the marketability gains found suggest engaging private sector food companies in partnerships around sustainable sourcing can incentivize adoption. Market linkages are important to motivate farmer technology adoption. Finally, the research highlights the need for interdisciplinary collaboration. Engineers, agronomists, marketers and policymakers all play key roles in supporting new technology implementation. Multidisciplinary partnerships can ensure an integrated approach to sustainable agricultural development.

Limitations and Future Recommendations

While providing valuable insights, this study has some limitations that suggest avenues for future research. The cross-sectional survey data provides only a snapshot, so longitudinal studies could enrich understanding of adoption and diffusion processes over time. The sample also focused specifically on Zimbabwe, so expanding the research to other developing country contexts could improve generalizability. Additionally, the sustainability measures were based on farmer perceptions, so incorporating objective indicators could be beneficial. Future studies could also compare additional technologies like nanotechnology or animal biotech. Finally, qualitative research through interviews and focus groups could provide more in-depth understanding of farmer decision-making and perspectives. Building on this research, future studies have an opportunity to generate an even more nuanced and holistic perspective on leveraging technologies to improve sustainability in agriculture.

REFERENCES

Aburumman, O. J., Omar, K., Al Shbail, M., & Aldoghan, M. (2022). How to Deal with the Results of PLS-SEM? *International Conference on Business and Technology*, 1196-1206.

Aguirre-Urreta, M. I., & Hu, J. (2019). Detecting common method bias: Performance of the Harman's single-factor test. *The Data Base for Advances in Information Systems*, *50*(2), 45–70. doi:10.1145/3330472.3330477

Akter, S., Fosso Wamba, S., & Dewan, S. (2017). Why PLS-SEM is suitable for complex modelling? An empirical illustration in big data analytics quality. *Production Planning and Control*, *28*(11-12), 1011–1021. doi:10.1080/09537287.2016.1267411

Ardani, W. 2021. Trust Drivers and Revisit Intention of Foreign Tourists in Bali. Budapest International Research and Critics Institute (BIRCI-Journal): Humanities and Social Sciences, 4, 287-297.

Asseng, S., Guarin, J. R., Raman, M., Monje, O., Kiss, G., Despommier, D. D., Meggers, F. M., & Gauthier, P. P. (2020). Wheat yield potential in controlled-environment vertical farms. *Proceedings of the National Academy of Sciences of the United States of America*, *117*(32), 19131–19135. doi:10.1073/pnas.2002655117 PMID:32719119

Avgoustaki, D. D., & Xydis, G. (2020a). *How energy innovation in indoor vertical farming can improve food security, sustainability, and food safety? In Advances in food security and sustainability*. Elsevier.

Avgoustaki, D. D., & Xydis, G. (2020b). Indoor vertical farming in the urban nexus context: Business growth and resource savings. *Sustainability (Basel)*, *12*(5), 1965. doi:10.3390/su12051965

Bakhar, M., Fajri, T. I., Ahmad, N. & Nuryanto, U. W. (2023). User Acceptance Analysis of My Pertamina Application Using UTAUT Model. *Jurnal Sistim Informasi dan Teknologi*, 182-186.

Bhat, S. A., & Huang, N.-F. (2021). Big data and ai revolution in precision agriculture: Survey and challenges. *IEEE Access : Practical Innovations, Open Solutions*, *9*, 110209–110222. doi:10.1109/ACCESS.2021.3102227

Blasch, J., Van Der Kroon, B., Van Beukering, P., Munster, R., Fabiani, S., Nino, P., & Vanino, S. (2022). Farmer preferences for adopting precision farming technologies: A case study from Italy. *European Review of Agriculture Economics*, *49*(1), 33–81. doi:10.1093/erae/jbaa031

Butturini, M. & Marcelis, L. F. (2020). Vertical farming in Europe: Present status and outlook. *Plant Factory*, 77-91.

Carrer, M. J., De Souza Filho, H. M., Vinholis, M. D. M. B., & Mozambani, C. I. (2022). Precision agriculture adoption and technical efficiency: An analysis of sugarcane farms in Brazil. *Technological Forecasting and Social Change, 177*, 121510. doi:10.1016/j.techfore.2022.121510

Chatterjee, A., Debnath, S., & Pal, H. (2020). *Implication of urban agriculture and vertical farming for future sustainability. In Urban horticulture-Necessity of the future*. IntechOpen.

Cheah, J.-H., Thurasamy, R., Memon, M. A., Chuah, F., & Ting, H. (2020). Multigroup analysis using SmartPLS: Step-by-step guidelines for business research. *Asian Journal of Business Research, 10*(3), I–XIX. doi:10.14707/ajbr.200087

Cho, E. (2021). Neither Cronbach's alpha nor McDonald's omega: A commentary on Sijtsma and Pfadt. *Psychometrika, 86*(4), 877–886. doi:10.1007/s11336-021-09801-1 PMID:34460069

Cisternas, I., Velásquez, I., Caro, A., & Rodríguez, A. (2020). Systematic literature review of implementations of precision agriculture. *Computers and Electronics in Agriculture, 176*, 105626. doi:10.1016/j.compag.2020.105626

Da Silveira, F., Lermen, F. H., & Amaral, F. G. (2021). An overview of agriculture 4.0 development: Systematic review of descriptions, technologies, barriers, advantages, and disadvantages. *Computers and Electronics in Agriculture, 189*, 106405. doi:10.1016/j.compag.2021.106405

Erickson, B., & Fausti, S. W. (2021). The role of precision agriculture in food security. *Agronomy Journal, 113*(6), 4455–4462. doi:10.1002/agj2.20919

Escobar-Pemberthy, N. & Calle Saldarriaga, M. A. (2020). International business, trade and the nagoya protocol: best practices and challenges for sustainability in access and benefit-sharing. *International Business, Trade and Institutional Sustainability*, 813-830.

Fornell, C., & Larcker, D. F. (1981). Evaluating structural equation models with unobservable variables and measurement error. *JMR, Journal of Marketing Research, 18*(1), 39–50. doi:10.1177/002224378101800104

Fox, M. P., Maclehose, R. F., & Lash, T. L. (2022). *Applying quantitative bias analysis to epidemiologic data*. Springer.

Gio, P. U. (2022). *Partial least squares structural equation modeling (pls-sem) dengan software smartpls*. Uwais Inspirasi Indonesia.

Glaze-Corcoran, S., Hashemi, M., Sadeghpour, A., Jahanzad, E., Afshar, R. K., Liu, X., & Herbert, S. J. (2020). Understanding intercropping to improve agricultural resiliency and environmental sustainability. *Advances in Agronomy, 162*, 199–256. doi:10.1016/bs.agron.2020.02.004

Godfrey, T. (2016). *The contribution of smallholder agriculture production to food security in rural Zimbabwe: A case study of Masvingo Province*. University of Fort Hare.

Gwanongodza, T. (2020). *Agricultural exports and economic growth in Zimbabwe*. Stellenbosch University.

Hair, J. F. Jr, Hult, G. T. M., Ringle, C. M., Sarstedt, M., Danks, N. P., & Ray, S. (2021). *Partial least squares structural equation modeling (PLS-SEM) using R: A workbook*. Springer Nature. doi:10.1007/978-3-030-80519-7

Hair, J. F. Jr, & Sarstedt, M. (2021). Data, measurement, and causal inferences in machine learning: Opportunities and challenges for marketing. *Journal of Marketing Theory and Practice*, *29*(1), 65–77. doi:10.1080/10696679.2020.1860683

Heo, H. K., & Lee, E. (2021). Types of vertical smart farms and awareness of their use in Korean cities types and feasibility analysis of vertical smart farms in Korean cities. *Journal of People, Plants, and Environment*, *24*(3), 257–266. doi:10.11628/ksppe.2021.24.3.257

Hermida, R. (2015). The problem of allowing correlated errors in structural equation modeling: Concerns and considerations. *Computational Methods in Social Sciences*, *3*, 5–17.

Karunathilake, E., Le, A. T., Heo, S., Chung, Y. S., & Mansoor, S. (2023). The path to smart farming: Innovations and opportunities in precision agriculture. *Agriculture*, *13*(8), 1593. doi:10.3390/agriculture13081593

Khan, M. (2020). *The fade-away phenomenon of initial non-response bias in panel surveys*. Freie Universitaet Berlin.

Krejcie, R. V., & Morgan, D. W. (1970). Determining sample size for research activities. *Educational and Psychological Measurement*, *30*(3), 607–610. doi:10.1177/001316447003000308

Lee, C.-L., Strong, R., & Dooley, K. E. (2021). Analyzing precision agriculture adoption across the globe: A systematic review of scholarship from 1999–2020. *Sustainability (Basel)*, *13*(18), 10295. doi:10.3390/su131810295

Lin, H., Wang, H., Deng, A., Rong, M., Ye, L., Shi, L., Wang, T., & Peng, X. (2021). *Applications of CRISPR/Cas Technology to Research the Synthetic Genomics of Yeast. In Synthetic Genomics-From BioBricks to Synthetic Genomes*. IntechOpen.

Lleo, A., Bastons, M., Rey, C., & Ruiz-Perez, F. (2021). Purpose implementation: Conceptualization and measurement. *Sustainability (Basel)*, *13*(4), 1921. doi:10.3390/su13041921

Long, C., Chen, Z., Zhou, Y. & Long, B. (2018). The role of biodiversity and plant conservation for ornamental breeding. *Ornamental Crops*, 1-12.

Maheshwari, S. (2021). Vertical Farming: Resilience Towards Climate Change. *Urban growth and environmental issues in India*, 207-221.

Mall, T., Gupta, M., Dhadialla, T. S. & Rodrigo, S. (2019). Overview of biotechnology-derived herbicide tolerance and insect resistance traits in plant agriculture. *Transgenic Plants: Methods and Protocols*, 313-342.

Martinez, M. (2023). Vertical Farming: A Bottom-Up Approach. *Seattle Journal of Technology, Environmental & Innovation Law*, *13*, 5.

Matandare, M. A. (2017). An analysis of the role of the Agriculture sector: Case of Zimbabwe. *Internafional Journal of Scienfific Research in Science and Technology*, *3*, 1255–1263.

Mendes, J., Pinho, T. M., Neves Dos Santos, F., Sousa, J. J., Peres, E., Boaventura-Cunha, J., Cunha, M., & Morais, R. (2020). Smartphone applications targeting precision agriculture practices—A systematic review. *Agronomy (Basel)*, *10*(6), 855. doi:10.3390/agronomy10060855

Meyer, R. (2019). Modern biotechnology and sustainable intensification: chances and limitations. *Agricultural Resilience*, 159.

Mir, M. S., Naikoo, N. B., Kanth, R. H., Bahar, F., Bhat, M. A., Nazir, A., Mahdi, S. S., Amin, Z., Singh, L., & Raja, W. (2022). Vertical farming: The future of agriculture: A review. *The Pharma Innovation Journal*, *11*, 1175–1195.

Mujeyi, A., Mudhara, M., & Munyaradzi, M. (2020). A review of current climate-smart agriculture technol-ogies: adoption and impact in integrated smallholder farming systems of Zimbabwe. *Biennial Africa Climate Smart Agriculture Stakeholders Conference*, 33.

Musara, J. P., Chimvuramahwe, J., & Borerwe, R. (2012). Adoption and efficiency of selected conservation farming technologies in Madziva Communal Area, Zimbabwe: A transcendental production function approach. *Bulletin of Environment. Pharmacology and Life Sciences*, *1*, 27–38.

Musungwini, S., Gavai, P. V., Munyoro, B., & Chare, A. (2023). *Emerging ICT Technologies for Agriculture, Training, and Capacity Building for Farmers in Developing Countries: A Case Study in Zimbabwe. In Applying Drone Technologies and Robotics for Agricultural Sustainability*. IGI Global. doi:10.4018/978-1-6684-6413-7.ch002

Ndhlovu, E., & Mhlanga, D. (2023). *Smart technologies, climate change, and smallholder farmer production in Zimbabwe. The fourth industrial revolution in Africa: Exploring the development implications of smart Technologies in Africa*. Springer.

Pek, J., Wong, O. & Wong, A. (2019). Data transformations for inference with linear regression: Clarifications and recommendations. *Practical Assessment, Research, and Evaluation, 22*, 9.

Price, C. (2022). *Agriculture 4.0: Bioinformationalism and postdigital hybrid assemblages. In Bioinformational Philosophy and Postdigital Knowledge Ecologies*. Springer.

Raj, E. F. I., Appadurai, M., & Athiappan, K. (2022). *Precision farming in modern agriculture. In Smart Agriculture Automation Using Advanced Technologies: Data Analytics and Machine Learning, Cloud Architecture, Automation and IoT*. Springer.

Roberts, J. M., Bruce, T. J., Monaghan, J. M., Pope, T. W., Leather, S. R., & Beacham, A. M. (2020). Vertical farming systems bring new considerations for pest and disease management. *Annals of Applied Biology*, *176*(3), 226–232. doi:10.1111/aab.12587

Rose, D. C., Bhattacharya, M., De Boon, A., Dhulipala, R. K., Price, C., & Schillings, J. (2022). *The fourth agricultural revolution: technological developments in primary food production. In A Research Agenda for Food Systems*. Edward Elgar Publishing.

Runganga, R. & Mhaka, S. (2021). *Impact of agricultural production on economic growth in Zimbabwe*. Academic Press.

Sanjeevi, P., Prasanna, S., Siva Kumar, B., Gunasekaran, G., Alagiri, I., & Vijay Anand, R. (2020). Precision agriculture and farming using Internet of Things based on wireless sensor network. *Transactions on Emerging Telecommunications Technologies*, *31*(12), e3978. doi:10.1002/ett.3978

Sanmukhiya, C. (2020). A pls–sem approach to the utaut model: The case of Mauritius. *Ann. Soc. Sci. Manag. Stud. ASM*, *6*(1), 8–16. doi:10.19080/ASM.2020.06.555677

Singh, R. K., Aernouts, M., De Meyer, M., Weyn, M., & Berkvens, R. (2020). Leveraging LoRaWAN technology for precision agriculture in greenhouses. *Sensors (Basel)*, *20*(7), 1827. doi:10.3390/s20071827 PMID:32218353

Singleton, H., Porter, S., Beavis, J., Falconer, L., Hernandez, J. P., & Holley, D. (2023). Accounting for complexity in critical realist trials: The promise of PLS-SEM. *Journal of Critical Realism*, *22*(3), 1–20. doi:10.1080/14767430.2023.2217652

Sishodia, R. P., Ray, R. L., & Singh, S. K. (2020). Applications of remote sensing in precision agriculture: A review. *Remote Sensing (Basel)*, *12*(19), 3136. doi:10.3390/rs12193136

Tawfik, E., Hammad, I., & Bakry, A. (2022). Production of transgenic Allium cepa by nanoparticles to resist Aspergillus niger infection. *Molecular Biology Reports*, *49*(3), 1–8. doi:10.1007/s11033-021-06988-5 PMID:34837626

Ugwu, U. F. (2021). Maximizing the differentiation principle in regional IP treaties to advance food security: Limitations in West Africa's regional IP and trade regime. *The Journal of World Intellectual Property*, *24*(5-6), 310–343. doi:10.1111/jwip.12193

Van Delden, S., Sharathkumar, M., Butturini, M., Graamans, L., Heuvelink, E., Kacira, M., Kaiser, E., Klamer, R., Klerkx, L., Kootstra, G., Loeber, A., Schouten, R. E., Stanghellini, C., van Ieperen, W., Verdonk, J. C., Vialet-Chabrand, S., Woltering, E. J., van de Zedde, R., Zhang, Y., & Marcelis, L. F. M. (2021). Current status and future challenges in implementing and upscaling vertical farming systems. *Nature Food*, *2*(12), 944–956. doi:10.1038/s43016-021-00402-w PMID:37118238

Van Gerrewey, T., Boon, N., & Geelen, D. (2021). Vertical farming: The only way is up? *Agronomy (Basel)*, *12*(1), 2. doi:10.3390/agronomy12010002

Yin, H., Cao, Y., Marelli, B., Zeng, X., Mason, A. J., & Cao, C. (2021). Soil sensors and plant wearables for smart and precision agriculture. *Advanced Materials*, *33*(20), 2007764. doi:10.1002/adma.202007764 PMID:33829545

Zengeya, T., Sambo, P., & Mabika, N. (2021). The Adoption of the Internet of Things for SMART Agriculture in Zimbabwe. *5th Conference on Networks and Communications (NET 2021)*, 99-106. 10.5121/csit.2021.111208

Compilation of References

Abdelsalam, S., Salim, N., Alias, R. A., & Husain, O. (2020). Understanding online impulse buying behavior in social commerce: A systematic literature review. *IEEE Access : Practical Innovations, Open Solutions*, *8*, 89041–89058. doi:10.1109/ACCESS.2020.2993671

Abioye, E. A., Abidin, M. S. Z., Mahmud, M. S. A., Buyamin, S., Ishak, M. H. I., Abd Rahman, M. K. I., ... Ramli, M. S. A. (2020). A review on monitoring and advanced control strategies for precision irrigation. *Computers and Electronics in Agriculture*, *173*, 105441. doi:10.1016/j.compag.2020.105441

Aburumman, O. J., Omar, K., Al Shbail, M., & Aldoghan, M. (2022). How to Deal with the Results of PLS-SEM? *International Conference on Business and Technology*, 1196-1206.

Abutabenjeh, S., & Jaradat, R. (2018). Clarification of research design, Research Methods, and research methodology. *Teaching Public Administration*, *36*(3), 237–258. doi:10.1177/0144739418775787

Acharya, S. S. (2006). Agricultural marketing and rural credit for strengthening. *Indian Agriculturist*. INRM Policy Brief No. 3,Asian Development BankbIndia Resident Mission (INRM), https://casi.sas.upenn.edu/sites/default/files/iit/ADB%20-%20INRM%20Policy%20Brief%20No.%203.pdf. Accessed on 29 March 2024

Acharyya, A. (2022). Climate-Smart Agriculture in Developing Economies. *Environmental Economics in Developing Countries: Issues and Challenges*, 231.

Adams, D. W. (2016). Microfinance and the poor: The evidence to date. *Journal of Economic Literature*, *54*(2), 436–472.

Adefare, T., Adeola, O., Mogaji, E., Nguyen, N. P., & Mogaji, S. A. (2024). Empowering women agriculture entrepreneurs: Banks' role in achieving sustainable development goals. *International Journal of Bank Marketing*. Advance online publication. doi:10.1108/IJBM-03-2023-0128

Adetiloye, K. A. (2012). Agricultural Financing in Nigeria: An Assessment of the Agricultural Credit Guarantee Scheme Fund (ACGSF) for Food Security in Nigeria (1978-2006). *Journal of Economic Literature*, *3*(1), 39–48.

Adeyemi, O., Grove, I., Peets, S., & Norton, T. (2017). Advanced monitoring and management systems for improving sustainability in precision irrigation. *Sustainability (Basel)*, *9*(3), 353. doi:10.3390/su9030353

Adeyemi, S. O. (2023). A Critical Review of Rural Agricultural Development Innovative Programmes in Nigeria. *Plant Biol Soil Health J*, *1*(1), 11–17.

Affandi, M. I. (2023). Linkages and Multipliers of the Agricultural Sector in the Regional Economy in Tanggamus Regency: Input-Output Analysis. *Eduvest-Journal of Universal Studies*, *3*(3), 594–603. doi:10.59188/eduvest.v3i3.775

Africa, A. S. R. (2013). Focus on staple crops. Alliance for a Green Revolution in Africa.

Afrin, S., Khan, A. T., Mahia, M., Ahsan, R., Mishal, M. R., Ahmed, W., & Rahman, R. M. (2018). Analysis of Soil Properties and Climatic Data to Predict Crop Yields and Cluster Different Agricultural Regions of Bangladesh. *2018 IEEE/ACIS 17th International Conference on Computer and Information Science (ICIS).* 10.1109/ICIS.2018.8466397

Agarwal, B. (2018). Gender equality, food security and the sustainable development goals. *Current Opinion in Environmental Sustainability*, *34*, 26–32. doi:10.1016/j.cosust.2018.07.002

Agbamu, J.U. (2000). Agricultural Research-Extension Linkage Systems: An International Perspective. *Agricultural Research and Extension Network*, Paper 106a.

Agbenyo, W., Jiang, Y., Jia, X., Wang, J., Ntim-Amo, G., Dunya, R., Siaw, A., Asare, I., & Twumasi, M. A. (2022). Does the Adoption of Climate-Smart Agricultural Practices Impact Farmers' Income? Evidence from Ghana. *International Journal of Environmental Research and Public Health*, *19*(7), 3804. doi:10.3390/ijerph19073804 PMID:35409488

Agriculture and Climate Change : An Introduction. (2012). World Bank. https://openknowledge.worldbank.org/server/api/core/bitstreams/93d709b7-4b38-548b-bd0a-d5431b34f296/content. Accessed on March 28 2024

Agriculture: Key to a Sustainable Future. (2021). The World Resources Institute (WRI). https://www.wri.org/research/agriadapt-strategies-participatory-design-and-development-climate-risk-tool. Accessed on 28 March 2024

Aguirre-Urreta, M. I., & Hu, J. (2019). Detecting common method bias: Performance of the Harman's single-factor test. *The Data Base for Advances in Information Systems*, *50*(2), 45–70. doi:10.1145/3330472.3330477

Ahmed, A., Isaac, W., Varshney, S., & Khan, E. (2016). An IoT based system for remote monitoring of soil characteristics. *2016 International Conference on Information Technology (InCITe) - the next Generation IT Summit on the Theme - Internet of Things: Connect Your Worlds.* 10.1109/INCITE.2016.7857638

Ahumada, O., & Villalobos, J. R. (2009). Application of planning models in the agri-food supply chain: A review. *European Journal of Operational Research*, *196*(1), 1–20. doi:10.1016/j.ejor.2008.02.014

Ajayi, A., Nageri, I., & Akolo, S. (2017). Impact of Agricultural Financing Policy and Deposit Money Bank Loan on Agricultural Sector Productivity in Nigeria. *Amity Journal of Agribusiness*, *2*(1), 1–11.

Ajayi, F. (2023). Strengthening Agriculture for Food and Nutrition Security, Market Development and Export in a Climate Change Environment. *Agricultural Society of Nigeria (ASN) 57th Annual Conference Proceedings.*

Aji, H. M., & Sutikno, B. (2015). The Extended Consequence of Greenwashing: Perceived Consumer Skepticism. *International Journal of Business and Information*, *10*(4), 433–463.

Aker, J. C. (2011). Dial "A" for agriculture: A review of information and communication technologies for agricultural extension in developing countries. *Agricultural Economics*, *42*(6), 631–647. doi:10.1111/j.1574-0862.2011.00545.x

Akter, S., Fosso Wamba, S., & Dewan, S. (2017). Why PLS-SEM is suitable for complex modelling? An empirical illustration in big data analytics quality. *Production Planning and Control*, *28*(11-12), 1011–1021. doi:10.1080/09537287.2016.1267411

Aldrich, D. G. (1966). Agricultural Research, a Key to Understanding our Environment. Symposium on Research in Agriculture. *USDA, Agricultural Research Service*, 50-53.

Alene, A.D. (n.d.). A Framework for Conceptualizing Impact Assessment and Promoting Impact Culture in Agricultural Research. *International Institute of Tropical Agriculture (IITA).*

Alfaro, L., & Chor, D. (2023). Global Supply Chains: The Looming "Great Reallocation" (No. w31661). *National Bureau of Economic Research.*

Ali, S., Hussain, T., Zhang, G., Nurunnabi, M., & Li, B. (2018). The Implementation of Sustainable Development Goals in "BRICS" Countries. *Sustainability (Basel)*, *10*(7), 2513. doi:10.3390/su10072513

Ali, S., Liu, Y., Ishaq, M., Shah, T., Abdullah, A., Ilyas, A., & Din, I. (2017). Climate change and its impact on the yield of major food crops: Evidence from Pakistan. *Foods*, *6*(6), 1–19. doi:10.3390/foods6060039 PMID:28538704

Altieri, M. A., Nicholls, C. I., Henao, A., & Lana, M. A. (2015). Agroecology and the design of climate change-resilient farming systems. *Agronomy for Sustainable Development*, *35*(3), 869–890. doi:10.1007/s13593-015-0285-2

Amos, C. C., Rahman, A., Karim, F., & Gathenya, J. M. (2018). A scoping review of roof harvested rainwater usage in urban agriculture: Australia and Kenya in focus. *Journal of Cleaner Production*, *202*, 174–190. doi:10.1016/j.jclepro.2018.08.108

Anderson, J. R., & Feder, G. (2004). Agricultural extension: Good intentions and hard realities. *The World Bank Research Observer*, *19*(1), 41–60. doi:10.1093/wbro/lkh013

Anderson, J. R., & Masters, W. A. (2017). Transforming market access for small-scale farmers: Value chains, standards, and collective action. *Annual Review of Environment and Resources*, *42*(1), 421–448.

Andiç, E., Yurt, Ö., & Baltacıo ˘glu, T. (2012). Green supply chains: Efforts and potential applications for the Turkish market. *Resources, Conservation and Recycling*, *2012*(58), 50–68. doi:10.1016/j.resconrec.2011.10.008

Andure, M. W., Rajiv, I., Nikesh, J., Vaibhav, K., & Rahul, K. (2018). Design and fabrication of manual seed sowing machine with fertilizer. *International Journal for Research in Applied Science and Engineering Technology*, *6*(2), 515–519. doi:10.22214/ijraset.2018.2073

Anibaldi, R., Rundle-Thiele, S., David, P., & Roemer, C. (2021). Theoretical Underpinnings in Research Investigating Barriers for Implementing Environmentally Sustainable Farming Practices: Insights from a Systematic Literature Review. *Land (Basel)*, *10*(4), 386. doi:10.3390/land10040386

Annapurna, B., & Anusha, B. (2019). Automated Seed Sowing Machine Using Atmega2560. *International Journal of Innovative Technology and Exploring Engineering*, *8*, 300–302.

Anthony, E. (2010). Agricultural Credit and Economic Growth in Nigeria: An Empirical Analysis. *Business and Economics Journal*, 14.

Antwi-Agyei, P., Abalo, E. M., Dougill, A. J., & Baffour-Ata, F. (2021). Motivations, enablers and barriers to the adoption of climate-smart agricultural practices by smallholder farmers: Evidence from the transitional and savannah agroecological zones of Ghana. *Regional Sustainability*, *2*(4), 375–386. doi:10.1016/j.regsus.2022.01.005

Ardani, W. 2021. Trust Drivers and Revisit Intention of Foreign Tourists in Bali. Budapest International Research and Critics Institute (BIRCI-Journal): Humanities and Social Sciences, 4, 287-297.

Arifin, S., Mohan, T., & Goh, Y. (2017). Influence of consumer's perceived risk on consumer's online purchase intention. *Journal of Research in Interactive Marketing*, 2040–7122.

Arora, N. K., & Mishra, I. (2022). Current scenario and future directions for sustainable development goal 2: A roadmap to zero hunger. *Environmental Sustainability*, *5*(2), 129–133. doi:10.1007/s42398-022-00235-8 PMID:37521584

Arora, N. K., & Mishra, I. (2023). Responsible consumption and production: A roadmap to sustainable development. *Environmental Sustainability*, *6*(1), 1–6. doi:10.1007/s42398-023-00266-9

Ashby, A. (2018). Developing closed loop supply chains for environmental sustainability. *Journal of Manufacturing Technology Management*, *29*(4), 699–722. doi:10.1108/JMTM-12-2016-0175

Asnawati & Wahyuni, S. (2018). The Influence of Hedonic Shopping Motivation on the Impulse Buying Of Online-Shopping Consumers on Instagram. *Russian Journal of Agricultural and Socio-Economic Sciences*, *74*(2), 99-107.

Asplund, N. M., Forster, D. L., & Stout, T. T. (1989). Farmers' use of forward contracting and hedging. *Review of Futures Markets*, *8*(1), 24–37.

Asseng, S., Guarin, J. R., Raman, M., Monje, O., Kiss, G., Despommier, D. D., Meggers, F. M., & Gauthier, P. P. (2020). Wheat yield potential in controlled-environment vertical farms. *Proceedings of the National Academy of Sciences of the United States of America*, *117*(32), 19131–19135. doi:10.1073/pnas.2002655117 PMID:32719119

Åstebro, T., & Serrano, C. J. (2015). Business partners: Complementary assets, financing, and invention commercialization. *Journal of Economics & Management Strategy*, *24*(2), 228–252. doi:10.1111/jems.12095

ATA. (2011). Federal Ministry of Agriculture and Rural Development. *Agricultural Transformation Agenda*. http://www.unaab.edu.ng/attachments/Agricultural%20Transformation%20Blue%20Print.pdf

Atapattu, A. A. A. J., Senarathne, S. H. S., Raveendra, S. A. S. T., Egodawatte, W. C. P., & Mensah, S. (2017). Effect of short-term agroforestry systems on soil quality in marginal coconut lands in Sri Lanka. *Agricultural Research Journal*, *54*(3), 324. doi:10.5958/2395-146X.2017.00060.6

Athwal, N., Wells, V. K., Carrigan, M., & Henninger, C. E. (2019). Sustainable Luxury Marketing: A synthesis and research agenda. *International Journal of Management Reviews*, *21*(4), 405–426. doi:10.1111/ijmr.12195

Attanasio, G., Preghenella, N., De Toni, A. F., & Battistella, C. (2022). Stakeholder engagement in business models for sustainability: The stakeholder value flow model for sustainable development. *Business Strategy and the Environment*, *31*(3), 860–874. doi:10.1002/bse.2922

Avgoustaki, D. D., & Xydis, G. (2020a). *How energy innovation in indoor vertical farming can improve food security, sustainability, and food safety? In Advances in food security and sustainability*. Elsevier.

Avgoustaki, D. D., & Xydis, G. (2020b). Indoor vertical farming in the urban nexus context: Business growth and resource savings. *Sustainability (Basel)*, *12*(5), 1965. doi:10.3390/su12051965

Avi, D. (2012). *"The 10 most Important Marketing Trends According to Sir Martin Sorrel", Baltzer, K. & Hansen, H. 2011*. Agricultural Input Subsidies in Sub-Saharan Africa.

Ayodele, O. S. (2013). Challenges facing the achievement of the Nigeria Vision 20:2020. *Global Advanced Research Journal of Social Science*, *2*(7), 143–157.

Ayunku, P. E., & Etale, L. M. (2015). Effect of agriculture spending on economic growth in Nigeria: Empirical evidence. *Research Journal of Finance and Accounting*, *6*(2), 138–143.

Azadi, H., Moghaddam, S. M., Burkart, S., Mahmoudi, H., Van Passel, S., Kurban, A., & Lopez-Carr, D. (2021). Rethinking resilient agriculture: From climate-smart agriculture to vulnerable-smart agriculture. *Journal of Cleaner Production*, *319*, 128602. doi:10.1016/j.jclepro.2021.128602

Bacon, C. M., Getz, C., Kraus, S., Montenegro, M., & Holland, K. (2012). The Social Dimensions of Sustainability and Change in Diversified Farming Systems. *Ecology and Society*, *17*(4), 41. doi:10.5751/ES-05226-170441

Baker, B. P., Green, T. A., & Loker, A. J. (2020). Biological control and integrated pest management in organic and conventional systems. *Biological Control*, *140*, 104095. doi:10.1016/j.biocontrol.2019.104095

Bakhar, M., Fajri, T. I., Ahmad, N. & Nuryanto, U. W. (2023). User Acceptance Analysis of My Pertamina Application Using UTAUT Model. *Jurnal Sistim Informasi dan Teknologi*, 182-186.

Balafoutis, A., Beck, B., Fountas, S., Vangeyte, J., Van der Wal, T., Soto, I., ... Eory, V. (2017). Precision agriculture technologies positively contributing to GHG emissions mitigation, farm productivity and economics. *Sustainability (Basel)*, *9*(8), 1339. doi:10.3390/su9081339

Balaji, G. N., Nandhini, V., Mithra, S., Priya, N., & Naveena, R. (2018). IoT based smart crop monitoring in farm land. *Imperial Journal of Interdisciplinary Research*, *4*(1), 88–92.

Balasundram, S. K., Shamshiri, R. R., Sridhara, S., & Rizan, N. (2023). The Role of Digital Agriculture in Mitigating Climate Change and Ensuring Food Security: An Overview. *Sustainability (Basel)*, *15*(6), 5325. doi:10.3390/su15065325

Balderjahn, I., Peyer, M., Seegebarth, B., Wiedmann, K.-P., & Weber, A. (2018). The many faces of sustainability-conscious consumers: A category-independent typology. *Journal of Business Research*, *91*, 83–93. doi:10.1016/j.jbusres.2018.05.022

Baltzer, K., & Hansen, H. (2012). Evaluation Study Agricultural input subsidies in Sub-Saharan Africa. *University of Copenhagen Rolighedsvej*, *25*, 1958.

Balyan, S., Jangir, H., Tripathi, S. N., Tripathi, A., Jhang, T., & Pandey, P. (2024). Seeding a Sustainable Future: Navigating the Digital Horizon of Smart Agriculture. *Sustainability (Basel)*, *16*(2), 475. doi:10.3390/su16020475

Banful, A. B. (2010). *Old problems in the new solutions?* IFPRI Discussion Paper 01002, IFPRI.

Bantilan, M. C. S., Brar, D. S., & Fukao, T. (2020). Breeding crops for improved abiotic stress tolerance: Achievements and future directions. *Frontiers in Plant Science*, *11*, 1107486.

Barney, J. (1991). Firm Resources and Sustained Competitive Advantage. *Journal of Management*, *17*(1), 99–120. doi:10.1177/014920639101700108

Barragan, J., Bralts, V., & Wu, I. P. (2006). Assessment of emission uniformity for micro-irrigation design. *Biosystems Engineering*, *93*(1), 89–97. doi:10.1016/j.biosystemseng.2005.09.010

Barratt, N., Chitundu, D., Dover, O., Elsinga, J., Eriksson, S., Guma, L., Haggblade, M., Haggblade, S., Henn, T. O., Locke, F. R., O'Donnell, C., Smith, C., & Stevens, T. (2006). *Cassava as droughtinsurance: Food security implications of cassava trials in Central Zambia*. AGRA.

Basher, R. (2006). Global early warning systems for natural hazards: systematic and people-centred. *Philosophical Transactions of the Royal Society A: Mathematical, Physical and Engineering Sciences*, *364*(1845), 2167–2182. doi:10.1098/rsta.2006.1819

Batchelor, C. (1999). Improving water use efficiency as part of integrated catchment management. *Agricultural Water Management*, *40*(2–3), 249–263. doi:10.1016/S0378-3774(98)00125-5

Beck, B., Zimmermann, J., McBratney, A., Whelan, B., & O'Kennedy, M. (2019). Digital soil mapping: A review of methods and applications. *European Journal of Soil Science*, *70*(6), 1164–1185.

Beddington, J. R., Asaduzzaman, M., Fernandez, A., Clark, M. E., Guillou, M., Jahn, M. M., ... Scholes, R. J. (2012). *Achieving food security in the face of climate change*. Final report from the Commission on Sustainable Agriculture and Climate Change.

Beintema, N., & Elliott, H. (2009). *Setting Meaningful Investment Targets in Agricultural Research and Development: Challenges, Opportunities and Fiscal Realities. In Expert Meeting on How to feed the World in 2050, Food and Agriculture Organization (FAO) & the United Nations, Economic and Social Development Department*. UNDESA.

Ben Nasr, J., Chaar, H., Bouchiba, F., & Zaibet, L. (2021). Assessing and building climate change resilience of farming systems in Tunisian semi-arid areas. *Environmental Science and Pollution Research International*, *28*(34), 46797–46808. doi:10.1007/s11356-021-13089-0 PMID:33646545

Bentley, J. W. (2008). More than just words: The importance of action in participatory research. *Action Research*, *6*(2), 127–146.

Bernardi, E. (2014). Joining Outside and Inside: Planted Terraces in Multi-Level Buildings. In New Urban Configurations (pp. 944-949). IOS Press.

Bernhofen, D.M. & Brown, J.C. (2004). *Comparative Advantage: From Theory to Data*. Paper to be presented at the 2005 AEA Meetings in Philadelphia: On the 75th Anniversary of the Opportunity Cost Formulation of Comparative Advantage.

Berti, G., & Mulligan, C. (2016). Competitiveness of Small Farms and Innovative Food Supply Chains: The Role of Food Hubs in Creating Sustainable Regional and Local Food Systems. *Sustainability (Basel)*, *8*(7), 616. doi:10.3390/su8070616

Bezner Kerr, R., Pender, J., & Lipper, L. (2010). The economics of adaptation to climate change: Agriculture. *Agricultural Economics*, *41*(S3), 817–833.

Bezuidenhout, L. M., Leonelli, S., Kelly, A. H., & Rappert, B. (2017). Beyond the digital divide: Towards a situated approach to open data. *Science & Public Policy*, *44*(4), 464–475. doi:10.1093/scipol/scw036

Bezuneh, M., & Demeke, S. (2011). Market access and farm productivity in Ethiopia: A panel data analysis. *Agricultural Economics*, *42*(3), 393–403.

Bhat, S. A., & Huang, N.-F. (2021). Big data and ai revolution in precision agriculture: Survey and challenges. *IEEE Access : Practical Innovations, Open Solutions*, *9*, 110209–110222. doi:10.1109/ACCESS.2021.3102227

Bhatta, G. D., & Aggarwal, P. K. (2016). Coping with weather adversity and adaptation to climatic variability: A cross-country study of smallholder farmers in South Asia. *Climate and Development*, *8*(2), 145–157. doi:10.1080/17565529.2015.1016883

Bhatt, R., Arora, S., & Chew, C. C. (2016). Improving irrigation water productivity using Tensiometers. *Journal of Soil and Water Conservation*, *15*(2), 120–124.

Biggs, T. W., Mishra, P. K., & Turral, H. (2008). Evapotranspiration and regional probabilities of soil moisture stress in rainfed crops, southern India. *Agricultural and Forest Meteorology*, *148*(10), 1585–1597. doi:10.1016/j.agrformet.2008.05.012

Birks, M., & Mills, J. (2011). *Grounded Theory: A Practical Guide*. Sage Publications, Inc.

Birtalan, M., Szigeti, C., & Baráth, L. (2021). The Role of Community Supported Agriculture (CSA) in Food Systems Transformation. *Sustainability*, *13*(12), 6957.

Blackburn, J., & Scudder, G. (2009). Supply chain strategies for perishable products: The case of fresh produce. *Production and Operations Management*, *18*(2), 129–137. doi:10.1111/j.1937-5956.2009.01016.x

Blasch, J., Van Der Kroon, B., Van Beukering, P., Munster, R., Fabiani, S., Nino, P., & Vanino, S. (2022). Farmer preferences for adopting precision farming technologies: A case study from Italy. *European Review of Agriculture Economics*, *49*(1), 33–81. doi:10.1093/erae/jbaa031

Blesh, J., Hoey, L., Jones, A. D., Friedmann, H., & Perfecto, I. (2019). Development pathways toward "zero hunger.". *World Development*, *118*, 1–14. doi:10.1016/j.worlddev.2019.02.004

Blonskaja, V., Põldnurk, J., & Loigu, E. (2014). Utilisation options for biodegradable kitchen waste in Estonia. SWOT analysis. *Proceedings of the International Conference on Environmental Engineering. ICEE, 9*(1).

Bloomfield, G., Bucht, K., Martínez-Hernández, J. C., Ramírez-Soto, A. F., Sheseña-Hernández, I., Lucio-Palacio, C. R., & Ruelas Inzunza, E. (2018). Capacity building to advance the United Nations sustainable development goals: An overview of tools and approaches related to sustainable land management. *Journal of Sustainable Forestry*, *37*(2), 157–177. doi:10.1080/10549811.2017.1359097

Blum, A. (2011). Drought resistance in plants and its improvement for crop production: a comprehensive overview. *Plant stress tolerance: methods and protocols*, 741-772.

Boerman, S. C., Kruikemeier, S., & Zuiderveen Borgesius, F. J. (2017). Online behavioral advertising: A literature review and research agenda. *Journal of Advertising*, *46*(3), 363–376. doi:10.1080/00913367.2017.1339368

Brown, C., & Wekerle, A. (2014). The future of community supported agriculture: Lessons from a decade of empirical research. *Agriculture and Human Values*, *31*(4), 431–444.

Bruce, J., & Fortmann, L. (1989). *Land tenure issues in project design and strategies for agricultural development in sub-Saharan Africa*. University of Wisconsin-Madison, Land Tenure Center.

Bryant, A. (2019). *Grounded theory, Case study research, Theory*. SAGE Publications Ltd. doi:10.4135/9781529716542

Building C.-R. F. S. Actions for Transforming Our World. (2023). The Food and Agriculture Organization of the United Nations (FAO.). https://www.fao.org/3/I9900EN/i9900en.pdf

Busari, M. A., Kukal, S. S., Kaur, A., Bhatt, R., & Dulazi, A. A. (2015). Conservation tillage impacts on soil, crop and the environment. *International Soil and Water Conservation Research*, *3*(2), 119–129. doi:10.1016/j.iswcr.2015.05.002

Butturini, M. & Marcelis, L. F. (2020). Vertical farming in Europe: Present status and outlook. *Plant Factory*, 77-91.

Buying Behaviour Community Middle Class on Online Shopping. (n.d.). *Academica,* 20(8), 8-11.

Camille, F. (2018). Grow what you eat eat what you grow: Urban agriculture as middle class intervention in India. *Journal of Political Ecology*, *25*, 222–249.

Campbell, B. M., Hansen, J., Rioux, J., Stirling, C. M., & Twomlow, S., & Wollenberg, E. (2018). Urgent action to combat climate change and its impacts (SDG 13): Transforming agriculture and food systems. *Current Opinion in Environmental Sustainability*, *34*, 13–20. doi:10.1016/j.cosust.2018.06.005

Campbell, B. M., Vermeulen, S. J., Aggarwal, P. K., Corner-Dolloff, C., Girvetz, E., Loboguerrero, A. M., Ramirez-Villegas, J., Rosenstock, T., Sebastian, L., Thornton, P. K., & Wollenberg, E. (2016). Reducing risks to food security from climate change. *Global Food Security*, *11*, 34–43. doi:10.1016/j.gfs.2016.06.002

Campbell, H., Lahner, S., & Campbell, D. (2017). Building and maintaining social capital through farmers' markets. *Journal of Rural Studies*, *54*, 198–208.

Carleton, T. A. (2017). Crop-damaging temperatures increase suicide rates in India. *Proceedings of the National Academy of Sciences*, 201701354. 10.1073/pnas.1701354114

Carlsson-Kanyama, A., & Gonzalez, A. (2009). Greenhouse gas emissions from food production and consumption. *Environmental Science & Technology*, *43*(17), 6260–6266.

Caron, P., Ferrero y de Loma-Osorio, G., Nabarro, D., Hainzelin, E., Guillou, M., Andersen, I., Arnold, T., Astralaga, M., Beukeboom, M., Bickersteth, S., Bwalya, M., Caballero, P., Campbell, B. M., Divine, N., Fan, S., Frick, M., Friis, A., Gallagher, M., Halkin, J.-P., ... Verburg, G. (2018). Food systems for sustainable development: Proposals for a profound four-part transformation. *Agronomy for Sustainable Development*, *38*(4), 41. doi:10.1007/s13593-018-0519-1 PMID:30956691

Carrer, M. J., De Souza Filho, H. M., Vinholis, M. D. M. B., & Mozambani, C. I. (2022). Precision agriculture adoption and technical efficiency: An analysis of sugarcane farms in Brazil. *Technological Forecasting and Social Change*, *177*, 121510. doi:10.1016/j.techfore.2022.121510

Carvalho, F., Domingues, P., & Sampaio, P. (2019). Communication of commitment towards sustainable development of certified Portuguese organisations: Quality, environment and occupational health and safety. *International Journal of Quality & Reliability Management*, *36*(4), 458–484. doi:10.1108/IJQRM-04-2018-0099

CBN. (2018). *Annual Statistical Bulletin*. Central Bank of Nigeria.

Chahal, H., Dangwal, R., & Raina, S. (2014). Antecedents and consequences of strategic green marketing orientation. *Journal of Global Responsibility*, *5*(2), 338–362. doi:10.1108/JGR-09-2013-0012

Chalmin-Pui, L. S., Griffiths, A., Roe, J., Heaton, T., & Cameron, R. (2021). Why garden?–Attitudes and the perceived health benefits of home gardening. *Cities (London, England)*, *112*, 103118. doi:10.1016/j.cities.2021.103118

Chand, R. (2017). *Doubling Farmers' Income Rationale*. Srategy, Prospects and Action Plan.

Chand, R., Saxena, R., & Rana, S. (2015). Estimates and analysis of farm income in India, 1983–84 to 2011–12. *Economic and Political Weekly*, *50*(22), 139–145.

Chapagain, T., & Raizada, M. N. (2017). Agronomic challenges and opportunities for smallholder terrace agriculture in developing countries. *Frontiers in Plant Science*, *8*, 331. doi:10.3389/fpls.2017.00331 PMID:28367150

Chartres, C. J., & Noble, A. (2015). Sustainable intensification: Overcoming land and water constraints on food production. *Food Security*, *7*(2), 235–245. doi:10.1007/s12571-015-0425-1

Chatterjee, C., Schneider, S., Peter, M., & Mayer, T. (2010). *Risk management in agriculture*. Deutsche Bank Research.

Chatterjee, A., Debnath, S., & Pal, H. (2020). Implication of urban agriculture and vertical farming for future sustainability. In *Urban Horticulture - Necessity of the Future*. Intech Open. doi:10.5772/intechopen.91133

Chatterjee, A., Debnath, S., & Pal, H. (2020). *Implication of urban agriculture and vertical farming for future sustainability. In Urban horticulture-Necessity of the future*. IntechOpen.

Chauhan. (2020). An Analytical Study On Impulse Buying For Online Shopping during Covid-19. *Amity School of Communication*, *12*(6), 198-209.

Cheah, J.-H., Thurasamy, R., Memon, M. A., Chuah, F., & Ting, H. (2020). Multigroup analysis using SmartPLS: Step-by-step guidelines for business research. *Asian Journal of Business Research*, *10*(3), I–XIX. doi:10.14707/ajbr.200087

Chidambaram, C., Nath, S. S., Varshney, P., & Kumar, S. (2022). Assessment of terrace gardens as modifiers of building microclimate. *Energy and Built Environment*, *3*(1), 105–112. doi:10.1016/j.enbenv.2020.11.003

Chikulo, S., Hebinck, P., & Kinsey, B. (2020). 'Mbare Musika is ours': An analysis of a fresh produce market in Zimbabwe. *African Affairs*, *119*(476), 311–337. doi:10.1093/afraf/adaa003

Chiogor, O. H., Gylych, J., & Alabi, O. O. (2023). Macroeconomic Impact of Agricultural Financing Reforms: A Computable General Equilibrium Analysis of Nigeria. *Journal of Economics Finance and Accounting*, *10*(2), 98–107. doi:10.17261/Pressacademia.2023.1732

Chivenge, P., Pascucci, S., Mafurayamutse, A., & Manyangadze, T. (2019). Consumers' willingness to pay for local and organic vegetables in Zimbabwe: A case study of Harare Metropolitan Province. *Journal of Agriculture and Environmental Research Development*, *11*(7), 1336–1344.

Cho, E. (2021). Neither Cronbach's alpha nor McDonald's omega: A commentary on Sijtsma and Pfadt. *Psychometrika*, *86*(4), 877–886. doi:10.1007/s11336-021-09801-1 PMID:34460069

Choquette-Levy, N., Wildemeersch, M., Oppenheimer, M., & Levin, S. A. (2021). Risk transfer policies and climate-induced immobility among smallholder farmers. *Nature Climate Change*, *11*(12), 1046–1054. doi:10.1038/s41558-021-01205-4

Christen, R., & Menkhoff, L. (2019). The impact of microfinance on the well-being of borrowers: A meta-analysis of impact evaluation studies. *The Journal of Development Studies*, *55*(1), 1–33.

Cisternas, I., Velásquez, I., Caro, A., & Rodríguez, A. (2020). Systematic literature review of implementations of precision agriculture. *Computers and Electronics in Agriculture*, *176*, 105626. doi:10.1016/j.compag.2020.105626

Climate Change and Agriculture : Addressing the Challenges. (2012). The United Nations Climate Change (UNFCCC). https://unfccc.int/resource/docs/2012/smsn/ngo/163.pdf

Climate Change and Food Systems . Global Assessments. (2023). The Food and Agriculture Organization of the United Nations (FAO). https://www.fao.org/policy-support/tools-and-publications/resources-details/en/c/435192/. Accessed on 28 March 2024

Climate Change Impacts on Global Food Production . (2019). Food and Agriculture Organization of the United Nations (FAO). https://www.fao.org/fileadmin/templates/est/meetings/climate_2013/Wheeler_CC_impacts_on_FoodSecurity.pdf. As Accessed on 28 March 2024

Cline, W.R. (2008). Global Warming and Agriculture. *Finance and Development.*

Commonwealth. (2003). Public –Private Partnerships: A Review with Special Reference to Local Government. In *Commonwealth Local Government Handbook*. KPL.

Connolly, J., & Prothero, A. (2008). Green consumption: Life-politics, risk and contradictions. *Journal of Consumer Culture*, *8*(1), 117–145. doi:10.1177/1469540507086422

Cooper, M., Messina, C. D., Podlich, D. W., Toenniessen, G. H., & Varshney, R. K. (2020). Breeding for drought tolerance: A wide array of potential approaches. *Plant Breeding Reviews*, *44*, 29–111.

Corona, B., Shen, L., Reike, D., Carreón, J. R., & Worrell, E. (2019). Towards sustainable development through the circular economy—A review and critical assessment on current circularity metrics. *Resources, Conservation and Recycling*, *151*, 104498. doi:10.1016/j.resconrec.2019.104498

Cotula, L., & Myers, D. J. (2009). *Land tenure security: Addressing the challenges*. International Food Policy Research Institute.

Cromwell, E., Friis, H., & Turner, M. (1992). *The seed sector in developing countries: A framework for performance analysis*. Academic Press.

Cucho-Padin, G., Loayza, H., Palacios, S., Balcazar, M., Carbajal, M., & Quiroz, R. (2020). Development of low-cost remote sensing tools and methods for supporting smallholder agriculture. *Applied Geomatics*, *12*(3), 247–263. doi:10.1007/s12518-019-00292-5

Da Silveira, F., Lermen, F. H., & Amaral, F. G. (2021). An overview of agriculture 4.0 development: Systematic review of descriptions, technologies, barriers, advantages, and disadvantages. *Computers and Electronics in Agriculture*, *189*, 106405. doi:10.1016/j.compag.2021.106405

Danciu V (2013) The future of marketing: an appropriate response to the environment changes, *Theoretical and Applied Economics, 5*(582), 33-52.

Dangelico, R. M., & Vocalelli, D. (2017). "Green marketing": An analysis of definitions, strategy steps, and tools through a systematic review of the literature. *Journal of Cleaner Production*, *165*, 1263–1279. doi:10.1016/j.jclepro.2017.07.184

Danish, M. S. S., Senjyu, T., Urasaki, N., Rahmany, N. A., Ershad, A. M., Sabory, N. R., Zarabie, A. K., Anwarzai, M. A., Karimy, H., & Zaheb, H. (2021). Developing nations as a foremost allotment for achieving 2030 SDGs - A case study. *Journal of Engineering and Technology Revolution*, *2*(1), 1–10. doi:10.37357/1068/jetr.2.1.01

Dash, D., & Deole, S. (2020). Terrace gardening: An amazing step towards growing vegetables on roof. *Vigyan Varta*, *1*(7), 20–24.

Davies, M., Guenther, B., Leavy, J., Mitchell, T., & Tanner, T. (2009). Climate Change Adaptation, Disaster Risk Reduction and Social Protection: Complementary Roles in Agriculture and Rural Growth? *IDS Working Papers, 2009*(320), 1–37. doi:10.1111/j.2040-0209.2009.00320_2.x

Davit, W. (2017). How to deal with cross-loadings in Exploratory Factor Analysis? *Research Gate.*

De Gobbi, M. (2020). *Environmental integrity and doing business in Zimbabwe: Challenges and engagement of sustainable enterprises.* ILO Working Paper 2.

Deininger, K., & Feder, G. (2009). Land rights and agricultural development in sub-Saharan Africa: Issues and opportunities. *The World Bank Research Observer*, *24*(2), 243–270.

Delabre, I., Alexander, A., & Rodrigues, C. (2020). Strategies for tropical forest protection and sustainable supply chains: Challenges and opportunities for alignment with the UN sustainable development goals. *Sustainability Science*, *15*(6), 1637–1651. doi:10.1007/s11625-019-00747-z

DeLind, L., & Ferguson, M. (2006). Networks and communities: Exploring the social capital of the Internet. *Journal of Computer-Mediated Communication*, *11*(4), 783–811.

Demirguc-Kunt, A., Klapper, L., & Singer, D. (2017). Financial inclusion and development. *The World Bank Research Observer*, *32*(2), 225–258.

Deng, C., Guangye, Z., Yaojun, L., Xiaodong, N., Zhongwu, L., Junyu, L., & Damei, Z. (2021). Advantages and disadvantages of terracing: A comprehensive review. *International Soil and Water Conservation Research*, *9*(3), 344–359. doi:10.1016/j.iswcr.2021.03.002

Deshmukh, B., & Verma, D. (2018). Fabrication and implementation of automatic seed sowing machine. *International Journal of Engineering Sciences & Research Technology*, *3*, 274–281.

Dev, S. M. (2008). *Challenges for revival of Indian agriculture. In 1st Prof. Dayanand Jha Memorial Lecture.* National Centre for Agricultural Economics and Policy Research.

Dhewa, C. (2015). *Social Investors Needed for Agric Sector.* https://www.herald.co.zw/social-investors-needed-for-agric-sector/

Dickson, M. A., & Bantikyan, H. (2014). Community-supported agriculture: A sustainable alternative for food production and distribution. *Journal of Sustainable Agriculture*, *38*(2), 172–193.

Dilanchiev, A., & Taktakishvili, T. (2022). Determinants of Bilateral Trade Balance between Georgia and China. *Ekonom*, *101*(1), 6–19. doi:10.15388/Ekon.2022.101.1.1

Directory of Public Companies in Nigeria. (n.d.). *Credit Risk Monitor (CRMZ).* http://www.crmz.com/Directory/CountryNG.htm

Dissanayaka, N. S., Udumann, S. S., Nuwarapaksha, T. D., & Atapattu, A. J. (2023b). Agroforestry: An Avenue for Resilient and Productive Farming through Integrated Crops and Livestock Production. In *Transitioning to Zero Hunger*. MDPI Books. https://www.mdpi.com/books/edition/1311-transitioning-to-zero-hunger

Dissanayaka, D. M. N. S., Dissanayake, D. K. R. P. L., Udumann, S. S., Nuwarapaksha, T. D., & Atapattu, A. J. (2023a). Agroforestry—a key tool in the climate-smart agriculture context: A review on coconut cultivation in Sri Lanka. *Frontiers in Agronomy*, *5*, 1162750. Advance online publication. doi:10.3389/fagro.2023.1162750

Dissanayaka, D. M. N. S., Udumann, S. S., Nuwarapaksha, T. D., & Atapattu, A. J. (2024). *Harnessing the potential of Mucuna cover cropping: a comprehensive review of its agronomic and environmental benefits*. Circular Agricultural Systems. doi:10.48130/cas-0024-0001

Dissanayaka, D., Nuwarapaksha, T., Udumann, S., Dissanayake, D., & Atapattu, A. J. (2022). A sustainable way of increasing productivity of coconut cultivation using cover crops: A review. *Circular Agricultural Systems*, *2*(1), 1–9. doi:10.48130/CAS-2022-0007

Dondapat, R. S. (2011). Consumer perception towards eco-Friendly FMCG products and its influence on their purchase decision: An observed research. *International Journal of Innovative Technology and Exploring Engineering, 9*(2).

Dong, H., Campbell, B., & Rabinowitz, A. N. (2019). Factors impacting producer marketing through community-supported agriculture. *PLoS One*, *14*(7), e0219498. doi:10.1371/journal.pone.0219498 PMID:31287848

Donovan, C. A. (2012). Financial inclusion and development in rural sub-Saharan Africa. *The Journal of Development Studies*, *48*(8), 1201–1225.

Dorcas, A. T. (2023). Funding of Agricultural Programme in Tertiary Education in Nigeria. *Best Journal of Innovation in Science. Research for Development*, *2*(9), 338–346.

Dorward & Chirwa. (2011). The Malawi Agricultural Input Subsidy Programme:2005-6 to 2008-9. *International Journal of Agricultural Sustainability, 9*(1).

Dorward, A. (2009). *Rethinking Agricultural Input Subsidy Programmes in a Changing World*. Academic Press.

Dorward, A. (2009). *Rethinking Agricultural Input Subsidy Programmes in a Changing World*. School of Oriental and African Studies.

Dwivedy, N. (2011). Challenges faced by the Agriculture Sector in Developing Countries with special reference to India. *International Journal of Rural Studies*, *18*(2), 1–6.

Eakin, H., & L derer, A. M. (2015). Differential impacts of global change on ecosystem services and livelihoods in the eastern and central Andes. *Global Change Biology*, *21*(4), 1204–1217.

Echeverria, R. G., Trigo, E. J., & Byerlee, D. (1998). Financing Agricultural Research in Latin America. In *Financing Agricultural Research: A Sourcebook*. International Service for National Agricultural Research.

Egli, L., Rüschhoff, J., & Priess, J. (2023). A systematic review of the ecological, social and economic sustainability effects of community-supported agriculture. *Frontiers in Sustainable Food Systems*, *7*, 1136866. Advance online publication. doi:10.3389/fsufs.2023.1136866

ElHaffar, G., Durif, F., & Dubé, L. (2020). Towards closing the attitude-intention-behavior gap in Green Consumption: A narrative review of the literature and an overview of future research directions. *Journal of Cleaner Production*, *275*, 122556. doi:10.1016/j.jclepro.2020.122556

Elkington, J. (1997). Cannibals with Forks: The Triple Bottom Line. *Oxford Capstone*.

Ellis, F. (2000). *Rural livelihoods and diversity in developing countries*. Oxford university press. doi:10.1093/oso/9780198296959.001.0001

Elsalih, O., Sertoglu, K., & Besim, M. (2021). Comparative Advantage of Crude Oil Production: Evidence from 28 Oil-producing Countries. *The Journal of International Trade & Economic Development*, *30*(2), 275–294. doi:10.1080/09638199.2020.1850846

Eno, E. J., & Eze, F. O. (2023). Relationship between Agricultural Financing and Agricultural Output in Nigeria. *Global Journal of Finance and Business Review*.

Environmental Quality. (2019). *International Journal (Toronto, Ont.)*, *31*(2), 406–430. doi:10.1108/MEQ-09-2019-0207

Erickson, B., & Fausti, S. W. (2021). The role of precision agriculture in food security. *Agronomy Journal*, *113*(6), 4455–4462. doi:10.1002/agj2.20919

Escobar-Pemberthy, N. & Calle Saldarriaga, M. A. (2020). International business, trade and the nagoya protocol: best practices and challenges for sustainability in access and benefit-sharing. *International Business, Trade and Institutional Sustainability*, 813-830.

Etzkowitz, H., & Dzisah, J. (2008). Triple Helix Circulation: The Heart of Innovation and Development. *EASST Rev.*, *14*(1), 14–19.

Evans, R. G., & Sadler, E. J. (2008). Methods and technologies to improve efficiency of water use. *Water Resources Research*, *44*(7). Advance online publication. doi:10.1029/2007WR006200

Evans, R. G., & Waller, P. M. (2007). 8. Application of chemical materials. In *Developments in Agricultural Engineering* (Vol. 13, pp. 285–327). Elsevier.

Ewubare, D. B. & Eyitope, J. A. (2015). The Effects of Public Expenditure on Agricultural Production Output in Nigeria. *Journal of Research in Humanities and Social Sciences*, *11*(3), 7-23.

Fanzo, J. (2019). Healthy and Sustainable Diets and Food Systems: The Key to Achieving Sustainable Development Goal 2? *Food Ethics*, *4*(2), 159–174. doi:10.1007/s41055-019-00052-6

FAO. (2016). *Building a common vision for sustainable food and agriculture - Principles and approaches*. FAO.

FAO. (2017). *The Future of Food and Agriculture: Trends and Challenges*. Food and Agriculture Organization.

FAO. (2020). *Empowering farmers and their organizations through the creation of social capital - Bond learning guide for trainers*. doi:10.4060/cb1423en

FAO. (n.d.). *Rainwater Harvesting for Agriculture and Water Supply*. https://www.fao.org/water/en/

Farmers' Suicide in India. (n.d.). https://en.wikipedia.org/wiki/Farmers%27_suicides_in_India

Feliciano, D. (2019). A review on the contribution of crop diversification to Sustainable Development Goal 1 "No poverty" in different world regions. *Sustainable Development (Bradford)*, *27*(4), 795–808. doi:10.1002/sd.1923

Findlay, R. (1987). Comparative Advantage. In The New Palgrave: A Dictionary of Economic. MacMillan. doi:10.1057/978-1-349-95121-5_67-1

Firas, Y. (2011). *Agricultural Risks and Risk Management of Different Farming Systems in Syria*. Academic Press.

Fischer, D., Brettel, M., & Mauer, R. (2020). The Three Dimensions of Sustainability: A Delicate Balancing Act for Entrepreneurs Made More Complex by Stakeholder Expectations. *Journal of Business Ethics*, *163*(1), 87–106. doi:10.1007/s10551-018-4012-1

Font, X., & McCabe, S. (2017). Sustainability and marketing in tourism: Its contexts, Paradoxes, approaches, challenges and potential. *Journal of Sustainable Tourism, 25*(7), 869–883. doi:10.1080/09669582.2017.1301721

Food and Agriculture Organization of the United Nations (FAO). (2013). *Climate-Smart Agriculture Sourcebook*. https://www.fao.org/3/i3325e/i3325e.pdf

Food and Agriculture Organization of the United Nations (FAO). (2019). *The State of the World's Biodiversity for Food and Agriculture*. FAO.

Food and Agriculture Organization of the United Nations. (2023, January 31). *Family farming*. https://www.fao.org/family-farming/home/en/

Food Security, Nutrition and Climate Change. (2016). The International Food Policy Research Institute (IFPRI). https://www.ipcc.ch/site/assets/uploads/sites/4/2020/02/SPM_Updated-Jan20.pdf. Accessed on March 28 2024

Fornell, C., & Larcker, D. F. (1981). Evaluating structural equation models with unobservable variables and measurement error. *JMR, Journal of Marketing Research, 18*(1), 39–50. doi:10.1177/002224378101800104

Fox, M. P., Maclehose, R. F., & Lash, T. L. (2022). *Applying quantitative bias analysis to epidemiologic data*. Springer.

Freeman, R. E. (1984). *Strategic Management: A Stakeholder Approach*. Pitman.

Frempong, M. F., Mu, Y., Adu-Yeboah, S. S., Hossin, M. A., & Adu-Gyamfi, M. (2021). Corporate Sustainability and firm performance: The Role of Green Innovation Capabilities and sustainability-oriented supplier–buyer relationship. *Sustainability (Basel), 13*(18), 10414. doi:10.3390/su131810414

Friedemann-Sánchez, G. (2006). Assets in Intrahousehold Bargaining Among Women Workers in Colombia's Cut-Flower Industry. *Feminist Economics, 12*(1–2), 247–269. doi:10.1080/13545700500508551

Fróna, D., Szenderák, J., & Harangi-Rákos, M. (2019). The Challenge of Feeding the World. *Sustainability (Basel), 11*(20), 5816. doi:10.3390/su11205816

Gadenne, D., Sharma, B., Kerr, D., & Smith, T. (2011). The influence of consumers' environmental beliefs and attitudes on energy saving behaviours. *Energy Policy, 39*(12), 7684–7694. doi:10.1016/j.enpol.2011.09.002

Gaffney, O. (2014). Sustainable development goals: Improving human and planetary wellbeing. *Global Change, 1*(82), 20–23.

García, L., Parra, L., Jimenez, J. M., Lloret, J., & Lorenz, P. (2020). IoT-based smart irrigation systems: An overview on the recent trends on sensors and IoT systems for irrigation in precision agriculture. *Sensors (Basel), 20*(4), 1042. doi:10.3390/s20041042 PMID:32075172

Gautam, P., Bhatt, A. K., Rawat, A., Nautiyal, O. P., & Dhanik, S. (2022). Design and Implementation of IoT-Based Smart System for Monitoring Soil and Parameters in the Agricultural Scenario. *IGI Global*. doi:10.4018/978-1-6684-2443-8.ch006

Gebbers, R., Adamchuk, V. I., Dobermann, A., Seidel, J., & Zeinstra, A. (2019). Quantitative approaches to site-specific crop management. *Advances in Agronomy, 154*, 1-50. https://pubmed.ncbi.nlm.nih.gov/31480709/

George, G., Haas, M. R., Joshi, H., McGahan, A. M., & Tracey, P. (2022). Introduction to the Business of Sustainability: An Organizing Framework for Theory, Practice and Impact. In G. George & ... (Eds.), *Handbook on the Business of Sustainability: The Organization, Implementation, and Practice of Sustainable Growth* (pp. 2–22). Edward Elgar. doi:10.4337/9781839105340.00007

Gericke, N., Boeve-de Pauw, J., Berglund, T., & Olsson, D. (2018). The Sustainability Consciousness Questionnaire: The theoretical development and empirical validation of an evaluation instrument for stakeholders working with Sustainable Development. *Sustainable Development (Bradford)*, *27*(1), 35–49. doi:10.1002/sd.1859

Gio, P. U. (2022). *Partial least squares structural equation modeling (pls-sem) dengan software smartpls*. Uwais Inspirasi Indonesia.

Glaser, B. G., & Strauss, A. L. (1967). *The Discovery of Grounded Theory. Strategies for Qualitative Research*. Aldine.

Glazebrook, T., & Opoku, E. (2020). Gender and Sustainability: Learning from Women's Farming in Africa. *Sustainability (Basel)*, *12*(24), 10483. doi:10.3390/su122410483

Glaze-Corcoran, S., Hashemi, M., Sadeghpour, A., Jahanzad, E., Afshar, R. K., Liu, X., & Herbert, S. J. (2020). Understanding intercropping to improve agricultural resiliency and environmental sustainability. *Advances in Agronomy*, *162*, 199–256. doi:10.1016/bs.agron.2020.02.004

Global A. G. Media (2020). *Zimbabwe reopens fresh markets after pressure from farmers hit by drought*. https://www.thepigsite.com/news/2020/04/zimbabwe-reopens-fresh-markets-after-pressure-from-farmers-hit-by-drought

Global Food Policy Report. (2023). The International Food Policy Research Institute (IFPRI). https://www.ifpri.org/publication/global-food-policy-report-2023-rethinking-food-crisis-responses as retrieved on 28th March 2024.

Godfrey, T. (2016). *The contribution of smallholder agriculture production to food security in rural Zimbabwe: A case study of Masvingo Province*. University of Fort Hare.

Gomez-Trujillo, A. M., & Velez-Ocampo, J., & Gonzalez-Perez, M. A. (2020). A literature review on the causality between sustainability and corporate reputation. *Scientific Research Journal*, *7*(5). Advance online publication. doi:10.31364/SCIRJ/v7.i5.2019.P0519656

Goulson, D., Nicholls, E., Botías, C., & Rotheray, E. L. (2015). Bee Declines Driven by Combined Stress from Parasites, Pesticides, and Lack of Flowers. *Science*, *347*(6229), 1255957. doi:10.1126/science.1255957 PMID:25721506

Govender, J. P., & Govender, T. L. (2016). The influence of green marketing on consumer purchase behavior. *Environment and Ecology*, *7*(2), 77–85.

Grosso, G., Mateo, A., Rangelov, N., Buzeti, T., & Birt, C. (2020). Nutrition in the context of the Sustainable Development Goals. *European Journal of Public Health*, *30*(Supplement_1), i19–i23. doi:10.1093/eurpub/ckaa034 PMID:32391903

Gul, M. S., & Shahzad, H. (2014). The relationship of social media with fashion consciousness and consumer buying behavior. *Journal of Management Info*, *1*(2), 12–20. doi:10.31580/jmi.v2i1.4

Gupta, S.D. (n.d.). *Comparative Advantage and Competitive Advantage: An Economics Perspective and a Synthesis*. St. Thomas University.

Gupta, D., Gujre, N., Singha, S., & Mitra, S. (2022). Role of existing and emerging technologies in advancing climate-smart agriculture through modeling: A review. *Ecological Informatics*, *71*, 101805. doi:10.1016/j.ecoinf.2022.101805

Gürhan-Canli, Z., Sarial-Abi, G., & Hayran, C. (2018). Consumers and brands across the globe: Research synthesis and new directions. *Journal of International Marketing*, *26*(1), 96–117. doi:10.1509/jim.17.0063

Gwanongodza, T. (2020). *Agricultural exports and economic growth in Zimbabwe*. Stellenbosch University.

Gwatidzo, T., & Muyengwa, S. (2020). *Zimbabwe - Assessing the Impact of the Command Agriculture Programme Using the Propensity Score Matching Approach*. World Bank.

Haghnazari, F., Shahgholi, H., & Feizi, M. (2015). Factors affecting the infiltration of agricultural soils. *International Journal of Agronomy and Agricultural Research*, *6*(5), 21–35.

Haider, M., Shannon, R., & Moschis, G. P. (2022). Sustainable Consumption Research and the role of marketing: A review of the literature (1976–2021). *Sustainability (Basel)*, *14*(7), 3999. doi:10.3390/su14073999

Hair, J. F. Jr, Hult, G. T. M., Ringle, C. M., Sarstedt, M., Danks, N. P., & Ray, S. (2021). *Partial least squares structural equation modeling (PLS-SEM) using R: A workbook*. Springer Nature. doi:10.1007/978-3-030-80519-7

Hair, J. F. Jr, & Sarstedt, M. (2021). Data, measurement, and causal inferences in machine learning: Opportunities and challenges for marketing. *Journal of Marketing Theory and Practice*, *29*(1), 65–77. doi:10.1080/10696679.2020.1860683

Hajli, N. (2016). Social commerce constructs and consumer's intention to buy. *International Journal of Information Management*, *35*(2), 183–191. doi:10.1016/j.ijinfomgt.2014.12.005

Halkos, G., & Gkampoura, E.-C. (2021). Where do we stand on the 17 Sustainable Development Goals? An overview on progress. *Economic Analysis and Policy*, *70*, 94–122. doi:10.1016/j.eap.2021.02.001

Hall, B. H. (2002). *The Financing of Research and Development*. Oxford University Press. doi:10.3386/w8773

Ham, D.-H., & Park, J. (2020). Use of a big data analysis technique for extracting HRA data from event investigation reports based on the safety-II concept. Reliability Engineering &amp. *Reliability Engineering & System Safety*, *194*, 106232. doi:10.1016/j.ress.2018.07.033

Hammer, J. H., & Spiker, D. A. (2018). Dimensionality, reliability, and predictive evidence of validity for three help-seeking intention instruments: ISCI, GHSQ, and MHSIS. *Journal of Counseling Psychology*, *65*(3), 394–401. doi:10.1037/cou0000256 PMID:29672088

Hanson, H. I., Eckberg, E., Widenberg, M., & Olsson, J. A. (2021). Gardens' contribution to people and urban green space. *Urban Forestry & Urban Greening*, *63*, 127198. doi:10.1016/j.ufug.2021.127198

Harrower, J. (2023, January 25). *Community-based marketing: Creating a loyal community around your brand*. https://www.socialmediaexaminer.com/

Hart, S. L., & Dowell, G. (2010). Invited editorial: A natural-resource-based view of the firm. *Journal of Management*, *37*(5), 1464–1479. doi:10.1177/0149206310390219

Harvey, C. A., Martínez-Rodríguez, M. R., Cárdenas, J. M., Avelino, J., Rapidel, B., Vignola, R., Donatti, C. I., & Vilchez-Mendoza, S. (2017). The use of Ecosystem-based Adaptation practices by smallholder farmers in Central America. *Agriculture, Ecosystems & Environment*, *246*, 279–290. doi:10.1016/j.agee.2017.04.018

Heifner, R., Coble, K., Perry, J., & Somwaru, A. (1999). *Managing risk in farming: concepts, research, and analysis* (J. L. Harwood, Ed.). US Department of Agriculture, Economic Research Service.

Henderson, J., Brown, C., & Zagorski, K. (2012). Community food security and education: Lessons from the field. *Agriculture and Human Values*, *29*(2), 123–136.

Henry, B. K., Eckard, R. J., & Beauchemin, K. A. (2018). Review: Adaptation of ruminant livestock production systems to climate changes. *Animal*, *12*, s445–s456. doi:10.1017/S1751731118001301 PMID:30092851

Henseler, J., Hubona, G., & Ray, P. A. (2016). Using PLS path modeling in new technology research: Updated guidelines. *Industrial Management & Data Systems*, *116*(1), 2–20. doi:10.1108/IMDS-09-2015-0382

Henson, C., & Reardon, T. (2007). *Staples, seeds, and standards: How global forces affect food security and poverty in the developing world*. International Food Policy Research Institute.

Heo, H. K., & Lee, E. (2021). Types of vertical smart farms and awareness of their use in Korean cities types and feasibility analysis of vertical smart farms in Korean cities. *Journal of People, Plants, and Environment*, *24*(3), 257–266. doi:10.11628/ksppe.2021.24.3.257

Hermida, R. (2015). The problem of allowing correlated errors in structural equation modeling: Concerns and considerations. *Computational Methods in Social Sciences*, *3*, 5–17.

Hien, N. N., Phuong, N. N., Tran, T. V., & Thang, L. D. (2020). The effect of country-of-origin image on purchase intention: The mediating role of Brand Image and Brand Evaluation. *Management Science Letters*, 1205–1212. doi:10.5267/j.msl.2019.11.038

Hillocks, R. J., & Cooper, J. E. (2012). Integrated Pest Management – Can it Contribute to Sustainable Food Production in Europe with Less Reliance on Conventional Pesticides? *Outlook on Agriculture*, *41*(4), 237–242. doi:10.5367/oa.2012.0107

Hirschman, A. O. (1958). *The Strategy of Economic Development.* Yale University Press.

Hoffman, N. P. (n.d.). An Examination of the "Sustainable Competitive Advantage. *Concept: Past, Present, and Future. Academy of Marketing Science Review*, *2*(4).

Hove, G., & Rathaha, T. (2021). The extent green marketing has been embraced in the construction industry-employee perspective in Zimbabwe. *International Business Research*, *14*(6), 1–78. doi:10.5539/ibr.v14n6p78

Hur, E., & Cassidy, T. (2019). Perceptions and attitudes towards Sustainable Fashion Design: Challenges and opportunities for implementing sustainability in fashion. *International Journal of Fashion Design, Technology and Education*, *12*(2), 208–217. doi:10.1080/17543266.2019.1572789

Hurni, H., Giger, M., Liniger, H., Mekdaschi Studer, R., Messerli, P., Portner, B., Schwilch, G., Wolfgramm, B., & Breu, T. (2015). Soils, agriculture and food security: The interplay between ecosystem functioning and human well-being. *Current Opinion in Environmental Sustainability*, *15*, 25–34. doi:10.1016/j.cosust.2015.07.009

Hussain, S., Hussain, S., Guo, R., Sarwar, M., Ren, X., Krstic, D., Aslam, Z., Zulifqar, U., Rauf, A., Hano, C., & El-Esawi, M. A. (2021). Carbon Sequestration to Avoid Soil Degradation: A Review on the Role of Conservation Tillage. *Plants*, *10*(10), 2001. doi:10.3390/plants10102001 PMID:34685810

IMF. (2019). *Zimbabwe Staff-Monitored Program – Press Release and Staff Report of 2019.* IMF Country Report No. 19/14.

Intergovernmental Panel on Climate Change (IPCC). (2023). *Impact of the 2022 floods on agriculture in Pakistan's Sindh Province.* International Commission on Irrigation and Drainage (ICID): https://icid-ciid.org/

International Centre for Climate Change and Development. (2023). *Promoting Indigenous knowledge to strengthen community-led adaptation.* https://www.preventionweb.net/news/promoting-indigenous-knowledge-strengthen-community-led-adaptation

IPCC. (2014). *Impacts, adaptation, and vulnerability. Part A: global and sectoral aspects. Contribution of working group II to the fifth Intergovernmental Panel on Climate Change assessment report.* Cambridge University Press.

IPCC. (2022). *Sixth Assessment Report.* https://www.ipcc.ch/report/sixth-assessment-report-working-group-ii/

Jablonski, B. B. R., Sullins, M., & Thilmany McFadden, D. (2019). Community-Supported Agriculture Marketing Performance: Results from Pilot Market Channel Assessments in Colorado. *Sustainability (Basel)*, *11*(10), 2950. doi:10.3390/su11102950

Jacquet, F., Jeuffroy, M.-H., Jouan, J., Le Cadre, E., Litrico, I., Malausa, T., Reboud, X., & Huyghe, C. (2022). Pesticide-free agriculture as a new paradigm for research. *Agronomy for Sustainable Development*, *42*(1), 8. doi:10.1007/s13593-021-00742-8

Jamroen, C., Komkum, P., Fongkerd, C., & Krongpha, W. (2020). An intelligent irrigation scheduling system using low-cost wireless sensor network toward sustainable and precision agriculture. *IEEE Access : Practical Innovations, Open Solutions*, *8*, 172756–172769. doi:10.1109/ACCESS.2020.3025590

Jayaram, Riese, & Sanghvi. (2010). Africa's path to growth: Sector by sector - Agriculture: Abundant opportunities. *McKinsey Quarterly*.

Jayne, T. S., & Minten, B. (2014). Access to markets and the transformation of African agriculture. *Food Policy*, *48*, 17–27.

Jin, J., & Tang, L. (2011). Coverage path planning on three-dimensional terrain for arable farming. *Journal of Field Robotics*, *28*(3), 424–440. doi:10.1002/rob.20388

Joe Millward, D., & Garnett, T. (2010). Plenary Lecture 3 Food and the planet: Nutritional dilemmas of greenhouse gas emission reductions through reduced intakes of meat and dairy foods. *The Proceedings of the Nutrition Society*, *69*(1), 103–118. doi:10.1017/S0029665109991868 PMID:20003639

Johnson, S., & Steele, L. (2000). Microfinance and its relevance for poverty alleviation. *Entwicklungsländer*, *31*(2), 269–299.

Johnston, B. F., & Mellor, J. W. (1961). The role of agriculture in economic development. *The American Economic Review*, *51*(4), 566–593.

Jolly, C. M., Nyandat, B., Yang, Z., Ridler, N., Matias, F., Zhang, Z., Murekezi, P., & Menezes, A. (2023). Dynamics of aquaculture governance. *Journal of the World Aquaculture Society*, *54*(2), 427–481. doi:10.1111/jwas.12967

Jones, A., Breuning-Madsen, H., Brossard, M., Dampha, A., Deckers, J., Dewitte, O., Gallali, T., Hallett, S., Jones, R., & Kilasara, M. (2013). Soil atlas of Africa. Luxembourg. *European CommissionJournal of Agricultural Economics*, *3*(4), 76–82.

Joshi, P. A. (2015). Challenges of agriculture economy of India. *The Business & Management Review*, *5*(4), 211.

Josson Akkara, P., & Mohan, R. (2023). Perception and Acceptance of Terrace and Balcony Gardening for Sustainable Future of Bengaluru. In *AI and Business, and Innovation Research: Understanding the Potential and Risks of AI for Modern Enterprises* (pp. 787–796). Springer Nature Switzerland.

Jurišić, M., Plaščak, I., Barač, Ž., Radočaj, D., & Zimmer, D. (2021). Sensors and Their Application in Precision Agriculture. *Tehnički glasnik, 15*(4), 529-533.

Kabulika, M. L. (2023). *Peri-Urban Farmers Revolutionalise Hwerb Supplhy for Mass Martkets.* https://www.herald.co.zw/peri-urban-farmers-revolutionise-herb-supply-for-mass-markets/

Kahane, R., Hodgkin, T., Jaenicke, H., Hoogendoorn, C., & Hermann, M., Keatinge, J. D. H., d'Arros Hughes, J., Padulosi, S., & Looney, N. (2013). Agrobiodiversity for food security, health and income. *Agronomy for Sustainable Development*, *33*(4), 671–693. doi:10.1007/s13593-013-0147-8

Kakraliya, S. K., Jat, H. S., Singh, I., Gora, M. K., Kakraliya, M., Bijarniya, D., Sharma, P. C., & Jat, M. L. (2022). Energy and economic efficiency of climate-smart agriculture practices in a rice–wheat cropping system of India. *Scientific Reports*, *12*(1), 1–14. doi:10.1038/s41598-022-12686-4 PMID:35610271

Kartika, N. F. N., Suharno, S., & Indriastuti, H. (2019). Pengaruh Adventure Shopping, Value Shopping, dan Idea Shopping Terhadap Impulse Buying Pada Konsumen Online Shop Instagram di Samarinda. *Jurnal Ilmu Manajemen Mulawarman, 3*(4).

Karunathilake, E., Le, A. T., Heo, S., Chung, Y. S., & Mansoor, S. (2023). The path to smart farming: Innovations and opportunities in precision agriculture. *Agriculture, 13*(8), 1593. doi:10.3390/agriculture13081593

Kassaw, T. (2020). *Magnitude and Determinants of Rural Household Poverty in Ebinat District of Amhara National Regional State of Ethiopia* (Doctoral dissertation).

Kaur, G. (2009). *Socio-economic status of farmers during pre and post liberalization: A case study from Punjab* [M.Phil. Dissertation]. School of Management and Social Sciences, Thapar University, Patiala, India.

Kautish, P., & Sharma, R. (2018). Consumer values, fashion consciousness and behavioural intentions in the online fashion retail sector. International Journal of Retail &amp. *International Journal of Retail & Distribution Management, 46*(10), 894–914. doi:10.1108/IJRDM-03-2018-0060

Kemper, J. A., & Ballantine, P. W. (2019). What do we mean by sustainability marketing? *Journal of Marketing Management, 35*(3–4), 277–309. doi:10.1080/0267257X.2019.1573845

Khalil, M. A., Khalil, R., & Khalil, M. K. (2022). Environmental, social and governance (ESG)-augmented investments in innovation and firms' value: a fixed-effects panel regression of Asian economies. *China Finance Review International*, ahead-of-print. doi:10.1108/CFRI-05-2022-0067

Khalil, M. K. (2022). *Leveraging Buyers' Interest in ESG Investments through Sustainability Awareness.* doi:10.3390/su142114278

Khalil, M. K., & Muneenam, U. (2021). Total Quality Management Practices and Corporate Green Performance: Does Organizational Culture Matter? *Sustainability (Basel), 13*(19), 11021. doi:10.3390/su131911021

Khan, M. (2020). *The fade-away phenomenon of initial non-response bias in panel surveys.* Freie Universitaet Berlin.

Khatri-Chhetri, A., Poudel, B., & Shirsath, P. (2017). *Assessment of Climate-Smart Agriculture (CSA).* Options in Nepal.

Kimaro, J. (2019). A Review on Managing Agroecosystems for Improved Water Use Efficiency in the Face of Changing Climate in Tanzania. *Advances in Meteorology, 2019*, 1–12. doi:10.1155/2019/9178136

Kim, N., Chun, E., & Ko, E. (2017). Country of origin effects on brand image, Brand Evaluation, and purchase intention. *International Marketing Review, 34*(2), 254–271. doi:10.1108/IMR-03-2015-0071

Kimura, S., Antón, J., & LeThi, C. (2010). *Farm level analysis of risk and risk management strategies and policies.* Academic Press.

Kini, J. (2022). Gender-aware inclusive value chain: A theoretical perspective. *Frontiers in Sustainable Food Systems, 6*, 1047190. Advance online publication. doi:10.3389/fsufs.2022.1047190

Kishore, M. V., & Patil, M. S. (2015). a terrace garden from biodegradable waste. *Bionano Frontier, 8*(3), 1–5.

Klein, A. M., Vaissière, B. E., Cane, J. H., Steffan-Dewenter, I., Cunningham, S. A., Kremen, C., & Tscharntke, T. (2007). Importance of Pollinators in Changing Landscapes for World Crops. *Proceedings. Biological Sciences, 274*(1608), 303–313. doi:10.1098/rspb.2006.3721 PMID:17164193

Kluza, K., Zioło, M., Bąk, I., & Spoz, A. (2021). Achieving Environmental Policy Objectives through the Implementation of Sustainable Development Goals. The Case for European Union Countries. *Energies, 14*(8), 2129. doi:10.3390/en14082129

Knickel, K., Redman, M., Darnhofer, I., Ashkenazy, A., Calvão Chebach, T., Šūmane, S., Tisenkopfs, T., Zemeckis, R., Atkociuniene, V., Rivera, M., Strauss, A., Kristensen, L. S., Schiller, S., Koopmans, M. E., & Rogge, E. (2018). Between aspirations and reality: Making farming, food systems and rural areas more resilient, sustainable and equitable. *Journal of Rural Studies*, *59*, 197–210. doi:10.1016/j.jrurstud.2017.04.012

Kondoh, K. (2015). The alternative food movement in Japan: Challenges, limits, and resilience of the teikei system. *Agriculture and Human Values*, *32*(1), 143–153. doi:10.1007/s10460-014-9539-x

Kong, W., Harun, A., Sulong, R. S., & Lily, J. (2014). The influence of consumers' perception of green products on green purchase intention. *Int J Asian SocSci*, *4*(8), 924–939.

Kothari, C. R. (2004). Methodology, Methods and Techniques. New Age International Publishers.

Kotler, P. (2018). *Principles of Marketing*. Prentice Hall.

Kovács-Hostyánszki, A., Espíndola, A., Vanbergen, A. J., Settele, J., Kremen, C., & Dicks, L. V. (2017). Ecological intensification to mitigate impacts of conventional intensive land use on pollinators and pollination. *Ecology Letters*, *20*(5), 673–689. doi:10.1111/ele.12762 PMID:28346980

Krejcie, R. V., & Morgan, D. W. (1970). Determining sample size for research activities. *Educational and Psychological Measurement*, *30*(3), 607–610. doi:10.1177/001316447003000308

Krishna Kumar, K., Rupa Kumar, K., Ashrit, R. G., Deshpande, N. R., & Hansen, J. W. (2004). Climate impacts on Indian agriculture. *International Journal of Climatology*, *24*(11), 1375–1393. doi:10.1002/joc.1081

Krugman, P. (2023). Cooperating for a better future. *CIRIEC-España, revista de economía pública, social y cooperativa*, (107), 5-14.

Krugman, P. (1984). Import Protection as Export Promotion: International Competition in the Presence of Oligopoly and Economies of Scale. In H. Kierzkowski (Ed.), *Monopolistic Competition and International Trade* (pp. 180–193). Clarendon Press.

Kudejira, D. (2014). *An integrated approach towards moderating the effects of climate change on agriculture: A policy perspective for Zimbabwe*. www.future-agricultures.org

Kulkarni, S. R., Nyamagoud, R. V., Naik, H., &Futane, M. (2015). Fabrication of Portable Foot Operated Agricultural Fertilizers and Pesticides Spraying Pump. *International Journal of Engineering Research and Technology*.

Kumar, C., & Venkatesh, V. (2018). *Cloud based soil monitoring and smart irrigation system using IoT and precision farming*. Semantic Scholar. https://www.semanticscholar.org/paper/Cloud-based-soil-monitoring-and-smart-irrigation-Kumar-Venkatesh/ca23e0098316741b004de61b37d6edb2f5d5cc73

Kumar, A., Yadav, S. K., Kharga, S., & Jangid, R. (2022). Rooftop Gardening: A New Era of Gardening. *Agriculture & Food: E-Newsletter*, *4*(7), 599–603.

Kumar, H. V., Shivamurthy, M., Gowda, V. G., & Biradar, G. S. (2017). Assessing decision-making and economic performance of farmers to manage climate-induced crisis in Coastal Karnataka (India). *Climatic Change*, *142*(1-2), 143–153. doi:10.1007/s10584-017-1928-x

Kumari, S., & Patil, Y. (2017). Achieving Climate Smart Agriculture with a Sustainable Use of Water: A Conceptual Framework for Sustaining the Use of Water for Agriculture in the Era of Climate Change. In Reconsidering the Impact of Climate Change on Global Water Supply, Use, and Management (pp. 122-143). IGI Global. doi:10.4018/978-1-5225-1046-8.ch008

Kumari, S., Patil, Y., & Rao, P. (2017). An Approach to Sustainable Watershed Management: Case Studies on Enhancing Sustainability with Challenges of Water in Western Maharashtra. In Reconsidering the Impact of Climate Change on Global Water Supply, Use, and Management (pp. 252-271). IGI Global.

Kumari, S. (2017). Review on developing a conceptual framework for technology adoption towards sustainability in agro based industry. *SAMVAD*, *13*, 14–19.

Kumari, S., Lal, K., & Lata, R. (2020). Assessing sustainability of urban and Peri-Urban agricultural development in Mumbai. *Journal of Pharmacognosy and Phytochemistry*, *9*(6S), 114–118.

Kumari, S., & Patil, Y. B. (2018). Enablers of sustainable industrial ecosystem: Framework and future research directions. *Management of Environmental Quality*.

Kumari, S., Venkatesh, V. G., & Shi, Y. (2023). Assessment of Risks and Risk Management for Agriculture Supply Chain. In *Supply Chain Risk and Disruption Management: Latest Tools, Techniques and Management Approaches* (pp. 155–172). Springer Nature Singapore. doi:10.1007/978-981-99-2629-9_8

Kumar, R. R., & Alok, K. (2020). Adoption of electric vehicle: A literature review and prospects for sustainability. *Journal of Cleaner Production*, *253*, 119911. doi:10.1016/j.jclepro.2019.119911

Kumbhare, D. N., Singh, V., Waghmare, P., Ansari, A., Tiwari, V., & Gorle, R. D. (2016). Fabrication of automatic pesticides spraying machine. *International Research Journal of Engineering and Technology (IRJET)*, *3*(4).

Kumbhar, V., Singh, T. P., & Kumari, S. (2017). The study of developing a sustainable crop production patternthrough trend analysis in Southern Region of Maharashtra: A case study of Satara District. *International Journal of Applied Business and Economic Research*, *15*(2), 149–160.

Kumutha, S., Gayathri, D. N., & Mohanbabu, S. (2018). IoT Concept for Smart System Monitoring Agricultural Land. *International Journal of Engineering Research & Technology (Ahmedabad)*, *6*(8), 1–8.

Ladha, J. K., Jat, M. L., Stirling, C. M., Chakraborty, D., Pradhan, P., Krupnik, T. J., Sapkota, T. B., Pathak, H., Rana, D. S., Tesfaye, K., & Gerard, B. (2020). Achieving the sustainable development goals in agriculture: The crucial role of nitrogen in cereal-based systems. doi:10.1016/bs.agron.2020.05.006

Lakshmi, S. (2016). Consumer Buying Behavior Towards Online Shopping. International Journal of Research – Granthaalayah, 4(8), 60-65. doi:10.29121/granthaalayah.v4.i8(SE).2016.2588

Lal, R., Bouma, J., Brevik, E., Dawson, L., Field, D. J., Glaser, B., Hatano, R., Hartemink, A. E., Kosaki, T., Lascelles, B., Monger, C., Muggler, C., Ndzana, G. M., Norra, S., Pan, X., Paradelo, R., Reyes-Sánchez, L. B., Sandén, T., Singh, B. R., ... Zhang, J. (2021). Soils and sustainable development goals of the United Nations: An International Union of Soil Sciences perspective. *Geoderma Regional*, *25*, e00398. doi:10.1016/j.geodrs.2021.e00398

Lartey, A. (2015). End hunger, achieve food security and improved nutrition and promote sustainable agriculture. *UN Chronicle*, *51*(4), 6–8. doi:10.18356/5940d90a-en

Lasco, R. D., Delfino, R. J. P., & Espaldon, M. L. O. (2014). Agroforestry systems: Helping smallholders adapt to climate risks while mitigating climate change. *Wiley Interdisciplinary Reviews: Climate Change*, *5*(6), 825–833. doi:10.1002/wcc.301

Lawther, W. C. (2002). *Contracting for the 21st Century: A Partnership Model.* The PricewaterhouseCoopers Endowment for the Business of Government.

Lazzarini, S. (2022). *The Right Privatization: Why Private Firms in Public Initiatives Need Capable Governments.* Cambridge University Press. doi:10.1017/9781009024167

Le Riche, L. (2008). Bring on the green. *Your Business, 13*(2), 26–29.

Leakey, R., Mabhaudhi, T., & Gurib-Fakim, A. (2021). African Lives Matter: Wild Food Plants Matter for Livelihoods, Justice, and the Environment—A Policy Brief for Agricultural Reform and New Crops. *Sustainability (Basel), 13*(13), 7252. doi:10.3390/su13137252

Leal Filho, W., Tripathi, S. K., Andrade Guerra, J. B. S. O. D., Giné-Garriga, R., Orlovic Lovren, V., & Willats, J. (2019). Using the sustainable development goals towards a better understanding of sustainability challenges. *International Journal of Sustainable Development and World Ecology, 26*(2), 179–190. doi:10.1080/13504509.2018.1505674

Lee, C.-L., Strong, R., & Dooley, K. E. (2021). Analyzing precision agriculture adoption across the globe: A systematic review of scholarship from 1999–2020. *Sustainability (Basel), 13*(18), 10295. doi:10.3390/su131810295

Lee, E.-B., Lee, S.-G., & Yang, C.-G. (2017). The influences of advertisement attitude and brand attitude on purchase intention of smartphone advertising. Industrial Management &amp. *Industrial Management & Data Systems, 117*(6), 1011–1036. doi:10.1108/IMDS-06-2016-0229

Lee, J. E., & Chen-Yu, J. H. (2018). Effects of a price discount on consumers' perceptions of savings, quality, and value for apparel products: Mediating effect of price discount affect. *Fash Text, 5*(1), 13. doi:10.1186/s40691-018-0128-2

Lee, J. T., Li, G., Zhang, Y., Yang, Z., Xu, Y., & Gao, C. (2019). Genome editing of crops for improved flood tolerance. *Plant Biotechnology Journal, 17*(10), 1907–1917.

Lesk, C., Rowhani, P., & Ramankutty, N. (2016). Influence of extreme weather disasters on global crop production. *Nature, 529*(7584), 84–87. doi:10.1038/nature16467 PMID:26738594

Liang, T.-P., Ho, Y.-T., Li, Y.-W., & Turban, E. (2019). What drives social commerce: The role of social support and relationship quality. *International Journal of Electronics.*

Liberati, A., Altman, D. G., Tetzlaff, J., Mulrow, C., Gøtzsche, P. C., Ioannidis, J. P., Clarke, M., Devereaux, P. J., Kleijnen, J., & Moher, D. (2009, October). The PRISMA statement for reporting systematic reviews and meta-analyses of studies that evaluate health care interventions: Explanation and elaboration. *Journal of Clinical Epidemiology, 62*(10), e1–e34. doi:10.1016/j.jclinepi.2009.06.006 PMID:19631507

Lim, W. M. (2017). Inside the sustainable consumption theoretical toolbox: Critical concepts for sustainability, consumption, and marketing. *Journal of Business Research, 78*, 69–80. doi:10.1016/j.jbusres.2017.05.001

Lin, H., Wang, H., Deng, A., Rong, M., Ye, L., Shi, L., Wang, T., & Peng, X. (2021). *Applications of CRISPR/Cas Technology to Research the Synthetic Genomics of Yeast. In Synthetic Genomics-From BioBricks to Synthetic Genomes.* IntechOpen.

Linu, J. (2021). A study on consumer perception towards green marketing. IJCRT, 9(5).

Lipper, L., Thornton, P., Campbell, B. M., Baedeker, T., Braimoh, A., Bwalya, M., Caron, P., Cattaneo, A., Garrity, D., Henry, K., Hottle, R., Jackson, L., Jarvis, A., Kossam, F., Mann, W., McCarthy, N., Meybeck, A., Neufeldt, H., Remington, T., ... Torquebiau, E. (2014). Climate-smart agriculture for food security. *Nature Climate Change, 4*(12), 1068–1072. doi:10.1038/nclimate2437

Liu, W. (2023). Smart sensors, sensing mechanisms and platforms of sustainable smart agriculture realized through the big data analysis. *Cluster Computing, 26*(5), 2503–2517. doi:10.1007/s10586-021-03295-3

Lleo, A., Bastons, M., Rey, C., & Ruiz-Perez, F. (2021). Purpose implementation: Conceptualization and measurement. *Sustainability (Basel), 13*(4), 1921. doi:10.3390/su13041921

Lockie, S., Lyons, K., & Lawrence, G. (2004). *Consumering food: A geographical and cultural approach.* Routledge.

Loebenstein, G., & Thottappilly, G. (2007). *Agricultural Research Management.* Springer. https://www.springer.com/gp/book/9781402060564

Long, C., Chen, Z., Zhou, Y. & Long, B. (2018). The role of biodiversity and plant conservation for ornamental breeding. *Ornamental Crops*, 1-12.

Lo, S. C., Tung, J., Wang, K. Y., & Huang, K.-P. (2017). Country-of-origin and consumer ethnocentrism: Effect on brand image and product evaluation. *Journal of Applied Sciences (Faisalabad)*, *17*(7), 357–364. doi:10.3923/jas.2017.357.364

Luo, B., Sun, Y., Shen, J., & Xia, L. (2020). How does green advertising skepticism on social media affect consumer intention to purchase green products? *Journal of Consumer Behaviour*, *19*(4), 371–381. doi:10.1002/cb.1818

Luo, Y. (2000). Dynamic Capabilities in International Expansion. *Journal of World Business*, *35*(4), 355–378. doi:10.1016/S1090-9516(00)00043-2

Lûpez-Pereira & Garcìa. (1994). The Maize Seed Industries of Brazil and Mexico: Past Performance, Current Issues, and Future Prospects. *CIMMYT Economics Program Working Paper.* El Batn, Mexico: International Maize and Wheat Improvement Center.

M., L. (2020). Impulsive Buying Behavior on Online Shopping. *International Journal of Psychosocial Rehabilitation*, *24*(5), 4024-4030.

Maddison, D. (2017). Community Supported Agriculture (CSA) in sub-Saharan Africa: A review of existing models and their potential for development. *Agriculture and Human Values*, *34*(3), 831–843.

Mahendra Dev, S. (2014). *Small farmers in India: Challenges and opportunities. "Emerging Economies Research Dialogue".* Indian Council for Research on International Economic Relations.

Maheshwari, S. (2021). Vertical Farming: Resilience Towards Climate Change. *Urban growth and environmental issues in India*, 207-221.

Makate, C. (2019). Effective scaling of climate smart agriculture innovations in African smallholder agriculture: A review of approaches, policy and institutional strategy needs. *Environmental Science & Policy*, *96*, 37–51. doi:10.1016/j.envsci.2019.01.014

Makonese, T., & Chivenge, P. (2017). An assessment of the drought preparedness of smallholder farmers in Chiredzi District, Zimbabwe. *International Journal of Disaster Risk Science*, *8*(1), 127–139.

Makoni, M. (2013). *Reaching for new heights. Information and communications technology takes agriculture to a new level in Zimbabwe.* https://www.un.org/africarenewal/magazine/december-2013/reaching-new-heights

Mall, T., Gupta, M., Dhadialla, T. S. & Rodrigo, S. (2019). Overview of biotechnology-derived herbicide tolerance and insect resistance traits in plant agriculture. *Transgenic Plants: Methods and Protocols*, 313-342.

Mall, R. K., Lal, M., Bhatia, V. S., Rathore, L. S., & Singh, R. (2004). Mitigating climate change impact on soybean productivity in India: A simulation study. *Agricultural and Forest Meteorology*, *121*(1), 113–125. doi:10.1016/S0168-1923(03)00157-6

Mall, R. K., Singh, N., Patel, S., Singh, S., Arora, A., Bhatla, R., & Srivastava, P. K. (2023). Climate Changes over the Indian Subcontinent: Scenarios and Impacts. In *Science, Policies and Conflicts of Climate Change: An Indian Perspective* (pp. 27–52). Springer International Publishing.

Mall, R. K., Singh, R., Gupta, A., Srinivasan, G., & Rathore, L. S. (2006). Impact of climate change on Indian agriculture: A review. *Climatic Change*, *78*(2-4), 445–478. doi:10.1007/s10584-005-9042-x

Manjula, E. (2017). Data Mining Technique to Analyze Soil Nutrients Based on Hybrid Classification. *International Journal of Advanced Research in Computer Science*, *8*(8), 505–510. doi:10.26483/ijarcs.v8i8.4794

Manolis, E. N., & Manoli, E. N. (2021). Raising awareness of the Sustainable Development Goals through Ecological Projects in Higher Education. *Journal of Cleaner Production*, *279*, 123614. doi:10.1016/j.jclepro.2020.123614

Manyise, T., Dentoni, D. & Trienekens, J. (2023). A more grounded view of 'farmer entrepreneurship': How Zimbabwean smallholder farmers fundamentally differ in their entrepreneurial behaviours. *Journal of Entrepreneurship in Emerging Economies*. doi:10.1108/JEEE-02-2023-0044

Mariani, L., Trivellato, B., Martini, M., & Marafioti, E. (2022). Achieving Sustainable Development Goals Through Collaborative Innovation: Evidence from Four European Initiatives. *Journal of Business Ethics*, *180*(4), 1075–1095. doi:10.1007/s10551-022-05193-z PMID:35855697

Mark, A. (2023, February 1). *Community-based marketing: Benefits & examples for independent financial advisors*. https://www.assetmark.com/

Markelova, H., Meinzen-Dick, R., Hellin, J., & Dohrn, S. (2009). Collective action for smallholder market access. *Food Policy*, *34*(1), 1–7. doi:10.1016/j.foodpol.2008.10.001

Marongwe, N. S. (2019). Revitalizing Farmer Markets in Urban Zimbabwe: A Case Study of Mbare Musika Market in Harare. *Journal of Food Products Marketing*, *25*(3), 359–373.

Martinez, M. (2023). Vertical Farming: A Bottom-Up Approach. *Seattle Journal of Technology, Environmental & Innovation Law*, *13*, 5.

Martínez-Peláez, R., Ochoa-Brust, A., Rivera, S., Félix, V. G., Ostos, R., Brito, H., Félix, R. A., & Mena, L. J. (2023). Role of Digital Transformation for Achieving Sustainability: Mediated Role of Stakeholders, Key Capabilities, and Technology. *Sustainability (Basel)*, *15*(14), 11221. doi:10.3390/su151411221

Matandare, M. A. (2017). An analysis of the role of the Agriculture sector: Case of Zimbabwe. *Internafional Journal of Scienfific Research in Science and Technology*, *3*, 1255–1263.

Matteoli, F., Schnetzer, J., & Jacobs, H. (2021). Climate-Smart Agriculture (CSA): An Integrated Approach for Climate Change Management in the Agriculture Sector. In J. M. Luetz & D. Ayal (Eds.), *Handbook of Climate Change Management*. Springer. doi:10.1007/978-3-030-57281-5_148

Maxwell, D., Levin, C., & Csete, J. (1998). Does urban agriculture help prevent malnutrition? Evidence from Kampala. *Food Policy*, *23*(5), 411–424. doi:10.1016/S0306-9192(98)00047-5

Mayr-Dorn, K. (2023). *Trade Diversion and Labor Market Adjustment: Vietnam and the US-China Trade War. Trinity Economics Papers. Trinity College Dublin*. Department of Economics.

McAfee, N., & Brydon-Miller, H. (2016). Food sovereignty and community food security in Canada's north. *Journal of Rural Studies*, *43*, 150–161.

McBride, W., Koning, A. M., & McKibbin, A. (2013). Reframing the community food movement: Critical perspectives on food systems, social justice, and sustainability. *Journal of Sustainable Agriculture*, *37*(2), 173–190.

McElwee, P., Calvin, K., Campbell, D., Cherubini, F., Grassi, G., Korotkov, V., Le Hoang, A., Lwasa, S., Nkem, J., Nkonya, E., Saigusa, N., Soussana, J., Taboada, M. A., Manning, F., Nampanzira, D., & Smith, P. (2020). The impact of interventions in the global land and agri-food sectors on Nature's Contributions to People and the UN Sustainable Development Goals. *Global Change Biology*, *26*(9), 4691–4721. doi:10.1111/gcb.15219 PMID:32531815

McEvoy, J. (2002). *The paradox of empowerment: collective action, knowledge, and development.* Cornell University Press.

McLaughlin, D., & Kinzelbach, W. (2015). Food security and sustainable resource management. *Water Resources Research*, *51*(7), 4966–4985. doi:10.1002/2015WR017053

Medha, M., Lavanya, T., Sunandini, G. P., & Meena, A. (2022). Terrace Gardening: A Sustainable yet Economic Approach. *International Journal of Environment and Climate Change*, *12*(10), 925–932. doi:10.9734/ijecc/2022/v12i1030881

Mehar, M., Mittal, S., & Prasad, N. (2016). Farmers coping strategies for climate shock: Is it differentiated by gender? *Journal of Rural Studies*, *44*, 123–131. doi:10.1016/j.jrurstud.2016.01.001

Memon, M., T, R., Cheah, J.-H., Ting, H., Chuah, F., & Cham, T. H. (2021). PLS-SEM statistical programs: A Review. *Journal of Applied Structural Equation Modeling*, *5*(1), 1–14. doi:10.47263/JASEM.5(1)06

Mendes, J., Pinho, T. M., Neves Dos Santos, F., Sousa, J. J., Peres, E., Boaventura-Cunha, J., Cunha, M., & Morais, R. (2020). Smartphone applications targeting precision agriculture practices—A systematic review. *Agronomy (Basel)*, *10*(6), 855. doi:10.3390/agronomy10060855

Mensah, J. (2019). Sustainable development: Meaning, history, principles, pillars, and implications for human action: Literature review. *Cogent Social Sciences*, *5*(1), 1653531. doi:10.1080/23311886.2019.1653531

Meyer, R. (2019). Modern biotechnology and sustainable intensification: chances and limitations. *Agricultural Resilience*, 159.

Mghenyi, E. W., Dankers, C., Thurlow, J., & Anyiro, C. (2021). *Nigeria transforming agribusiness for inclusive recovery, jobs creation, and poverty reduction: policy reforms and investment priorities.* Academic Press.

Milestad, R., & Darnhofer, I. (2003). Building farm resilience: The prospects and challenges of organic farming. *Journal of Sustainable Agriculture*, *22*(3), 81–97. doi:10.1300/J064v22n03_09

Mir, M. S., Naikoo, N. B., Kanth, R. H., Bahar, F., Bhat, M. A., Nazir, A., Mahdi, S. S., Amin, Z., Singh, L., & Raja, W. (2022). Vertical farming: The future of agriculture: A review. *The Pharma Innovation Journal*, *11*, 1175–1195.

Mishra, A. K., & El-Osta, H. S. (2002). Managing risk in agriculture through hedging and crop insurance: What does a national survey reveal? *Agricultural Finance Review*, *62*(2), 135–148. doi:10.1108/00214930280001134

Mittelman, R. (2012). Green consumer behaviour in emerging markets: A review of research. *Proc Ann Market DivConf*, *33*(3), 392–441.

Miyazoe, K. (2022). The concept of community-based marketing. *Annals of Business Administrative Science.*, *21*(3), 61–73. Advance online publication. doi:10.7880/abas.0220420a

Mladenović, E., Lakićević, M., Pavlović, L., Hiel, K., & Padejčev, J. (2017). Opportunities and benefits of green balconies and terraces in urban conditions. *Contemporary Agriculture*, *66*(3-4), 38–45. doi:10.1515/contagri-2017-0017

Mohan, S. V., Hemalatha, M., Amulya, K., Velvizhi, G., Chiranjeevi, P., Sarkar, O., Kumar, A. N., Krishna, K. V., Modestra, J. A., Dahiya, S., Yeruva, D. K., Butti, S. K., Sravan, J. S., Chatterjee, S., & Kona, R. (2020). Decentralized urban farming through keyhole garden: A case study with circular economy and regenerative perspective. *Materials Circular Economy*, *2*(1), 1–12. doi:10.1007/s42824-020-00011-1

Montagnini, F., Ewel, J. J., & Rheingans, R. (2018). *Agroforestry for soil management in Latin America.* CRC Press.

Morgan, K., & Murdoch, J. (2000). Taste of the future? Consumers' attitudes to new food products in the UK. *Sociology, 34*(4), 605–624.

Morrison, M. (2009). Encouraging the adoption of decision support systems by irrigators. *Rural Society, 19*(1), 17–31. doi:10.5172/rsj.351.19.1.17

Moyo, S. (2013). Exploring the potential of farmer markets in enhancing food security and income generation in Zimbabwe: A case study of Chitungwiza and Mbare Musika farmer markets. *International Journal of African Rural Development, 1*(1), 59–73.

Moyo, S., & Chamboko, T. (2012). Cooperatives as drivers of rural development in Zimbabwe: Challenges and opportunities. *Journal of Southern African Studies, 38*(3), 801–815.

Mugabe, F. T., Nhamo, N., & Pieterse, E. (2013). Smallholder farmers' perceptions of climate change and variabilities in the SADC region: Effects and coping strategies in Zimbabwe. *African Journal of Environmental Science and Technology, 7*(5), 274–286.

Muhie, S. H. (2022). Novel approaches and practices to sustainable agriculture. *Journal of Agriculture and Food Research, 10*, 100446. doi:10.1016/j.jafr.2022.100446

Muhtar, K. (2012). *The five mega-trends shaping tomorrow's customers.* BBC News Business.

Mujeyi, A., Mudhara, M., & Munyaradzi, M. (2020). A review of current climate-smart agriculture technol-ogies: adoption and impact in integrated smallholder farming systems of Zimbabwe. *Biennial Africa Climate Smart Agriculture Stakeholders Conference*, 33.

Mujtaba, G., Shah, M. U. H., Hai, A., Daud, M., & Hayat, M. (2024). A holistic approach to embracing the United Nation's Sustainable Development Goal (SDG-6) towards water security in Pakistan. *Journal of Water Process Engineering, 57*, 104691. doi:10.1016/j.jwpe.2023.104691

Mukonza & Swarts. (2019). The influence of green marketing strategies on business performance and corporate image in the retail sector. In *Bus StratEnv*. John Wiley & Sons, Ltd.

Mukurambwa, P. (2017). *Determinants of smallholder vegetable farmers' participation in post-harvest practices and market access: Evidence from Mashonaland East Province of Zimbabwe* [Doctoral Dissertation]. University of Fort Hare. https://core.ac.uk/download/pdf/224300014.pdf

Mulla, D. J. (2013). Twenty-five years of remote sensing in precision agriculture: Key advances and remaining knowledge gaps. *Biosystems Engineering, 114*(4), 358–371. doi:10.1016/j.biosystemseng.2012.08.009

Mupangwa, W., Nyika, D., & Tirivangasi, H. P. (2018). The status of land degradation in communal areas of Zimbabwe. *Journal of Sustainable Agriculture and Environment, 18*(3), 100–113.

Musara, J. P., Chimvuramahwe, J., & Borerwe, R. (2012). Adoption and efficiency of selected conservation farming technologies in Madziva Communal Area, Zimbabwe: A transcendental production function approach. *Bulletin of Environment. Pharmacology and Life Sciences, 1*, 27–38.

Musungwini, S., Gavai, P. V., Munyoro, B., & Chare, A. (2023). *Emerging ICT Technologies for Agriculture, Training, and Capacity Building for Farmers in Developing Countries: A Case Study in Zimbabwe. In Applying Drone Technologies and Robotics for Agricultural Sustainability*. IGI Global. doi:10.4018/978-1-6684-6413-7.ch002

Mutami, C. (2015). Smallholder Agriculture Production in Zimbabwe: A Survey. *Consilience, 14*, 140–157. https://www.jstor.org/stable/26188746

MuyangaM.NyirendaZ.LifeyoY.BurkeW. (2020). The Future of Smallholder Farming in Malawi. doi:10.13140/RG.2.2.33903.87201

Muzammil, H., Zaman, M., Shahid, M. A., Safdar, M., Majeed, M. D., & Sabir, R. M. (2023). The Impacts of Climate Change on Monsoon Flood Situations in Pakistan. *Environmental Sciences Proceedings*, *25*(1), 63.

Nair, P. K. R., Nair, V. D., Kumar, B. M., & Haile, S. G. (2009). Soil carbon sequestration in tropical agroforestry systems: A feasibility appraisal. *Environmental Science & Policy*, *12*(8), 1099–1111. doi:10.1016/j.envsci.2009.01.010

Narula, R., & Van Huylenbroeck, G. (2015). Empowering smallholder farmers through collective action: A review of the literature. *Journal of Rural Studies*, *39*, 1–11.

NASCIO-CLC. (2006). Keys to Collaboration: Building Effective Public-Private Partnerships. *The National Association of State Chief Information Officers (NASCIO) Corporate Leadership Council (CLC) Issue Brief.*

Naveena, K. P., & Sahana, R. T. (2017). Organic terrace gardening, a new vista in bringing back to traditional food system: Economic issues. *The Journal of Rural and Agricultural Research*, *17*(2), 11–14.

Nazari, B., Liaghat, A., Akbari, M. R., & Keshavarz, M. (2018). Irrigation water management in Iran: Implications for water use efficiency improvement. *Agricultural Water Management*, *208*, 7–18. doi:10.1016/j.agwat.2018.06.003

Ndhlovu, E., & Mhlanga, D. (2023). *Smart technologies, climate change, and smallholder farmer production in Zimbabwe. The fourth industrial revolution in Africa: Exploring the development implications of smart Technologies in Africa*. Springer.

Nikolaou, G., Neocleous, D., Christou, A., Kitta, E., & Katsoulas, N. (2020). Implementing Sustainable Irrigation in Water-Scarce Regions under the Impact of Climate Change. *Agronomy (Basel)*, *10*(8), 1120. doi:10.3390/agronomy10081120

Nin-Pratt, A., & Stads, G. (2023). Innovation Capacity, Food System Development, and the Size of the Agricultural Research System. *Frontiers in Sustainable Food Systems*, *7*, 1051356. doi:10.3389/fsufs.2023.1051356

NISER. (2005). Review of Public Private Partnership in Agricultural Production in Nigeria. In Public-Private Partnership in Nigerian Development. NISER Review of Nigerian Development, 2003/2004, 235-258.

Njuki, J., Mwangi, A., & Ouma, E. (2014). Challenges and opportunities for smallholder farmers' access to markets in Kenya: A review of existing literature. *CGIAR Climate Change, Agriculture and Food Security (CCAFS) Working Paper No. 112.*

Nuchhi, S., Bagali, V., & Annigeri, S. (2020). IOT Based Soil Testing Instrument For Agriculture Purpose. *2020 IEEE Bangalore Humanitarian Technology Conference (B-HTC)*. 10.1109/B-HTC50970.2020.9297897

Nunes, A. N., Gonçalves, J. P., & Figueiredo, A. (2023). Soil Erosion in Extensive versus Intensive Land Uses in Areas Sensitive to Desertification: A Case Study in Beira Baixa, Portugal. *Land (Basel)*, *12*(8), 1591. doi:10.3390/land12081591

Nuwarapaksha, T. D., Udumann, S. S., Dissanayaka, N. S., & Atapattu, A. J. (2023c). Coconut-Based Livestock Farming: A Sustainable Approach to Enhancing Food Security in Sri Lanka. In *Transitioning to Zero Hunger*. MDPI Books. https://www.mdpi.com/books/edition/1311-transitioning-to-zero-hunger

Nuwarapaksha, T. D., Dissanayaka, N. S., Udumann, S. S., & Atapattu, A. J. (2023a). Gliricidia as a Beneficial Crop in Resource-Limiting Agroforestry Systems in Sri Lanka. *Indian Journal of Agroforestry*, *25*, 12–18.

Nuwarapaksha, T. D., Rajapaksha, U. N., Ekanayake, J., Weerasooriya, S. A., & Atapattu, A. J. (2023b). Exploring the Economic Viability of Integrating Jamnapari Goat into Underutilized Pastures under Coconut Cultivations in Coconut Research Institute, Sri Lanka. *IECAG*, *27*, 27. Advance online publication. doi:10.3390/IECAG2023-14990

Nuwarapaksha, T., Udumann, S., Dissanayaka, D., Dissanayake, D., & Atapattu, A. J. (2022). Coconut based multiple cropping systems: An analytical review in Sri Lankan coconut cultivations. *Circular Agricultural Systems*, *2*(1), 1–7. doi:10.48130/CAS-2022-0008

Nyagadza, B. (2021). Fostering green economies in Africa through green marketing strategies for environmental sustainability: An overview. *Journal of Environmental Medicine*, *2*(1), 17–22. doi:10.1386/jem_00037_1

Nyagumbo, I., & Munyaradzi, J. (2014). An assessment of the impacts of climate change on livelihoods in the communal lands of Zimbabwe. *Climate and Development*, *6*(2), 207–216.

Nyangau, P., Muriithi, B., Diiro, G., Akutse, K. S., & Subramanian, S. (2022). Farmers' knowledge and management practices of cereal, legume, and vegetable insect pests and willingness to pay for biopesticides. *International Journal of Pest Management*, *68*(3), 204–216. doi:10.1080/09670874.2020.1817621

O'Mara, F. P. (2011). The significance of livestock as a contributor to global greenhouse gas emissions today and in the near future. *Animal Feed Science and Technology*, *166–167*, 7–15. doi:10.1016/j.anifeedsci.2011.04.074

Obaideen, K., Yousef, B. A. A., AlMallahi, M. N., Tan, Y. C., Mahmoud, M., Jaber, H., & Ramadan, M. (2022). An overview of smart irrigation systems using IoT. *Energy Nexus*, *7*, 100124. doi:10.1016/j.nexus.2022.100124

Obansa, S. A. J., & Madueke, I. M. (2013). Agriculture Financing and Economic Growth in Nigeria. *European Scientific Journal*.

Octavia, D., Suharti, S., Murniati, Dharmawan, I. W. S., Nugroho, H. Y. S. H., Supriyanto, B., Rohadi, D., Njurumana, G. N., Yeny, I., Hani, A., Mindawati, N., Suratman, Adalina, Y., Prameswari, D., Hadi, E. E. W., & Ekawati, S. (2022). Mainstreaming Smart Agroforestry for Social Forestry Implementation to Support Sustainable Development Goals in Indonesia: A Review. *Sustainability (Basel)*, *14*(15), 9313. doi:10.3390/su14159313

Okoji, J. (2023). *Agric allocation relative to national budget lowest in 7yrs*. https://businessday.ng/agriculture/article/agric-allocation-relative-to-national-budget-lowestin7yrs/#:~:text=The%20highest%20proportion%20of%20the,to%201.25%20percent%20in%202022

Oktarina, S., Sumardjo, S., Purnaningsih, N., & Hapsari, D. R. (2020). Activities of Farmer Women Groups in Utilizing Digital Communication Media in Urban Farming Activities in Bogor City. *International Journal of Progressive Sciences and Technologies*, *19*(1), 241–249.

Olsson, D., & Gericke, N. (2017). The effect of gender on students' sustainability consciousness: A nationwide Swedish study. *The Journal of Environmental Education*, *48*(5), 357–370. doi:10.1080/00958964.2017.1310083

Olsson, D., Gericke, N., Sass, W., & Boeve-de Pauw, J. (2020). Self-perceived action competence for sustainability: The theoretical grounding and empirical validation of a novel research instrument. *Environmental Education Research*, *26*(5), 742–760. doi:10.1080/13504622.2020.1736991

Onah, H. C., & Nwodo, A. J. (2023). Backlash of "Uncamped" Internally Displaced Persons Fleeing Insurgency in Northern Nigeria and Its Correlation to Fuelling Insecurity and Violence in the South East Region, Nigeria. *African Journal of Current Research*, *4*(1), 1–19.

Onakoya, A. B. (2013). Agriculture and Intersectoral Linkages and their Contribution to Nigerian Economic Growth. *Economics*, *2*(5), 38–54.

Onyimadu, C. O. (2023). Climate change adaptation and wellbeing among smallholder women farmers in Gwagwalada and Kokona, Nigeria. *Futures*, *153*, 103238. doi:10.1016/j.futures.2023.103238

Oukaira, A., Benelhaouare, A. Z., Kengne, E., & Lakhssassi, A. (2021). FPGA-embedded smart monitoring system for irrigation decisions based on soil moisture and temperature sensors. *Agronomy (Basel)*, *11*(9), 1881. doi:10.3390/agronomy11091881

Oyeyinka, B. O. (2020). SME: Issues, Challenges and Prospects. *Financial System Strategy (FSS) International Conference.* http://www.cenbank.org/fss/wed/sme_issues,%20challenges%20and%20prospects_oyeyinka%20banji.pdf

Pacheco, D. C., Moniz, A. I. D. D. S. A., Caldeira, S. N., & Silva, O. D. L. (2022). Online Impulse Buying—Integrative Review of Psychological Factors. *Perspectives and Trends in Education and Technology*, 527-536.

Pacheco, D. F., Dean, T. J., & Payne, D. S. (2010). Escaping the green prison: Entrepreneurship and the creation of opportunities for sustainable development. *Journal of Business Venturing*, *25*(5), 464–480. doi:10.1016/j.jbusvent.2009.07.006

Page, K. L., Dang, Y. P., & Dalal, R. C. (2020). The Ability of Conservation Agriculture to Conserve Soil Organic Carbon and the Subsequent Impact on Soil Physical, Chemical, and Biological Properties and Yield. *Frontiers in Sustainable Food Systems*, *4*, 31. Advance online publication. doi:10.3389/fsufs.2020.00031

Pagliai, M., Vignozzi, N., & Pellegrini, S. (2004). *Soil structure and the effect of management practices*. Academic Press.

Palm, H. W., Knaus, U., Appelbaum, S., Goddek, S., Strauch, S. M., Vermeulen, T., Haïssam Jijakli, M., & Kotzen, B. (2018). Towards commercial aquaponics: A review of systems, designs, scales and nomenclature. *Aquaculture International*, *26*(3), 813–842. doi:10.1007/s10499-018-0249-z

Pal, S., Rahija, M. A., & Beintema, N. M. (2012). *India: Recent Developments in Agricultural Research. ASTI Country Note*. IFPRI and Indian Council of Agricultural Research.

Palusuk, N., Koles, B., & Hasan, R. (2019). 'All you need is brand love': A critical review and comprehensive conceptual framework for brand love. *Journal of Marketing Management*, *35*(1–2), 97–129. doi:10.1080/0267257X.2019.1572025

Pandey, P. C., & Pandey, M. (2023). Highlighting the role of agriculture and geospatial technology in food security and sustainable development goals. *Sustainable Development (Bradford)*, *31*(5), 3175–3195. doi:10.1002/sd.2600

Pardey, P. G., & Roseboom, J. (1998). Trend in Financing African Agricultural Research. In *Financing Agricultural Research: A Sourcebook*. International Service for National Agricultural Research.

Pardey, P. G., Roseboom, J., & Fan, S. (1998). Trend in Financing Asian and Australian Agricultural Research. In *Financing Agricultural Research: A Sourcebook*. International Service for National Agricultural Research.

Parmar, B. L., Freeman, R., & Harrison, J. S. (2010). Stakeholder Theory: The State of the Art. Management Faculty Publications.

Parot, J., Wahlen, S., & Schryro, J. (2023). *Food justice in community supported agriculture – differentiating charitable and emancipatory social support actions*. Agric Hum Values. doi:10.1007/s10460-023-10511-w

Partey, S. T., Zougmoré, R. B., Ouédraogo, M., & Campbell, B. M. (2018). Developing climate-smart agriculture to face climate variability in West Africa: Challenges and lessons learnt. *Journal of Cleaner Production*, *187*, 285–295. doi:10.1016/j.jclepro.2018.03.199

Patel, J. D., Gadhavi, D. D., & Shukla, Y. S. (2016). Consumers' responses to cause related marketing: Moderating influence of cause involvement and skepticism on attitude and purchase intention. *International Review on Public and Nonprofit Marketing*, *14*(1), 1–18. doi:10.1007/s12208-016-0151-1

Patil, A. R., Patil, A., & Mail, I. D. (2020). The role of the food and fertilizers in antimicrobial resistance in human and its preventive measures. *International Journal of Innovative Science, Engineering and Technology, 1*, 31-37.

Paull, J. (2011). Nanomaterials in food and agriculture: The big issue of small matter for organic food and farming. *Proceedings of the Third Scientific Conference of ISOFAR (International Society of Organic Agriculture Research),* 96–99.

Paustian, K., Lehmann, J., Ogle, S., Reay, D., Robertson, G. P., & Smith, P. (2016). Climate-smart soils. *Nature, 532*(7597), 49–57. doi:10.1038/nature17174 PMID:27078564

Paustian, M., & Theuvsen, L. (2017). Adoption of precision agriculture technologies by German crop farmers. *Precision Agriculture, 18*(5), 701–716. doi:10.1007/s11119-016-9482-5

Pek, J., Wong, O. & Wong, A. (2019). Data transformations for inference with linear regression: Clarifications and recommendations. *Practical Assessment, Research, and Evaluation, 22*, 9.

Petrescu-Mag, R. M., Petrescu, D. C., & Reti, K.-O. (2019). My land is my food: Exploring social function of large land deals using food security–land deals relation in five Eastern European countries. *Land Use Policy, 82*, 729–741. doi:10.1016/j.landusepol.2019.01.003

Petway, J. R., Lin, Y.-P., & Wunderlich, R. F. (2019). Analyzing Opinions on Sustainable Agriculture: Toward Increasing Farmer Knowledge of Organic Practices in Taiwan-Yuanli Township. *Sustainability (Basel), 11*(14), 3843. doi:10.3390/su11143843

Piao, S., Ciais, P., Huang, Y., Shen, Z., Peng, S., Li, J., ... Friedlingstein, P. (2010). The impacts of climate change on water resources and agriculture in China. *Nature, 467*(7311), 43–51. doi:10.1038/nature09364 PMID:20811450

Pingali, P. (2010). Agriculture renaissance: making "Agriculture for development" work in the 21st century. Handbook of agricultural economics, 4, 3867-3894. doi:10.1016/S1574-0072(09)04074-2

Pise, A. A., Sawalkar, R. S., & Saxena, V. D. (2021). An analytical study on online consumer buying behaviour concerning e-shopping portals (B2C). *International Journal of Management, 12*(3).

Place, F., & De Janvry, A. (2012). Land tenure security and agricultural development: Evidence from a global perspective. *American Economic Journal. Applied Economics, 4*(1), 1–32.

Platteau, J. P. (1996). The evolutionary theory of agrarian institutions. *The Journal of Development Studies, 32*(6), 775–810.

Pokharel, A., Hennessy, D. A., & Wu, F. (2023). Health burden associated with tillage-related PM2.5 pollution in the United States, and mitigation strategies. *The Science of the Total Environment, 903*, 166161. doi:10.1016/j.scitotenv.2023.166161 PMID:37574060

Pomering, A. (2017). Marketing for sustainability: Extending the conceptualisation of the marketing mix to drive value for individuals and society at large. *Australasian Marketing Journal, 25*(2), 157–165. doi:10.1016/j.ausmj.2017.04.011

Poornima, C. P., & Suryawanshi, P. D. (2020). Vegetable Consumption Pattern and Health Consciousness of Urban Terrace Garden Farmers. *Journal of Extension Education, 32*(1), 6471. doi:10.26725/JEE.2020.1.32.6471-6474

Pop, C., & Khampirat, B. (2019). Self-assessment instrument to measure the competencies of Namibian graduates: Testing of validity and reliability. *Studies in Educational Evaluation, 60*, 130–139. doi:10.1016/j.stueduc.2018.12.004

Porter, M. E. (1990). *The Competitive Advantage of Nations. Harvard Business Review.* doi:10.1007/978-1-349-11336-1

Power, A. G. (2010). Ecosystem services and agriculture: Tradeoffs and synergies. *Philosophical Transactions of the Royal Society of London. Series B, Biological Sciences, 365*(1554), 2959–2971. doi:10.1098/rstb.2010.0143 PMID:20713396

Prahalad, C. K., & Hamel, G. (1990). *The Core Competence of the Corporation. Harvard Business Review.*

Pray, C., & Nagarajan, L. (2012). Innovation and Research by Private Agribusiness in India. *IFPRI Discussion Paper 1181.*

Pretty, J. (2018). *The future of food: Sustainability and security*. Routledge.

Pretty, J., Noble, J., Bossio, D., Taylor, J., Morison, J., & Pilgrim, L. (2011). Resource-conserving agriculture increases food security in sub-Saharan Africa. *Journal of Experimental Agriculture, 49*(1), 401–417.

Pretty, J., & Smith, A. (2004). Social capital and the collective management of resources. *Journal of Environmental Management, 71*(1), 89–102.

Pretty, J., Toulmin, C., & Bharucha, Z. P. (2011). Paying for biodiversity conservation services in the agricultural landscape. *Environmental Science & Policy, 14*(3), 260–270.

Pretty, J., & Ward, H. (2001). Social capital and the environment. *World Development, 29*(2), 209–227. doi:10.1016/S0305-750X(00)00098-X

Price, C. (2022). *Agriculture 4.0: Bioinformationalism and postdigital hybrid assemblages. In Bioinformational Philosophy and Postdigital Knowledge Ecologies*. Springer.

Production Behavior—From the Perspective of the Market Structure. (2023). *International Journal of Environmental Research and Public Health, 20*, 506. doi:10.3390/ijerph20010506

Purnhagen, K. P., Clemens, S., Eriksson, D., Fresco, L. O., Tosun, J., Qaim, M., Visser, R. G. F., Weber, A. P. M., Wesseler, J. H. H., & Zilberman, D. (2021). Europe's Farm to Fork Strategy and Its Commitment to Biotechnology and Organic Farming: Conflicting or Complementary Goals? *Trends in Plant Science, 26*(6), 600–606. doi:10.1016/j.tplants.2021.03.012 PMID:33893049

Putri, V. P. (2020). A shopping lifestyle mediates between hedonic shopping on impulse buying online at the marketplace. *Jurnal Manajemen Bisnis, 10*(2), 1–8.

Quaye, D., & Mensah, I. (2019). Marketing innovation and sustainable competitive advantage of manufacturing smes in Ghana. *Management Decision, 57*(7), 1535–1553. doi:10.1108/MD-08-2017-0784

Quélin, B., Cabral, S., Lazzarini, S., & Kivleniece, I. (2019). The Private Scope in Public–Private Collaborations: An Institutional and Capability-based Perspective. *Organization Science, 30*(4), 831–846. doi:10.1287/orsc.2018.1251

Quélin, B., Kivleniece, I., & Lazzarini, S. (2017). Public-Private Collaboration, Hybridity and Social Value: Towards New Theoretical Perspectives. *Journal of Management Studies, 54*(6), 763–792. doi:10.1111/joms.12274

Quested, E., Thøgersen-Ntoumani, C., Uren, H., Hardcastle, S. J., & Ryan, R. M. (2018). Community gardening: Basic psychological needs as mechanisms to enhance individual and community well-being. *Ecopsychology, 10*(3), 173–180. doi:10.1089/eco.2018.0002

Quisumbing, A., Cole, S., Elias, M., Faas, S., Galiè, A., Malapit, H., Meinzen-Dick, R., Myers, E., Seymour, G., & Twyman, J. (2023). Measuring Women's Empowerment in Agriculture: Innovations and evidence. *Global Food Security, 38*, 100707. doi:10.1016/j.gfs.2023.100707 PMID:37752898

Qureshi, A., Sarwar, A., & Qureshi, A. S. (2022). Water Management in the Indus Basin in Pakistan. *Challenges and Opportunities Water Management in the Indus Basin in Pakistan: Challenges and Opportunities., 31*, 252–260.

Qureshi, M. R. N. M., Almuflih, A. S., Sharma, J., Tyagi, M., Singh, S., & Almakayeel, N. (2022). Assessment of the Climate-Smart Agriculture Interventions towards the Avenues of Sustainable Production–Consumption. *Sustainability (Basel), 14*(14), 8410. doi:10.3390/su14148410

Ragab, M. A., & Arisha, A. (2017). Research methodology in business: A starter's guide. *Management and Organizational Studies*, *5*(1), 1. doi:10.5430/mos.v5n1p1

Rahi, S., & Ishaq, M. (2020). Factors Influencing Online Shopping behavior With Mediating Role of Customer Perceived Value. *Economic and Social Development: Book of Proceedings,* 128-141.

Rahman, M. (2022). Is co-management a double-edged sword in the protected areas of Sundarbans mangrove? *Biology & Philosophy*, *37*(1), 4. doi:10.1007/s10539-022-09836-3

Rajagopal, A., Jha, S., Khari, M., Ahmad, S., Alouffi, B., & Alharbi, A. (2021). A Novel Approach in Prediction of Crop Production Using Recurrent Cuckoo Search Optimization Neural Networks. *Applied Sciences (Basel, Switzerland)*, *11*(21), 9816. doi:10.3390/app11219816

Rajak, P., Ganguly, A., Adhikary, S., & Bhattacharya, S. (2023). Internet of Things and smart sensors in agriculture: Scopes and challenges. *Journal of Agriculture and Food Research*, *14*, 100776. doi:10.1016/j.jafr.2023.100776

Raj, E. F. I., Appadurai, M., & Athiappan, K. (2022). *Precision farming in modern agriculture. In Smart Agriculture Automation Using Advanced Technologies: Data Analytics and Machine Learning, Cloud Architecture, Automation and IoT*. Springer.

Rajeev, A., Pati, R. K., Padhi, S. S., & Govindan, K. (2017). Evolution of sustainability in Supply Chain Management: A literature review. *Journal of Cleaner Production*, *162*, 299–314. doi:10.1016/j.jclepro.2017.05.026

Rajendran, S., Afari-Sefa, V., Shee, A., Bocher, T., Bekunda, M., dominick, I., & Lukumay, P. J. (2017). Does crop diversity contribute to dietary diversity? Evidence from integration of vegetables into maize-based farming systems. *Agriculture & Food Security*, *6*(1), 50. doi:10.1186/s40066-017-0127-3

Ram, A., Dev, I., Uthappa, A. R., Kumar, D., Kumar, N., Chaturvedi, O. P., Dotaniya, M. L., & Meena, B. P. (2017). *Reactive Nitrogen in Agroforestry Systems of India. In The Indian Nitrogen Assessment*. Elsevier. doi:10.1016/B978-0-12-811836-8.00014-8

Ramayah, T., Lee, J. W. C., & Mohamad, O. (2010). Green product purchase intention: Some insights from a developing country. *Resources, Conservation and Recycling*, *54*(12), 1419–1427. doi:10.1016/j.resconrec.2010.06.007

Ramirez-Andreotta, M. D., Tapper, A., Clough, D., Carrera, J. S., & Sandhaus, S. (2019). Understanding the intrinsic and extrinsic motivations associated with community gardening to improve environmental public health prevention and intervention. *International Journal of Environmental Research and Public Health*, *16*(3), 494–504. doi:10.3390/ijerph16030494 PMID:30754611

Rana, J., & Paul, J. (2017). Consumer behavior and purchase intention for Organic Food: A Review and Research Agenda. *Journal of Retailing and Consumer Services*, *38*, 157–165. doi:10.1016/j.jretconser.2017.06.004

Rao, C. S., Lal, R., Prasad, J. V., Gopinath, K. A., Singh, R., Jakkula, V. S., & Virmani, S. M. (2015). Potential and challenges of rainfed farming in India. In *Advances in agronomy* (Vol. 133, pp. 113–181). Academic Press.

Rao, P. B., Inampudi, M. P., & Srikumar, M. S. (2022). Consumer Attitude towards Organic Terrace Gardening-A Study in Bangalore City. *International Journal of All Research Education and Scientific Methods*, *10*(9), 1–6.

Rao, V. V., Mathapati, S., & Amarapur, B. (2013). Multiple power supplied fertilizer sprayer. *International Journal of Scientific and Research Publications*, *3*(8), 1–5.

Raveendra, S. A. S. T., Atapattu, A. A. A. J., Senarathne, S. H. S., Ranasinghe, C. S., & Weerasinghe, K. W. L. K. (2017). Evaluation of the carbon sequestration potential of intercropping systems under coconut in Sri Lanka. *International Journal of Geology, Earth & Environmental Sciences*, *7*, 1–7.

Ravi Kumar, P., Jayaram, A., & Somashekar, R. K. (2009). Assessment of the performance of different compost models to manage urban household organic solid wastes. *Clean Technologies and Environmental Policy*, *11*(4), 473–484. doi:10.1007/s10098-009-0204-9

Rayhana, R., Xiao, G., & Liu, Z. (2021). RFID sensing technologies for smart agriculture. *IEEE Instrumentation & Measurement Magazine*, *24*(3), 50–60. doi:10.1109/MIM.2021.9436094

Razzaq, A., Ansari, N., Razzaq, Z., & Awan, H. M. (2018). The impact of fashion involvement and pro-environmental attitude on sustainable clothing consumption: The moderating role of Islamic religiosity. *SAGE Open*, *8*(2), 1–17. doi:10.1177/2158244018774611

Reddy, K. P., Chandu, V., & Osei, B. (2023). Consumers perception on green marketing towards eco-friendly fast moving consumer goods. *International Journal of Engineering Business Management*, *15*, 18479790231170962. doi:10.1177/18479790231170962

Reddy, R. V. S. K., Omprasad, J., & Janakiram, T. (2022). Technological innovations in commercial high-tech horticulture, vertical farming and landscaping. *International Journal of Innovative Horticulture*, *11*(1), 78–91. doi:10.5958/2582-2527.2022.00008.2

Redlich, S., Martin, E. A., & Steffan-Dewenter, I. (2021). Sustainable landscape, soil and crop management practices enhance biodiversity and yield in conventional cereal systems. *Journal of Applied Ecology*, *58*(3), 507–517. doi:10.1111/1365-2664.13821

Rehman, S. U., Bhatti, A., Kraus, S., & Ferreira, J. J. M. (2020). The role of Environmental Management Control Systems for ecological sustainability and sustainable performance. *Management Decision*, *59*(9), 2217–2237. doi:10.1108/MD-06-2020-0800

Reicosky, D. (2020). Conservation agriculture systems: Soil health and landscape management. In *Advances in Conservation Agriculture* (pp. 87–154). Burleigh Dodds Science Publishing.

Reith, J., Ghazaryan, G., Muthoni, F., & Dubovyk, O. (2021). Assessment of Land Degradation in Semiarid Tanzania—Using Multiscale Remote Sensing Datasets to Support Sustainable Development Goal 15.3. *Remote Sensing (Basel)*, *13*(9), 1754. doi:10.3390/rs13091754

Richard, B., Qi, A., & Fitt, B. D. (2022). Control of crop diseases through integrated crop management to deliver climate-smart farming systems for low-and high-input crop production. *Plant Pathology*, *71*(1), 187–206. doi:10.1111/ppa.13493

Roberts, J. M., Bruce, T. J., Monaghan, J. M., Pope, T. W., Leather, S. R., & Beacham, A. M. (2020). Vertical farming systems bring new considerations for pest and disease management. *Annals of Applied Biology*, *176*(3), 226–232. doi:10.1111/aab.12587

Rodrigues, C. I. D., Brito, L. M., & Nunes, L. J. R. (2023). Soil Carbon Sequestration in the Context of Climate Change Mitigation: A Review. *Soil Systems*, *7*(3), 64. doi:10.3390/soilsystems7030064

Rodriguez, S. I., Roman, M. S., Sturhahn, S. C., & Terry, E. H. (2002). *Sustainability assessment and reporting for the University of Michigan's Ann Arbor Campus*. Center for Sustainable Systems, Report No. CSS02-04. University of Michigan.

Rogers, E. M. (2003). *Diffusion of innovations* (5th ed.). Free Press.

Rondhi, M., Pratiwi, P. A., Handini, V. T., Sunartomo, A. F., & Budiman, S. A. (2018). Agricultural land conversion, land economic value, and sustainable agriculture: A case study in East Java, Indonesia. *Land (Basel)*, *7*(4), 148. doi:10.3390/land7040148

Rose, D. C., Bhattacharya, M., De Boon, A., Dhulipala, R. K., Price, C., & Schillings, J. (2022). *The fourth agricultural revolution: technological developments in primary food production. In A Research Agenda for Food Systems*. Edward Elgar Publishing.

Rossi, V., Caffi, T., Salotti, I., & Fedele, G. (2023). Sharing decision-making tools for pest management may foster implementation of Integrated Pest Management. *Food Security*, *15*(6), 1459–1474. doi:10.1007/s12571-023-01402-3

Rotunno, L. (2023). Trade Policy and Jobs in Vietnam: The Unintended Consequences of Trump's Trade War. *QPE Working Paper*, Center for Open Science.

Ruan, J.H., Wang, Y.X., Chan, F.T.S., Hu, X.P., Zhao, M.J., Zhu, F.W., Shi, B.F., Shi, Y., & Lin, F. (2019). *A life cycle framework of green IoT-based agriculture and its finance, operation, and management issues*. IEEE.

Runganga, R. & Mhaka, S. (2021). *Impact of agricultural production on economic growth in Zimbabwe*. Academic Press.

Safdar, M., Shahid, M. A., Sarwar, A., Rasul, F., Majeed, M. D., & Sabir, R. M. (2023). Crop Water Stress Detection Using Remote Sensing Techniques. *Environmental Sciences Proceedings*, *25*(1), 20.

Sagar, M., Hullur, P., Ankush, S., Kannolli, V., & Rayakar, V. (2020). *Design and Fabrication of Semi-Automatic Seed Sowing Vehicle*. Academic Press.

Sailer, A., Wilfing, H., & Straus, E. (2022). Greenwashing and bluewashing in Black Friday-related sustainable fashion marketing on Instagram. *Sustainability (Basel)*, *14*(3), 1494. doi:10.3390/su14031494

Samoggia, A., Perazzolo, C., Kocsis, P., & Del Prete, M. (2019). Community Supported Agriculture Farmers' Perceptions of Management Benefits and Drawbacks. *Sustainability (Basel)*, *11*(12), 3262. doi:10.3390/su11123262

Samus, A., Freeman, C., Dickinson, K. J., & van Heezik, Y. (2023). An examination of the factors influencing engagement in gardening practices that support biodiversity using the theory of planned behavior. *Biological Conservation*, *286*, 110252. doi:10.1016/j.biocon.2023.110252

Sanjeevi, P., Prasanna, S., Siva Kumar, B., Gunasekaran, G., Alagiri, I., & Vijay Anand, R. (2020). Precision agriculture and farming using Internet of Things based on wireless sensor network. *Transactions on Emerging Telecommunications Technologies*, *31*(12), e3978. doi:10.1002/ett.3978

Sanmukhiya, C. (2020). A pls–sem approach to the utaut model: The case of Mauritius. *Ann. Soc. Sci. Manag. Stud. ASM*, *6*(1), 8–16. doi:10.19080/ASM.2020.06.555677

Saravanan, K., Sivam, S. S. S., Kumar, S. R., & Moorthy, K. S. (2018). Design and fabrication of automatic seed sowing robot for agricultural field. *International Journal of Pure and Applied Mathematics*, *120*(6), 11749–11766.

Sarker, R. P., & Biswas, B. C. (1978). Agricultural Meteorology in India: A Status Report Agroclimatological Research Needs of the Semi-Arid Tropics. In *Proceedings of the International Workshop on the Agroclimatological Research Needs of the Semi-Arid Tropics*, International Crops Research Institute for the Semi-Arid Tropics.

Sarker, A., Bornman, J., & Marinova, D. (2019). A Framework for Integrating Agriculture in Urban Sustainability in Australia. *Urban Science (Basel, Switzerland)*, *3*(2), 50. doi:10.3390/urbansci3020050

Sarstedt, M., & Cheah, J.-H. (2019). Partial least squares structural equation modeling using SmartPLS: A software review. *Journal of Marketing Analytics*, *7*(3), 196–202. doi:10.1057/s41270-019-00058-3

Satpathy, S., Rao, K. R., Charyulu, G. M., Naidu, T. P., Saikumar, K., & Rajesh, V. (2022). Online shopping and the estimation of their effecting parameters using Consumer Attitudes: Banking sector. *Journal of Positive School Psychology*, *6*(2), 2671–2677.

Scherr, S. J. (2003). A decade of change in the global food system: Research for a hungry world. *Food Policy*, *28*(4), 341–355.

Seck, P. A. (2013). Repackaging Agricultural Research for Greater Impact on Agricultural Growth in Africa. *Journal of Food Security*, *1*(3), 30–14.

Selvan, T., Panmei, L., Murasing, K. K., Guleria, V., Ramesh, K. R., Bhardwaj, D. R., Thakur, C. L., Kumar, D., Sharma, P., Digvijaysinh Umedsinh, R., Kayalvizhi, D., & Deshmukh, H. K. (2023). Circular economy in agriculture: Unleashing the potential of integrated organic farming for food security and sustainable development. *Frontiers in Sustainable Food Systems*, *7*, 1170380. Advance online publication. doi:10.3389/fsufs.2023.1170380

Sengodan, P., & Jbara, Y. H. F. (2017). Development of IoT controlled agri-rover for automatic seeding. *International Journal of Pure and Applied Mathematics*, *114*(11), 241–251.

Sexsmith, K., & McMichael, P. (2015). Formulating the SDGs: Reproducing or Reimagining State-Centered Development? *Globalizations*, *12*(4), 581–596. doi:10.1080/14747731.2015.1038096

Shafiee, M., & Hasan, S. (2020). The effect of price promotions on impulse buying: The mediating role of service innovation in fast-moving consumer goods. *International Journal of Business Information Systems*, *33*(3).

Shafiq, M. N., Gillani, S., & Shafiq, S. (2021). Climate Change and Agricultural Production in Pakistan. *IRASD Journal of Energy & Environment*, *2*(2), 47–54. doi:10.52131/jee.2021.0202.0016

Shahbaz, P., ul Haq, S., Abbas, A., Batool, Z., Alotaibi, B. A., & Nayak, R. K. (2022). Adoption of Climate Smart Agricultural Practices through Women Involvement in Decision Making Process: Exploring the Role of Empowerment and Innovativeness. *Agriculture*, *12*(8), 1161. doi:10.3390/agriculture12081161

Shahmohamadloo, R. S., Febria, C. M., Fraser, E. D. G., & Sibley, P. K. (2022). The sustainable agriculture imperative: A perspective on the need for an agrosystem approach to meet the United Nations Sustainable Development Goals by 2030. *Integrated Environmental Assessment and Management*, *18*(5), 1199–1205. doi:10.1002/ieam.4558 PMID:34821459

Sharma, V. P. (2011). *India's Agricultural Development under the New Economic Regime: Policy Perspective and Strategy for the 12th Five Year Plan*. Indian Institute of Management.

Shaw, E., & Iazzolino, G. (2014). Building a community-based marketing strategy. *Journal of Marketing Management*, *30*(1-2), 143–164.

Shelef, O., Weisberg, P. J., & Provenza, F. D. (2017). The value of native plants and local production in an era of global agriculture. *Frontiers in Plant Science*, *8*, 2069. doi:10.3389/fpls.2017.02069 PMID:29259614

Shennan, C., Krupnik, T. J., Baird, G., Cohen, H., Forbush, K., Lovell, R. J., & Olimpi, E. M. (2017). Organic and Conventional Agriculture: A Useful Framing? *Annual Review of Environment and Resources*, *42*(1), 317–346. doi:10.1146/annurev-environ-110615-085750

Siebrecht, N. (2020). Sustainable Agriculture and Its Implementation Gap—Overcoming Obstacles to Implementation. *Sustainability (Basel)*, *12*(9), 3853. doi:10.3390/su12093853

Siedlecki, S. L. (2020). Understanding descriptive research designs and methods. *Clinical Nurse Specialist CNS*, *34*(1), 8–12. doi:10.1097/NUR.0000000000000493 PMID:31789957

Singh, R. K., Aernouts, M., De Meyer, M., Weyn, M., & Berkvens, R. (2020). Leveraging LoRaWAN technology for precision agriculture in greenhouses. *Sensors (Basel)*, *20*(7), 1827. doi:10.3390/s20071827 PMID:32218353

Singleton, H., Porter, S., Beavis, J., Falconer, L., Hernandez, J. P., & Holley, D. (2023). Accounting for complexity in critical realist trials: The promise of PLS-SEM. *Journal of Critical Realism, 22*(3), 1–20. doi:10.1080/14767430.2023.2217652

Sinha, S. K., Singh, G. B., & Rai, M. (1998). *Decline in crop productivity in Haryana and Punjab: Myth or reality.* Indian Council of Agricultural Research.

Sishodia, R. P., Ray, R. L., & Singh, S. K. (2020). Applications of remote sensing in precision agriculture: A review. *Remote Sensing (Basel), 12*(19), 3136. doi:10.3390/rs12193136

Slayi, M., Zhou, L., & Jaja, I. F. (2023). Constraints Inhibiting Farmers' Adoption of Cattle Feedlots as a Climate-Smart Practice in Rural Communities of the Eastern Cape, South Africa: An In-Depth Examination. *Sustainability (Basel), 15*(20), 14813. doi:10.3390/su152014813

SMEs as Gateway to Economic Prosperity. (n.d.). *ThisDayLive.* http://www.thisdaylive.com/articles/smes-as-gateway-to-economic-prosperity/166328/

Smith, R., & Baillie, J. (2009, January). Defining precision irrigation: A new approach to irrigation management. In *Irrigation Australia 2009: Irrigation Australia Irrigation and Drainage Conference: Proceedings.* University of Southern Queensland.

Sneha, K. (2017). *Exploration and development of a sustainable agro based industrial ecosystem model with special reference to sugar industry.* Academic Press.

Sokol, J., Amrose, S., Nangia, V., Talozi, S., Brownell, E., Montanaro, G., Abu Naser, K., Bany Mustafa, K., Bahri, A., Bouazzama, B., Bouizgaren, A., Mazahrih, N., Moussadek, R., Sikaoui, L., & Winter, A. G. (2019). Energy reduction and uniformity of low-pressure online drip irrigation emitters in field tests. *Water (Basel), 11*(6), 1195. doi:10.3390/w11061195

Solanki, S. N., Ramteke, R. T., & Dhande, P. S. (2022). Agricultural Automatic Seed Sowing Machine: A Review. *International Journal of Research in Engineering. Science and Management, 5*(2), 114–119.

Spielman, D. J., Davis, K., Negash, M., & Ayele, G. (2009). Rural innovation systems and networks: Findings from a study of Ethiopian smallholders. *Agriculture and Human Values, 27*(2), 195–212. doi:10.1007/s10460-010-9273-y

Sri, A. N. S., & Reddy, P. A. (2017). A Smart Framework for Agriculture Production Improvement using Web of Things. *International Journal of Engineering Trends and Technology, 50*(5), 251–259. doi:10.14445/22315381/IJETT-V50P241

Srinivasa. (2015). Potential and Challenges of Rainfed Farming in India. *Advances in Agronomy, 133*, 113–181. doi:10.1016/bs.agron.2015.05.004

Strand, L. (2017). Decomposing the Triple-Helix Synergy into the Regional Innovation Systems of Norway: Firm Data and Patent Networks. *Qual. Quantity, 51*, 963–988.

Sudirjo, F. (2021). Hedonic Shopping Motives, Sales Promotion, and Impulse Buying at Matahari Department Store Consumers in Central Java Province. *Review of International Geographical Education Online, 11*(10), 925–932.

Sugandh, U., Nigam, S., & Khari, M. (2023). Ecosystem of Technologies for Smart Agriculture to Improve the Efficiency and Profitability of Indian Farmers. *2023 10th International Conference on Computing for Sustainable Global Development (INDIACom).* https://ieeexplore.ieee.org/document/10112573

Sujatha, R., & Nithya. (2017). A Survey on Soil Monitoring and Testing In Smart Farming Using IoT And Cloud Platform. *Int. Journal of Engineering Research and Application, 7*(11), 55–59. https://doi.org/ doi:10.9790/9622-0711065559

Sumberg, J., & Evelyne, D. (2010). The social capital of markets: Collective action and market outcomes in Africa. *World Development, 38*(5), 694–706.

Sumetha, M. (2019). Factors Influencing Online Impulsive Buying Behaviour. *Academia (Caracas), 2*(2), 2393–9419.

Sustainable Agriculture Research & Education (SARE). (n.d.). https://www.sare.org/

Swapnil, T., Kasturi, M. L., Girish, P., & Rajkumar, P. (2017). Design and fabrication of seed sowing machine. *International Research Journal of Engineering and Technology, 4*(09), 704–707.

Tăbăraşu, A. M., Anghelache, D., & Dumitru, D. (2021). Benefits Regarding the Implementation of Agriculture 4.0 In the Current Context. Annals of the University of Craiova-Agriculture, Montanology. *Cadastre Series, 50*(2), 544–551.

Tawfik, E., Hammad, I., & Bakry, A. (2022). Production of transgenic Allium cepa by nanoparticles to resist Aspergillus niger infection. *Molecular Biology Reports, 49*(3), 1–8. doi:10.1007/s11033-021-06988-5 PMID:34837626

The State of Food Security and Nutrition in the World. (2022). The Food and Agriculture Organization of the United Nations (FAO). https://www.fao.org/documents/card/en?details=cc0639en. Accessed on 28 march 2024

Thierfelder, C., & Wall, P. C. (2014). Climate variability and agricultural production in communal areas of southern Africa. *Regional Environmental Change, 14*(6), 1839–1851.

Thirathon, U., Wieder, B., Matolcsy, Z., & Ossimitz, M.-L. (2017). Big Data, analytic culture and analytic-based decision making evidence from Australia. *Procedia Computer Science, 121*, 775–783. doi:10.1016/j.procs.2017.11.100

Tian, J., Ni, L., Song, T., Olson, J., & Zhao, J. (2018). An overview of operating parameters and conditions in hydrocyclones for enhanced separations. *Separation and Purification Technology, 206*, 268–285. doi:10.1016/j.seppur.2018.06.015

Titton, M., & Gijsbers, H. J. (2014). The dynamics of land use change in tropical regions and the challenges of sustainable development. *Environmental Science & Policy, 41*, 43–48.

Todeschini, B. V., Cortimiglia, M. N., Callegaro-de-Menezes, D., & Ghezzi, A. (2017). Innovative and sustainable business models in the fashion industry: Entrepreneurial drivers, opportunities, and challenges. *Business Horizons, 60*(6), 759–770. doi:10.1016/j.bushor.2017.07.003

Tomich, T. P., Lidder, P., Coley, M., Gollin, D., Meinzen-Dick, R., Webb, P., & Carberry, P. (2019). Food and agricultural innovation pathways for prosperity. *Agricultural Systems, 172*, 1–15. doi:10.1016/j.agsy.2018.01.002

Tran, P. T., Nguyen, T. Q., Van Huynh, C., Pham, T. H., & Schinkel, U. (2023). A Nuanced Analysis on Livelihood Resilience of Vietnamese Upland Households: An Intersectional Lens of Ethnicity and Gender. *Sustainability (Basel), 15*(4), 3510. doi:10.3390/su15043510

Trendov, M., Varas, S., & Zeng, M. (2019). Digital technologies in agriculture and rural areas: status report. *Digital technologies in agriculture and rural areas: status report.*

Tsani, S., Koundouri, P., & Akinsete, E. (2020). Resource management and sustainable development: A review of the European water policies in accordance with the United Nations' Sustainable Development Goals. *Environmental Science & Policy, 114*, 570–579. doi:10.1016/j.envsci.2020.09.008

Tshuma, S. (2022). Embrace green marketing or lose competitive advantage: An insight on the performance of small and medium-sized enterprises in the foundry sector in Zimbabwe. *Economics. Management and Sustainability, 7*(2), 63–69. doi:10.14254/jems.2022.7-2.5

Udumann, S. S., Dissanayaka, N. S., Nuwarapaksha, T. D., Thelwadana, E. P., & Atapattu, A. J. (2023). Assessing the growth potential of Sunn hemp (Crotalaria juncea L.) as a cover crop for major coconut-growing soils. *Trends in Horticulture*, *6*(2), 3579. doi:10.24294/th.v6i2.3579

Ugwu, U. F. (2021). Maximizing the differentiation principle in regional IP treaties to advance food security: Limitations in West Africa's regional IP and trade regime. *The Journal of World Intellectual Property*, *24*(5-6), 310–343. doi:10.1111/jwip.12193

Umaru, A., & Zubairu, A. A. (2012). An Empirical Analysis of the Contribution of Agriculture and Petroleum Sector to the Growth and Development of the Nigerian Economy from 1960-2010. *The International Journal of Social Sciences (Islamabad)*, *2*(4).

UN. (2019). *World population prospects 2019: Data booklet*. UN.

UNFCCC. (2016) Intended Nationally Determined Contributions (INDCs). http://www4.unfccc.int/Submissions/INDC/Submission%20Pages/submissions.aspx. Accessed on 29 March 2024

UNGC. (n.d.). https://www.unglobalcompact.org/howtoparticipate/business_participation/index.html

United Nations. (2015). *Sustainable development goals*. United Nations Development Programme. https://sdgs.un.org/goals

USDA Natural Resources Conservation Service. (2022, April). *Windbreaks and hedge rows for erosion control*. United States Department of Agriculture.

Ustanti, M. (2018). *Effect of Shopping Lifestyle*. Hedonic Shopping On Impulse.

Valkila, J. (2009). Fair Trade organic coffee production in Nicaragua — Sustainable development or a poverty trap? *Ecological Economics*, *68*(12), 3018–3025. doi:10.1016/j.ecolecon.2009.07.002

Van Delden, S., Sharathkumar, M., Butturini, M., Graamans, L., Heuvelink, E., Kacira, M., Kaiser, E., Klamer, R., Klerkx, L., Kootstra, G., Loeber, A., Schouten, R. E., Stanghellini, C., van Ieperen, W., Verdonk, J. C., Vialet-Chabrand, S., Woltering, E. J., van de Zedde, R., Zhang, Y., & Marcelis, L. F. M. (2021). Current status and future challenges in implementing and upscaling vertical farming systems. *Nature Food*, *2*(12), 944–956. doi:10.1038/s43016-021-00402-w PMID:37118238

Van den Ban, A.W. (2000). Different Ways of Financing Agricultural Extension. *Agricultural Research and Extension Network*, Paper 106a.

Van Gerrewey, T., Boon, N., & Geelen, D. (2021). Vertical farming: The only way is up? *Agronomy (Basel)*, *12*(1), 2. doi:10.3390/agronomy12010002

Varadarajan, R. (2015). Innovating for sustainability: A framework for sustainable innovations and a model of sustainable innovations orientation. *Journal of the Academy of Marketing Science*, *45*(1), 14–36. doi:10.1007/s11747-015-0461-6

Varma, S., & Namara, R. E. (2006). Promoting micro irrigation technologies that reduce poverty. *Water Policy Briefing*, *23*, 1–6.

Vasquez, A., Sherwood, N. E., Larson, N., & Story, M. (2017). Community-Supported Agriculture as a Dietary and Health Improvement Strategy: A Narrative Review. *Journal of the Academy of Nutrition and Dietetics*, *117*(1), 83–94. doi:10.1016/j.jand.2016.09.029 PMID:27847311

Vasu, Singh, S. K., Ray, S. K., Duraisami, V. P., Tiwary, P., Chandran, P., Nimkar, A. M., & Anantwar, S. G. (2016). Soil quality index (SQI) as a tool to evaluate crop productivity in semi-arid Deccan plateau, India. *Geoderma*, *282*, 70–79. doi:10.1016/j.geoderma.2016.07.010

Vazhacharickal, P. J. (2014). Balcony and terrace gardens in urban greening and local food production, scenarios from Mumbai Metropolitan Region, India. *International Journal of Food. Agriculture and Veterinary Sciences*, *4*(2), 149–162.

Veenadhari, S., Misra, B., & Singh, C. (2014). *Machine learning approach for forecasting crop yield based on climatic parameters*. IEEE Xplore. doi:10.1109/ICCCI.2014.6921718

Velandia, M., Rejesus, R. M., Knight, T. O., & Sherrick, B. J. (2009). Factors affecting farmers' utilization of agricultural risk management tools: The case of crop insurance, forward contracting, and spreading sales. *Journal of Agricultural and Applied Economics*, *41*(1), 107–123. doi:10.1017/S1074070800002583

Venezian, L. (1995). *El financiamiento de las actividades de I&D en el ambito de la universidad*. Paper presented at the workshop "Enfoques Innovativos para el Financiamiento de la Investigacion Agropecuaria en America Latinaî". Fundacion ArgenINTA, Buenos Aires.

Venkateswarlu. (2011). *ICAR sponsored Training Course on "Sustainable Agriculture Production through Innovative Approaches for Enhanced Livelihoods" 2-15th September, 2011*. Academic Press.

Viana, C. M., Freire, D., Abrantes, P., Rocha, J., & Pereira, P. (2022). Agricultural land systems importance for supporting food security and sustainable development goals: A systematic review. *The Science of the Total Environment*, *806*, 150718. doi:10.1016/j.scitotenv.2021.150718 PMID:34606855

Virmani, S. M., Sivakumar, M. V. K., & Reddy, S. J. (1978). Climatological Features of the Semi-Arid Tropics in Relation to the Farming Systems. *International Workshop on the Agroclimatological Research Needs of the Semi-Arid Tropics*, 5-16.

Viviers, S. (2009). Going green: A SMME perspective. *The Southern African Journal of Entrepreneurship and Small Business Management*, *2*(1), 30–49. doi:10.4102/sajesbm.v2i1.18

Voss, J. F. (2005). Toulmin's model and the solving of ill-structured problems. *Argumentation*, *19*(3), 321–329. doi:10.1007/s10503-005-4419-6

Wahyuni, D. F., & Rachmawati, I. (2018). Hedonic Shopping Motivation Terhadap Impulse Buying. *Journal Riset Bisnis Dan Management*, *11*(2), 59–65.

Wall, E., & Smit, B. (2005). Climate change adaptation in light of sustainable agriculture. *Journal of Sustainable Agriculture*, *27*(1), 113–123. doi:10.1300/J064v27n01_07

Wang, W.B., Lin, W.Y., Cai, J.H., & Chen, M.X. (2021). *Impact of demand forecast information sharing on the decision of a green supply chain with government subsidy*. Academic Press.

Wang, X., Zhang, J., Ma, D., & Sun, H. (2023). Green Agricultural Products Supply Chain Subsidy Scheme with Green Traceability and Data-Driven Marketing of the Platform. *International Journal of Environmental Research and Public Health*, *20*(4), 3056. doi:10.3390/ijerph20043056 PMID:36833775

Wanzala, M., & Groote, R. (2013). *Fertilizer market development in Africa*. IFDC.

Weeks, J. B., Smith, K. M., & Hulland, J. (2021). Consumer brand curation on social shopping sites. *Journal of Business Research*, *133*, 399–408. doi:10.1016/j.jbusres.2021.05.010

Wen-Yong, W., Yan, H., Hong-Lu, L., Shi-Yang, Y., & Yong, N. (2015). Reclaimed water filtration efficiency and drip irrigation emitter performance with different combinations of sand and disc filters. *Irrigation and Drainage*, *64*(3), 362–369. doi:10.1002/ird.1909

Wezel, A., Casulli, A., & Pascual, U. (2014). Agrodiversity as a driver of ecosystem services and resilience in agriculture. *Ecology and Society*, *19*(3), 54.

Wezel, A., Herren, B. G., Kerr, R. B., Barrios, E., Gonçalves, A. L. R., & Sinclair, F. (2020). Agroecological principles and elements and their implications for transitioning to sustainable food systems. A review. *Agronomy for Sustainable Development*, *40*(6), 40. doi:10.1007/s13593-020-00646-z

Whitcraft, A. K., Becker-Reshef, I., Justice, C. O., Gifford, L., Kavvada, A., & Jarvis, I. (2019). No pixel left behind: Toward integrating Earth Observations for agriculture into the United Nations Sustainable Development Goals framework. *Remote Sensing of Environment*, *235*, 111470. doi:10.1016/j.rse.2019.111470

Williams, N., & Vorley, T. (2014). Economic resilience and entrepreneurship: Lessons from the Sheffield City Region. *Entrepreneurship and Regional Development*, *26*(3–4), 257–281. doi:10.1080/08985626.2014.894129

Winkler, B., Maier, A., & Lewandowski, I. (2019). Urban gardening in Germany: Cultivating a sustainable lifestyle for the societal transition to a bioeconomy. *Sustainability (Basel)*, *11*(3), 801. doi:10.3390/su11030801

Woiceshyn, J., & Daellenbach, U. (2018). Evaluating inductive vs deductive research in management studies: Implications for authors, editors, and reviewers. Qualitative Research in Organizations and Management: An International Journal.

Wong, G. (2013). Meta-narrative reviews. *BMC Medicine*, *11*(12). PMID:23360661

Wong, G., Greenhalgh, T., Westhorp, G., Buckingham, J., & Pawson, R. (2013). RAMESES publication standards: Meta-narrative reviews. *Journal of Advanced Nursing*, *69*(5), 987–1004. doi:10.1111/jan.12092 PMID:23356699

World Agroforestry Centre. (2017). *Corporate Strategy 2017-2020—Transforming Lives and Landscapes with Trees*. Author.

World Bank. (2010). *Zambia Impact Assessment of the Fertilizer Support Program, Analysis of Effectiveness and Efficiency*. Report No.54864, World Bank, Africa Region.

World Bank. (2021). *Climate-smart agriculture: A call to action.* https://www.worldbank.org/en/topic/climate-smart-agriculture

World Bank. (2023). *Climate-Smart Agriculture*. https://www.worldbank.org/en/topic/climate-smart-agriculture

Wulandari, G. A., Suryaningsih, I. B., & Abriana, R. M. (2021). Co-shopper, Mall Environment, Situational Factors Effects On Shopping Experience To Encourage Consumers Shopping Motivation. *Jurnal Aplikasi Manajemen*, *19*(3).

Xu, L., Liu, F., Fu, Z., Jin, J., & Zhang, H. (2020). Breeding for crop drought resistance: Strategies and progress. *Plant Methods*, *16*(1), 1–21. PMID:33292390

Xu, Y., Ferwerda, B., & Lee, M. J. (2020). A Qualitative Study of User Participation and Challenges in a Social Shopping Context. *Proceedings of the 4th HUMANIZE Workshop.*

Yang, S., Song, Y., & Tong, S. (2017). Sustainable retailing in the fashion industry: A Systematic Literature Review. *Sustainability (Basel)*, *9*(7), 1266. doi:10.3390/su9071266

Yee, H. P. (2016). *Green Marketing Practice and Its Effect on Small and Medium Enterprises Performance*. Bachelor of Technopreneurship dissertation. UniversitiTeknikal Malaysia Melaka.

Yigit, M. K., & Tıgli, M. (2018). The moderator role of brand awareness and brand loyalty on consumers' online impulse buying behavior. *International Journal of Research in Business and Social Science*, *7*(1), 35-48.

Yin, H., Cao, Y., Marelli, B., Zeng, X., Mason, A. J., & Cao, C. (2021). Soil sensors and plant wearables for smart and precision agriculture. *Advanced Materials*, *33*(20), 2007764. doi:10.1002/adma.202007764 PMID:33829545

Yin, S. J., Wu, L. H., Du, L., & Chen, M. (2010). Consumers' purchase intention of organic food in China. *Journal of the Science of Food and Agriculture*, *90*(8), 1361–1367. doi:10.1002/jsfa.3936 PMID:20474056

Yu, J. (2018). Consumer responses toward Green Advertising: The effects of gender, advertising skepticism, and Green Motive Attribution. *Journal of Marketing Communications*, *26*(4), 414–433. doi:10.1080/13527266.2018.1514317

Zengeya, T., Sambo, P., & Mabika, N. (2021). The Adoption of the Internet of Things for SMART Agriculture in Zimbabwe. *5th Conference on Networks and Communications (NET 2021)*, 99-106. 10.5121/csit.2021.111208

Zhang, H., Zhao, L., & Gupta, S. (2018). The role of online product recommendations on customer decision-making and loyalty in social shopping communities. *International Journal of Information Management*, *38*(1), 150–166. doi:10.1016/j.ijinfomgt.2017.07.006

Zhang, J., Guan, K., Peng, B., Jiang, C., Zhou, W., Yang, Y., Pan, M., Franz, T. E., Heeren, D. M., Rudnick, D. R., Abimbola, O., Kimm, H., Caylor, K., Good, S., Khanna, M., Gates, J., & Cai, Y. (2021). Challenges and opportunities in precision irrigation decision-support systems for center pivots. *Environmental Research Letters*, *16*(5), 053003. doi:10.1088/1748-9326/abe436

Zhao, J., Liu, D., & Huang, R. (2023). A Review of Climate-Smart Agriculture: Recent Advancements, Challenges, and Future Directions. *Sustainability (Basel)*, *15*(4), 3404. doi:10.3390/su15043404

Zhao, Z., Chen, M., & Zhang, W. (2019). Social community, personal involvement, and psychological processes: A study of impulse buying in the online shopping carnival. *Journal of Electronic Commerce Research*, *20*(4), 255–272.

Zimand-Sheiner, D., Kol, O., & Levy, S. (2022). *Help me if you can: the advantage of farmers' altruistic message appeal in generating engagement with social media posts during COVID-19*. Electron Commer Res. doi:10.1007/s10660-022-09637-6

About the Contributors

Mubashir Ahmed is a Pakistani Agricultural Engineer, M.Sc. (Hons), leveraging Python and deep learning to revolutionize seasonal monsoon prediction. Impacting agriculture, water management, and sustainability, establishing prominence in the field.

Ibukunoluwa Akinrinde is a member of the North West University (NWU) Trade and Development (TRADE) Research Entity, with specialization in trade competitiveness. Prior to now, he has worked as an economist with the Nigerian Economic Summit Group (NESG), National Competitiveness Council of Nigeria (NCCN), PricewaterhouseCoopers (PwC), and Nigerian Institute of Social and Economic Research (NISER).

Sodiq Arogundade holds a Ph.D. in economics from the School of Economics and Econometrics, University of Johannesburg (UJ). He has more than seven years of cognate experience in lecturing, the financial market, public policy advocacy, and management consulting. He worked as a lead economist at the Nigerian Stock Exchange, a senior economist at the Lagos Chamber of Commerce and Industry, and a research analyst at Kainos Edge Consulting, Nigeria. Sodiq currently works as a research fellow and assistant lecturer at the School of Economics, UJ. He has published several scientific papers in international peer-reviewed journals and has been involved in several professional consulting projects. Specifically, Sodiq has consulted for organizations such as the IAU-British Council, South African Reserve Bank, Gauteng Department of Economic Development, MTN Nigeria, Financial and Fiscal Commission (FFC), Nigerian Maritime Administration and Safety Agency, and Center for International Private Enterprise (CIPE). His expertise covers impact assessment studies, monitoring and evaluation, public policy advocacy, environmental scanning, and business intelligence. He is also very vast in the following quantitative methods: causal inference, time series analysis, panel data econometrics, data envelopment analysis, and machine learning econometric method.

Opeyemi Aromolaran is a Postdoctoral Research Fellow at the University of Johannesburg, College of Business and Economics, South Africa, and a lecturer at Dominion University, Ibadan, Nigeria. He primarily concentrates on macroeconomic studies and has co-authored some publications.

Kishore G. is currently pursuing undergraduate degree (B.E) in Mechanical Engineering at Saranathan College of Engineering since 2020.

Ashvin J. is currently pursuing undergraduate degree (B.E) in Mechanical Engineering at Saranathan College of Engineering since 2020.

Tendai Kaponda is a seasoned local governance practitioner with a passion for rural development He is a public policy scholar with a decade of experience working in local governance. His professional career has focused on empowering rural communities through sustainable practices and effective policy implementation. His research interests lie at the intersection of sustainable agriculture, rural governance, and public policy, with a particular focus on facilitating sustainable rural development. Currently pursuing a Ph.D. in Public Administration at Central China Normal University, his doctoral research delves into the complexities of rural governance, aiming to bridge the gap between theory and practice. This research aligns with his extensive experience working within local governance structures, where firsthand exposure to the challenges and opportunities faced by rural communities has fueled his academic pursuits. His dedication to sustainable development extends beyond academia. His experience as a local governance practitioner demonstrates a commitment to translating theory into action, fostering positive change for rural communities. This combination of academic rigour and practical expertise positions him to make significant contributions to the fields of rural governance and sustainable development.

Manju Khari is currently working as a Professor at Jawaharlal Nehru University, New Delhi, India. Prior to this, she worked as Assistant Professor at Netaji Subhas University of Technology, East Campus, formerly, the Ambedkar Institute of Advanced Communication Technologies and Research, under the Government of NCT Delhi, India. She received Ph.D. degree in computer science and engineering from the National Institute of Technology, Patna, India. She received Master of Technology degree in information security from the Ambedkar Institute of Advanced Communication Technologies and Research, affiliated with Guru Gobind Singh Indraprastha University, New Delhi, India. She has published 80 papers in referred national/international journals and conferences, such as IEEE, ACM, Springer, Inderscience, and Elsevier, ten book chapters in a Springer, CRC press, IGI Global, and Auerbach. She has coauthored of two books published by NCERT of XI and XII and co-editor in ten edited books. She has also organized five international conference sessions, three faculty development programme, one workshop, and one industrial meet in her experience. She delivered an expert talk, guest lecturers in international conference, and a member of reviewer/technical program committee in various international conferences. Besides this, she associated with many international research organizations as an Associate Editor/a Guest Editor of Springer, Wiley, and Elsevier books, and a reviewer of various international journals.

Pratyush Kumar is a M.Tech Student at Symbiosis Institute of Geoinformatics. He is also preparing for PhD. His research interests include GIS, Agriculture and GIS.

Sneha Kumari is an Assistant Professor at Symbiosis School of Economics, Symbiosis International (Deemed University), Pune. She has completed her graduation in Agriculture under Indian Council of Agriculture Research fellowship and her Masters in Agribusiness Management under Indian council of Agriculture Research merit.. She has published 25 plus research papers in the area of agriculture, sustainability, technology, climate change and attended severalnational and international conferences.

Vidya Kumbhar is an Assistant Professor and Deputy Director at Symbiosis Institute of Geoinformatics. She has 17+ years of research and teaching in the domain of Programing, Data mining, Machine Learning, Data Science, Geoinformatics, Geospatial technology, she has worked on various funded research projects at national level and Presently, she holds 10+ publications in reputed and peer reviewed international/national/ conferences/book chapters in the field of agriculture, spatial decision support system and Ge Intelligent techniques. Her research interests are Data Science, Machine Learning, Geospatial Technology, agroinformatics and education data mining.

Harish M. is currently pursuing undergraduate degree (B.E) in Mechanical Engineering at Saranathan College of Engineering since 2020.

Reason Masengu holds the position of Senior Lecturer of Marketing in the Department of Business Management at Middle East College, located in Muscat, Oman. In this role, he imparts knowledge in courses such as Global Marketing, Marketing Management, Digital Marketing, and other subjects pertaining to the business management domain. He undertakes various responsibilities, including student counseling, developing curricula, ensuring quality standards, and facilitating teaching and learning activities. Reason holds the distinction of being a Fellow of the Chartered Institute of Marketing, a Senior Member of the Marketers Association of Zimbabwe, and a Professional Member of the Australian Marketing Institute. As an ardent researcher in the fields of marketing, logistics, and supply management, he has acquired international exposure in administering research grants.

Peripaul Onuoha is master's holder in economics from the prestigious University of Lagos. He is the author of Job Creation Strategies for Nigeria. He has also published other journals.

Rajesh P. V. completed his under graduate programme (B.E) in Mechanical Engineering at M.A.M. College of Engineering and Technology, Trichy in 2014. He did his Masters' degree (M.E) in Manufacturing Engineering, Trichy in 2016 with an aggregate of 8.72. Since then, he is working as an Assistant Professor in the department of Mechanical Engineering at Saranathan College of Engineering, Trichy. He has so far presented nearly 10 papers at different National and International Conferences. He has also published various articles in 25 different journals of international reputation and indexing. In addition to this, he wrote 2 book chapters. His core areas of interest are: bio-ceramics, composites, optimization and MCDM.

Kailash Parihar is an M.Tech Student at Symbiosis Institute of Geoinformatics. He is also preparing for PhD. His research interests include GIS, Agriculture and GIS.

Samaya Pillai Iyengar is a full-time faculty in Symbiosis Institute of Digital and Telecom Management, Symbiosis International University, Pune (India). She holds MCA, MCM and MPhil degree under the Faculty of Management. She is Perusing her PhD with (Savitribai Phule Pune University) SPPU, Pune University under Computer-Management. She has 18 years of teaching experience with specialisation in databases. Her research interests focus around computer sciences and big data domains. She has taught subjects in programming -Java, Python, C. Databases –Oracle, MySQL, MongoDB, Operating Systems, MIS, Basic Statistics, and Networking etc. She has 20 publications in Scopus. 10 Papers in Conference proceedings.

Nidheesh S. is currently pursuing undergraduate degree (B.E) in Mechanical Engineering at Saranathan College of Engineering since 2020.

Yashi Tiwari, a B.Tech graduate in Computer Science from Banasthali, is a dynamic individual with diverse areas of interests and accomplishments. In addition to her academic achievements, She has been actively involved in organizing and participating in cultural events. She has also dedicated her time to working with NGOs and taking the lead in science seminars. A sports enthusiast, she brings a well-rounded approach to her pursuits. Her keen interest in current affairs is complemented by her passion for writing, often demonstrated through the creation of insightful news articles on current happenings.

Ayush Verma is a Ph.D scholar at the Jawaharlal Nehru University, Delhi, India. He received his M.Tech degree in Computer Science & Technology in 2022 from School of Computer & Systems Sciences, Jawaharlal Nehru University, Delhi, India and B.Tech degree in Computer Science & Engineering from Ambedkar Institute of Advanced Communication Technologies & Research in 2020 affiliated to Guru Gobind Singh Indraprastha University, Delhi, India.

Index

G

I

L

M

N

P

R

S

T

V

W

Z

www.ingramcontent.com/pod-product-compliance
Lightning Source LLC
LaVergne TN
LVHW060422300926
849162LV00006B/1500
9798369348642